T0340103

MUTAGENICITY: ASSAYS AND APPLICATIONS

MUTAGENICITY: ASSAYS AND APPLICATIONS

Edited by

ASHUTOSH KUMAR

VASILY N. DOBROVOLSKY

ALOK DHAWAN

RISHI SHANKER

ACADEMIC PRESS

An imprint of Elsevier

Academic Press is an imprint of Elsevier
125 London Wall, London EC2Y 5AS, United Kingdom
525 B Street, Suite 1800, San Diego, CA 92101-4495, United States
50 Hampshire Street, 5th Floor, Cambridge, MA 02139, United States
The Boulevard, Langford Lane, Kidlington, Oxford OX5 1GB, United Kingdom

Notices
Knowledge and best practice in this field are constantly changing. As new research and
experience broaden our understanding, changes in research methods, professional practices,
or medical treatment may become necessary.

Practitioners and researchers must always rely on their own experience and knowledge
in evaluating and using any information, methods, compounds, or experiments described
herein. In using such information or methods they should be mindful of their own safety
and the safety of others, including parties for whom they have a professional responsibility.

To the fullest extent of the law, neither the Publisher nor the authors, contributors, or
editors, assume any liability for any injury and/or damage to persons or property as a
matter of products liability, negligence or otherwise, or from any use or operation of any
methods, products, instructions, or ideas contained in the material herein.

Library of Congress Cataloging-in-Publication Data
A catalog record for this book is available from the Library of Congress

British Library Cataloguing-in-Publication Data
A catalogue record for this book is available from the British Library

ISBN: 978-0-12-809252-1

For information on all Academic Press publications visit our website at
https://www.elsevier.com/books-and-journals

  **Working together
to grow libraries in
developing countries**

www.elsevier.com • www.bookaid.org

Publisher: Mica Haley
Acquisition Editor: Rob Sykes
Editorial Project Manager: Tracy Tufaga
Production Project Manager: Anusha Sambamoorthy
Cover Designer: Victoria Pearson

Typeset by TNQ Books and Journals

CONTENTS

4. Chromosomal Aberrations   69

Abhishek K. Jain, Divya Singh, Kavita Dubey, Renuka Maurya,
and Alok K. Pandey

5. In Vivo Cytogenetic Assays   93

Pasquale Mosesso and Serena Cinelli

LIST OF CONTRIBUTORS

Diana Anderson
University of Bradford, Bradford, United Kingdom

Amit Bafana
CSIR-National Environmental Engineering Research Institute, Nagpur, India

Martin H. Brinkworth
University of Bradford, Bradford, United Kingdom

Serena Cinelli
Research Toxicology Centre, Roma, Italy

Ranjeet Prasad Dash
Auburn University, Auburn, AL, United States

Mukul Das
CSIR-Indian Institute of Toxicology Research, Lucknow, India

Jayant Dewangan
CSIR-Central Drug Research Institute, Lucknow, India

Alok Dhawan
CSIR-Indian Institute of Toxicology Research, Lucknow, India

Wei Ding
National Center for Toxicological Research, US Food and Drug Administration, Jefferson, AR, United States

Aman Divakar
CSIR-Central Drug Research Institute, Lucknow, India

Manisha Dixit
Institute of Toxicology Research (CSIR-IITR), Lucknow, India; Academy of Scientific and Innovative Research (AcSIR), Chennai, India

Vasily N. Dobrovolsky
National Center for Toxicological Research, US Food and Drug Administration, Jefferson, AR, United States

Kavita Dubey
CSIR-Indian Institute of Toxicology Research, Lucknow, India

Souvik Sen Gupta
Ahmedabad University, Ahmedabad, India

Khaled Habas
University of Bradford, Bradford, United Kingdom

Robert H. Heflich
National Center for Toxicological Research, US Food and Drug Administration, Jefferson, AR, United States

Abhishek K. Jain
CSIR-Indian Institute of Toxicology Research, Lucknow, India

Mukul R. Jain
Zydus Research Centre, Cadila Healthcare Limited, Ahmedabad, India

Krupa Kansara
Ahmedabad University, Ahmedabad, India

Kannan Krishnamurthi
CSIR-National Environmental Engineering Research Institute, Nagpur, India

Amit Kumar
CSIR-Indian Institute of Toxicology Research, Lucknow, India

Ashutosh Kumar
Ahmedabad University, Ahmedabad, India

Payal Mandal
CSIR-Indian Institute of Toxicology Research, Lucknow, India

Renuka Maurya
CSIR-Indian Institute of Toxicology Research, Lucknow, India

Sakshi Mishra
CSIR-Central Drug Research Institute, Lucknow, India

Pasquale Mosesso
Università degli Studi della Tuscia, Viterbo, Italy

Pravin K. Naoghare
CSIR-National Environmental Engineering Research Institute, Nagpur, India

Manish Nivsarkar
B.V. Patel Pharmaceutical Education and Research Development (PERD) Centre, Ahmedabad, India

Alok K. Pandey
CSIR-Indian Institute of Toxicology Research, Lucknow, India

Prabhash Kumar Pandey
CSIR-Central Drug Research Institute, Lucknow, India

Shraddha Pandit
CSIR-Indian Institute of Toxicology Research, Lucknow, India

Ramakrishnan Parthasarathi
CSIR-Indian Institute of Toxicology Research, Lucknow, India

Dayton M. Petibone
National Center for Toxicological Research, US Food and Drug Administration, Jefferson, AR, United States

Ankita Rai
CSIR-Indian Institute of Toxicology Research, Lucknow, India

Srikanta Kumar Rath
CSIR–Central Drug Research Institute, Lucknow, India

Rishi Shanker
CSIR–Indian Institute of Toxicology Research, Lucknow, India

Divya Singh
CSIR–Indian Institute of Toxicology Research, Lucknow, India

Saravanadevi Sivanesan
CSIR–National Environmental Engineering Research Institute, Nagpur, India

Sonal Srivastava
CSIR–Central Drug Research Institute, Lucknow, India

Rajesh Sundar
Zydus Research Centre, Cadila Healthcare Limited, Ahmedabad, India

Anurag Tripathi
CSIR–Indian Institute of Toxicology Research, Lucknow, India

Darshan Valani
Zydus Research Centre, Cadila Healthcare Limited, Ahmedabad, India

FOREWORD

Human ingenuity in creating and developing new materials knows no bounds, and the demand and appetite for them continues unabated. Whether these are medicines, novel materials, consumer products, or food components, their impact on living organisms, humans, animals, or vegetables and their release into the environment can have consequences on the health of the planet. There is a continuing need for testing new materials for toxicity and especially carcinogenicity. Hence the importance of the study of genetic toxicology, whereby the potential of materials to be mutagenic, can be assessed.

There is no perfect single test for mutagenicity—none has 100% selectivity or specificity. Therefore the concept of a battery of tests has evolved over several decades. Review of these core tests is continuous and influenced by the emergence of new tests and the modification of established ones. New insights into the mechanisms of mutagenicity open up new possibilities for testing and predicting mutagenicity, and the data generated are essential for accurate and well-informed assessment of risk to human health and the environment.

This book is timely and important for students, researchers, regulators, and all those who are concerned with the safety of medicines and consumer products. It presents the state-of-the-art methods to determine the mutagenicity of materials and their potential to cause cancer or heritable mutations to the germline of species. It is both a practical guide on how to assess the mutagenicity of a compound and a textbook on the mechanistic understanding that underlies the tests.

Professor David H. Phillips
King's College London
London, United Kingdom

PREFACE

In the last decade, population growth and increased industrial activity has led to both intentional and inadvertent release of artificial chemicals into the environment. Therefore, large number of chemicals exist in the environment for which toxicity data are not sufficient due to limited regulatory framework and constraints in methodology. The improvement in our understanding of interaction of artificial chemicals with macromolecules in the post–human genome era has led to development, modification, and validation of test systems to evaluate the risk and predict health effects on low-level long-term exposures.

The creation of "Environmental Mutagen Society" in 1969 by geneticists and researchers under the leadership of Alexander Hollaender led to recognition of the discipline of genetic toxicology. Genetic toxicology is the study of the toxic effects of agents, which interact with the hereditary material, resulting in alterations of the deoxyribonucleic acid or its components, leading to inactivation and/or modification in its structure and/or function. Genetic toxicology initially used for hazard identification is now an important component of integrated risk assessment. It is also an intrinsic component of safety assessment for the approval of drugs and animal health products. The US Environmental Protection Agency uses genetic toxicology as a regulatory tool for environmental exposures.

A range of tests need to be conducted to cover the possible wide category of damages possible in response to a chemical in living systems. The workgroups of the European Union, Organization for Economic Co-operation and Development, and International Conference on Harmonization of Technical Requirements for Registration of Pharmaceuticals for Human Use have delineated standard battery of in vitro as well as in vivo *tests*, which need to be carried out for a compound. International workgroups on genetic toxicology, COMNET for the Comet assay, and HuMN for micronucleus assay in human cells have a pivotal role in conducting interlaboratory validations of test systems for their acceptance by the regulatory agencies.

This book incorporates comprehensive protocols and reviews, which will serve as a useful and ready resource for students and scientists working in regulatory toxicology as well as biomedical and pharmaceutical sciences. The authors have actively contributed to peer-reviewed scientific literature in the area of genetic toxicology. The book covers a range of methodologies in safety assessment of chemicals for mutagenesis potential of chemicals. The book comprises 14 chapters that address issues ranging from detection of mutagenic potential of a chemical in bacterial or mammalian cells to chromosomal aberrations and in vivo testing. The book attempts to address the issues of immense relevance to human and environmental health such as mutagens in food and need for mutagen determination in environmental pollution control. The book also has component chapters on emerging methodologies and regulatory guidelines. In essence, the book intends to provide an understanding of mutagenesis, protocols, methodologies, and new approaches that should serve as tools for protection of human health and environment.

This book epitomizes the long-term scientific association that the editors have enjoyed with the authors and is a culmination of their collaborative effort.

ACKNOWLEDGMENTS

The editors wish to acknowledge with thanks the unstinted support and contributions of all the authors in the book. They also acknowledge for the generous funding from the Council of Scientific and Industrial Research, India, under its network projects NWP 34, NWP 35, INDEPTH (BSC 0111), and NanoSHE (BSC 0112); Department of Biotechnology for the project NanoToF; and Gujarat Institute of Chemical Technology, India, for CENTRA (Centre for Nanotechnology and Research Application); risk assessment project and DST SERB for the project EMR/2016/005286.

CHAPTER ONE

Mutagenesis, Genetic Disorders and Diseases

Manisha Dixit[1,2], Amit Kumar[3]

[1]Institute of Toxicology Research (CSIR-IITR), Lucknow, India; [2]Academy of Scientific and Innovative Research (AcSIR), Chennai, India; [3]CSIR-Indian Institute of Toxicology Research, Lucknow, India

1. INTRODUCTION

We are subjected to continuous genotoxic insults every day by environmental factors that include harmful UV rays from sunlight or carcinogens in our food consumables, which potentially could be harmful to the human body. Even in the presence of multiple mechanisms devised to efficiently cope with these insults, very often failures occur in the exhaustive defensive strategies of the cell leading to a number of diseases such as cancers, developmental disorders, and neurodegeneration. The changes in DNA, due to exposure to certain insults (such as harmful radiations, chemicals, or biological agents) or the inability of the cells to repair damaged chromosomes, is referred as mutagenesis; while the physical, chemical, or biological entities that are able to mutagenize are called mutagens (Fig. 1.1). There are around 20,000 genes in the human body [1], and they often undergo alterations due to endogenous or exogenous DNA damage as stated above. Although the efficient repair system is competent enough to protect the cells from such damage, at times it fails to repair the chromosomes efficiently, and this leads to changes in the DNA sequence and exome readout through phenotypic and genotypic alterations, which are referred as mutations.

Although mammalian DNA replication is meticulous and tightly controlled owing to its complexity (although all DNA replications are tightly regulated, the mammalian one is more complex), there remains a possibility of 10^{-8} error per base pair (bp) [2]. Mutations are imperative to evolution and form the basis of incorporation of new alleles, which pass to subsequent generations. They serve as the reason for variation among individuals. These new alleles or variants could be envisioned as the reason behind how eukaryotes have survived so far in the course of time and fought to cope

Mutagenicity: Assays and Applications
ISBN 978-0-12-809252-1
http://dx.doi.org/10.1016/B978-0-12-809252-1.00001-8

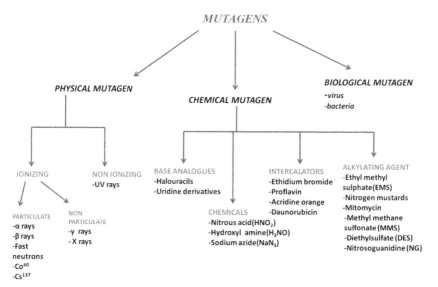

Figure 1.1 Type of mutagens.

against diseases that were once considered deadly. In this chapter, we will elaborate on various mutagens, their role in causing mutations, molecular mechanisms of repair pathways, and some rare and common diseases arising from distinct mutations.

2. MUTAGENS

2.1 Physical Mutagens

Physical mutagens include various kinds of ionizing and nonionizing radiations such as ultraviolet (UV) rays, infrared (IR) rays, X-rays, gamma rays, alpha particles, beta particles, or fast moving neutrons (Fig. 1.1) [3], which can be emitted either by natural or artificial sources. These particles possess enough energy to free atoms or molecules from their electronic orbits and induce reactive oxygen species (ROS) formation. Cobalt^{60}, a sterilizing agent, and cesium^{137} are few examples of ionizing agents.

2.2 Chemical Mutagens

Chemical mutagens react with DNA and lead to faulty base pairing that can be classified into different categories (Fig. 1.1). Base analogues mimic a particular nucleobase in nucleic acid and are misread by the replicating machinery as a normal base. Following incorporation into DNA, they form non–Watson pairing with the DNA. The most studied base analogues are halouracils and uridine derivatives that include 5-bromodeoxyuridine (BrdU). This potent mutagen is capable of inducing point mutations by substituting thymine residues and pairing with guanine instead of adenine. Such base pair changes are deleterious and are more prone to alterations by physical mutagens such as X-rays, UV rays, or gamma rays relative to normal DNA [4]. Certain chemicals such as nitrous acid, hydroxyl amine, and sodium azide can modify the bases by deamination, thus modifying the regular base pairing. Nitrous acid deaminates adenine, guanine, and cytosine substituting adenine to hypoxanthine, guanine to xanthine, and cytosine to uracil [5]. These substitutions induce AT to GC transitions leading to faulty base pairing. Chemical mutagens are also comprised of alkylating agents and DNA intercalating agents. Ethyl methyl sulfate, nitrogen mustards, mitomycin, methyl methane sulfonate (MMS), diethyl sulfate, and nitroso-guanidine (NTG, NG, MNNG) are few examples of alkylating agents as well as DNA-intercalating agents, which include acridine orange, ethidium bromide (EtBr), proflavin, daunorubicin, among others.

Acridine orange or EtBr introduces frameshift mutations either by addition or deletion of base pairs. Certain pesticides such as rotenone, paraquat, or maneb have the capacity to induce mutations such as base pair changes in genes and could lead to neurodegenerative diseases such as Parkinson disease (PD) [6].

2.3 Biological Mutagens

Biological mutagens include agents such as transposons, viruses, or bacteria (Fig. 1.1). Both transposons and viral DNA may integrate into the human genome and change the genetic composition during cell division [7]. Upon incorporation into DNA, they can cause change in normal functioning of gene and can lead to several disorders.

3. MUTATIONS

Mutations can be spontaneous or induced; spontaneous mutations arise de novo and include mutations due to DNA replication or lesions, whereas induced mutations occur upon interaction with mutagens. Mutations are classified into three broad categories: gene mutation, chromosomal mutation, and genome mutation (Fig. 1.2). Gene mutations include *transition, transversion, frameshift mutation* [8], *neutral mutation, silent mutation, and missense or nonsense mutation* [9]. Chromosomal mutations involve *deletion, duplication,*

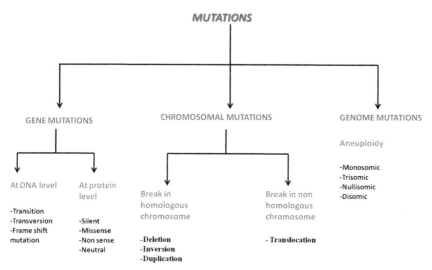

Figure 1.2 Classification of mutation.

inversion, or *translocation* of a group of genes. Genome mutations include aneuploidy such as *monosomic, nullisomic, double monosomic, trisomic, double trisomic, and tetrasomic* [10]. All mutation subtypes are discussed in the following section.

Gene mutations in general are permanent changes in the DNA and normally include changes in a single nucleotide to a large segment of chromosomes having multiple genes. Transitions and transversions both occur due to changes in a single nucleotide leading to a point mutation. Transition involves the replacement of either purine with purine or pyrimidine with pyrimidine, whereas transversion comprises the exchange of purines with pyrimidines and vice versa. Transitions occur due to a tautomeric shift of either an amino group to an imino group or a keto to an enol form of base pairs or due to oxidative deamination and are more common than transversions [11].

Not all mutations could produce visible changes in the phenotype and the genotype of an organism; such mutations are termed **silent mutations,** wherein a single base pair mutation does not change the amino acid product [12]. Since a single amino acid can be coded by multiple codons, such mutations are considered ineffective. For example, a change from the codon TTT to TTC has no effect since both code for phenylalanine. **Neutral mutations** include substitutions wherein a basic amino acid is replaced by another basic amino acid or an acidic amino acid by another acidic amino acid such that there is no apparent change in the physiology of the formed protein [12]. For example, when an arginine is replaced by a lysine or an aspartate is replaced by a glutamate.

A **missense mutation** reflects change in a base pair that could alter the coding sequence and results in coding for a different amino acid, thus affecting the function of a given protein that may alter physiology of the cell entirely. For example, in sickle cell anemia, change of GAG to GTG replaces valine (hydrophobic) instead of glutamic acid (charged) leading to sickle cell anemia. Such types of disorders are discussed in Section "Genetic Disorders and Diseases."

Mutations where a normal chain is often terminated due to point mutation of a normal codon to a stop codon, UAA, UAC, UGA, are classified as **nonsense mutations**. Premature termination of a growing peptide chain results in formation of truncated protein products [13]. Thalassemia, Duchenne muscular dystrophy, and cystic fibrosis are few examples occurring due to nonsense mutations. Mutations that alter protein function are clearly more deleterious relative to other types of gene mutations.

Loss of certain genome regions, gene amplification, and alteration in genetic code or translocation among genes are classified as "chromosomal mutations."

3.1 Break in Homologous Chromosome

During DNA replication, often the chromosomes break, but the broken ends may soon rejoin through a process called restitution so that the chromosome remains intact. However, if they fail to ligate, there is often generation of a centric fragment with the centromere and an acentric fragment without the centromere (Fig. 1.3A). Breaks can occur either in homologous or nonhomologous chromosomes. Breaks in homologous chromosomes include deletions, inversions, and duplications (Fig. 1.3B), whereas breaks in nonhomologous chromosome include translocation. The acentric fragment is soon lost and eventually degraded, whereas, the centric fragment is retained and participates normally in the DNA replication process; however, part of it has undergone a deletion making it shorter than the parental copy; this is classified as a **deletion mutation** (Fig. 1.3B). Deletion often causes pseudodominance, a phenomenon where due to deletion of certain segment of gene in a homologue, the recessive gene in the other homologue is unmasked and is expressed. This leads to unexpected expression of certain deleterious genes which till now were masked by the dominant gene and could lead to several disorders [14].

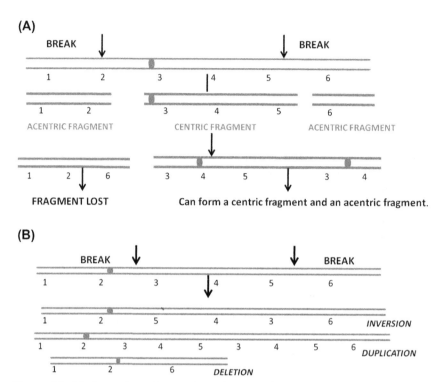

Figure 1.3 (A): Formation of acentric and centric fragments during chromosomal breakage (*red line* (gray in print versions) represents the chromosome, whereas the *red dot* (gray in print versions) represents the centromere). (B): Break in homologous chromosome showing inversion, deletion, and duplication. *Red lines* (gray in print versions) represent the chromosome while the *red dot* (gray in print versions) depicts the centromere.

Often two breaks in the same chromosome could join in an inverted fashion, setting the phenomenon of **inversion** (Fig. 1.3B). The outcome of such changes includes new linkage relations on chromosomal mapping and leads to "position effect," change expression of a gene due to change in the linkage arrangement. Different products can form during inversion; the two non–sister chromatids that were not involved in crossing-over would form normal gametes. Crossovers would form a dicentric and an acentric chromatid. Thus, the product of inversion would be chromosomes with deletions and duplications. Inversions could either include centromeres within the loop (**pericentric inversion**) or the centromere could be outside the loop (**paracentric inversion**); the most common outcome of both scenarios is formation of nonviable gametes.

Often during DNA damage, the telomeres are lost making DNA ends sticky, which fuse together, and during anaphase upon migration to respective poles, there occurs unequal breakage of chromosome generating acentric and dicentric fragments. The acentric fragment is lost, while the dicentric fragment is pulled to opposite ends, leading to breakage of the fragment. This yields an imbalanced product as the breakage does not necessarily occur at the center of the chromosome and creates duplications in one strand called **duplication mutation** and deletions in the other (Fig. 1.3B) [15].

3.2 Breaks in Nonhomologous Chromosomes

Often nonhomologous chromosomes can break and join to produce different gene products. When the ends of nonhomologous chromosomes are translocated, it is referred as reciprocal translocation (Fig. 1.4A). The outcome of translocation could be the same as that occurring during inversion and meiosis: a cross-shaped pattern is observed (Fig. 1.4B). Reciprocal translocation heterozygotes fail to produce viable progeny [16]. In meiotic divisions, separation of centromeres during recombination may result in various types of segregations such as "alternate," "adjacent one," or "adjacent two" (Fig. 1.4C).

For alternate segregation, the first centromere is segregated with the fourth centromere allowing the second and third centromeres to move to opposite pole, thus producing one normal gamete and another with reciprocal translocation [17].

In Adjacent 1 type segregation, the first and third centromeres segregate together and move in a direction opposite to the second and fourth. Here, both types of gametes have either deletion or duplication and thus are rendered lethal. A third outcome would be Adjacent 2 type segregation, wherein the homologous chromosomes move to the same pole, meaning the first with second and the third with fourth. Thus, reciprocal translocations are set in new linkage arrangements and often semisterility [17].

Cytologist W. Robertson described that often two acrocentric chromosomes join closely to their centromeres via a process named "Robertsonian fusion." The result of such translocation is the decreased number of chromosomes without affecting the amount of genetic content. Changes in chromosomes may also occur via variation in chromosome number. Such anomalies could give rise to **euploidy** that entails changes in the whole set of chromosome or **aneuploidy** that involves

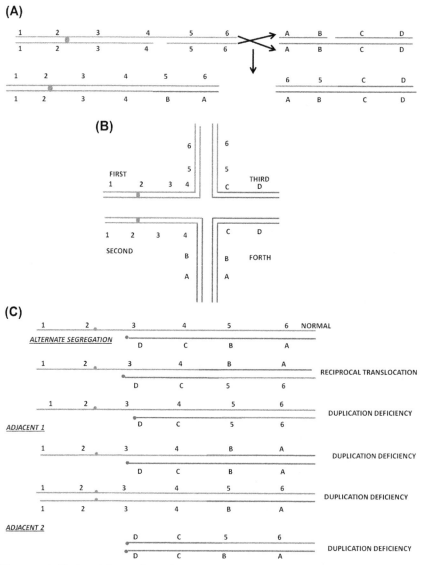

Figure 1.4 (A): Break in nonhomologous chromosome showing reciprocal translocation. *Red lines* (light gray in print versions) and *blue lines* (gray in print versions) depict the nonhomologous chromosomes and gaps depict break in chromosomes (above) and exchange of nonhomologous chromosomes (below). (B): Synapsis at meiosis yields a cross-shaped figure. (C): The possible outcome of various translocation types during break in nonhomologous chromosome (*red* (light gray in print versions) and *blue lines* (gray in print versions) show the two nonhomologous chromosomes).

changes in chromosome number by addition or deletion of less than a whole set of chromosomes (Fig. 1.2).

Aneuploidy is the most common form of chromosomal abnormality in humans and is responsible for the highest number of congenital birth defects or miscarriages. Intense research work has been done in this field to identify the possible reasons for such outcomes [18]. The first aneuploid disorder was identified almost six decades ago [19]. Aneuploidy occurs due to nondisjunction of chromosomes at first division or second division of the meiotic cycle, during which the sister chromatids tend to adhere together and do not separate during the normal disjunction process. The various types of aneuploidies include nullisomy $(2n - 2)$, monosomy $(2n - 1)$, disomy $(n + 1)$, and trisomy $(2n + 1)$. Diseases or human abnormalities due to aneuploidy will be discussed later in the chapter.

4. DNA DAMAGE RESPONSE AND REPAIR

Whenever damage is encountered in DNA strand, cells activate the DNA damage response to guard its genome integrity, which includes sensing the damage (sensors), transducing the signals (transducers), and enabling the effectors to initiate the DNA damage repair (effectors). The transducers involve ATM (ataxia-telangiectasia mutated), ATR (ATM- and Rad3-related), and DNA-PKcs (DNA-dependent protein kinase), all belonging to the phosphatidylinositol 3 kinase like kinase family [20] (Fig. 1.9).

Once DNA is damaged by any intrinsic or extrinsic insults, several proteins are phosphorylated at Ser/Thr-Glu motif in an ATM- or ATR-dependent manner [21]; while DNA-PKcs regulate protein targets that play a role in nonhomologous end joining (NHEJ) repair as described below [20].

On detection of DNA damage, both ATR and ATM, along with other partners, initiate a signaling cascade involving Chk1 and Chk2 ser/thr kinases along with Cdc25 phosphatase.

These transducers then activate p53 (which in a normal cell cycle is inactivated by Mdm2) that binds the regulatory region of p21 gene, activating it. This activated p21 is an inhibitor protein, which inhibits cell cycle progression by preventing cells from dividing and arresting them at checkpoints (regulatory points which ensure normal cell division) either at G1 to S transition (at G1/S checkpoint) or G2 to mitosis transition (at G2/M checkpoint) [22]. If these protein sense that the damage is large enough to be repaired, they can even target the cell to apoptosis.

5. DNA REPAIR PATHWAYS

DNA repair pathways are majorly classified into direct repair, single strand DNA damage repair, and DNA double strand repair (Fig. 1.5). Due to space limitation, we present an overview of the pathways involved in brief in following section while we refer for more detailed description of DNA repair pathways.

5.1 Direct Repair

DNA adducts (thymine dimers), formed by UV irradiation, are repaired by the direct repair mechanism that involves two processes: reversal of UV-induced pyrimidine dimers by *DNA photolyase* and removal of O^6 methyl group from O^6 methylguanine in DNA by *methylguanine DNA methyltransferase* (MGMT). DNA photolyase is, however, absent in humans while MGMT is evolutionary conserved [22]. *Photolyase* is a monomeric protein of 55–65 kDa consisting two chromophore cofactors: a methyltetrahydrofolate and a flavin in the form

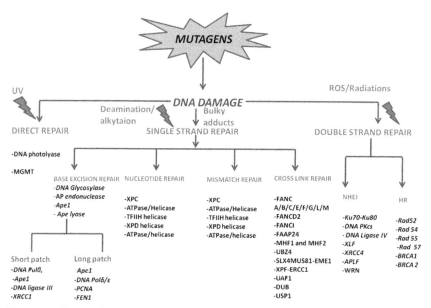

Figure 1.5 Types of mutagens causing DNA damage, the respective repair pathways, and various enzymes associated with these pathways.

of flavin adenine dinucleotide. To repair such dimers, it uses blue light that is absorbed by the folate group [23] and transferred to the flavin molecule, which then transfers the electron to the pyrimidine dimer forming a dimer radical. The dimer radical is cleaved into monomeric units, and electrons return to the flavin radical regenerating it to its original state (Fig. 1.6A). MGMT is a small protein of 20 kDa without any cofactors and diffuses among the strand of DNA to form a low-stability complex with the damaged site and flips out the O^6MeGua (formed due to attachment of an oxygen atom to the methyl group rendering a faulty base pairing of thymine rather than cytosine) base into the active site of enzyme [24], thus transferring the methyl group to the active cysteine molecule (Fig. 1.6B). After repairing DNA damage, MGMT dissociates but the C—S bond in the complex is stable enough that it fails to break and renders the enzyme inactive after one cycle.

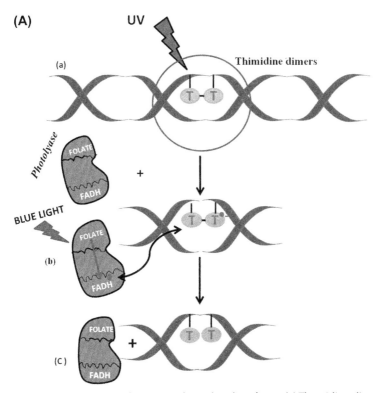

Figure 1.6A Direct repair mechanism pathway by photolyase. (a) Thymidine dimers are formed, upon UV exposure, which are resolved by photolyase. (b) Photolyase uses blue light (gray in print versions) as their sole source of energy and transfers electron from folate to flavin adenine dinucleotide to the dimer. (c) This process repairs the DNA and photolyase enzyme is recycled for next set of reaction.

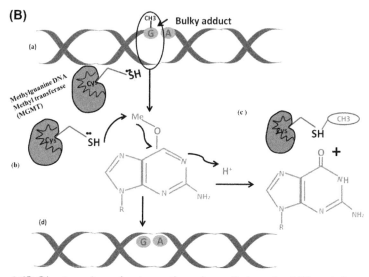

Figure 1.6B Direct repair mechanism pathway by *methylguanine DNA methyltransferase* (MGMT). (a) The methylation of base pairs is resolved by MGMT enzyme, (b) The cysteine residue interacts with the methyl group of the base pair. (c) The methyl group is transferred to the cysteine residue. (d) The DNA is repaired, but the enzyme becomes inactive and cannot be reused further.

5.2 Single-Strand Break Repair

5.2.1 Base Excision Repair

Base excision repair (BER) is the major pathway that repairs point mutations in DNA caused through chemical modifications such as products of oxidized or reduced base, alkylated bases, deaminated bases, or mismatched bases [22]. This pathway is essential for maintaining the genome integrity and prevents premature aging or cancer. The mechanism has been explained in detail elsewhere [25]. In brief (Fig. 1.7A), the major enzyme of this pathway is *DNA glycosylase,* which has 11 isoforms and each type has a specific role in particular type of lesions. *DNA glycosylase* hydrolyze the N-glycosidic bond between deoxyribose of DNA and base, forming an AP site to be cleaved by an AP *endonuclease,* which hydrolyzes the phosphodiester bond present at the 5′ of the AP site, thereby removing the altered base and generating a single-strand break [26]. If the cleaved site generates a single nucleotide, the repair occurs via "short patch base excision repair" and is often initiated by glycosylases, while if the cleaved site generates multinucleotides, repair occurs via "long patch excision repair" wherein AP sites are generated by oxidative base loss. "Short patch" utilizes *DNA Polβ, Ape1,* and *DNA ligase*

III-XRCC1, while "long patch" requires *Ape1, DNA Polδ/ε, PCNA, and FEN1.* As the "long patch" generates a flap of 2–10 nucleotides, the junction of single to double strand displacement is cut by FEN1 endonuclease. Another patch of equivalent size is synthesized by Polδ/ε, aided by PCNA and is later ligated by DNA ligase (Fig. 1.7A).

5.2.2 Nucleotide Excision Repair

Nucleotide excision repair (NER) is mainly involved in removal of bulky DNA distortions such as dipyrimidinic photolesions that are formed after

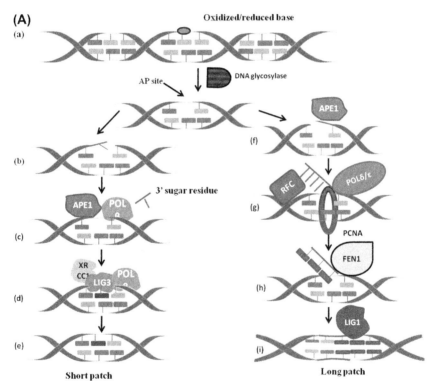

Figure 1.7A Base excision repair mechanism in mammals. (a) The damaged base (oxidized/reduced) is removed by DNA glycosylase generating AP site. The repairing mechanism could be processed by two pathways: short patch (single nucleotide) repair pathway (b–e) and long patch (2–10 nucleotide) repair pathway (f–i). (b) After removal of damaged base, APE1 cleaves 5′ bond, (c) pol β is recruited to fill in the gap, (d) ligation is done via Lig3/XRCC1 complex. (e) Repaired DNA by short patch repair pathway. (f) Cleavage is done by APE1 endonuclease. (g) The RFC-PCNA and pol δ/ε complex completes the repair synthesis along with nick translation removing several nucleotides. (h) Flap is processed by FEN1 nuclease. (i) The patch is ligated by Ligase 1.

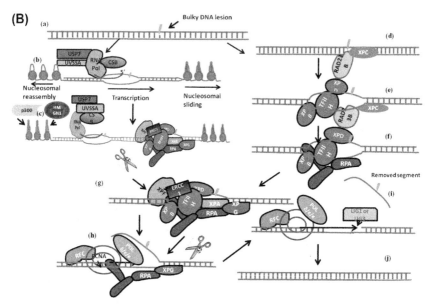

Figure 1.7B Mechanism for nucleotide excision repair (NER). (a) The bulky DNA lesions destabilize the DNA duplex. The lesions if detected during transcription leads to transcription arrest and is repaired via transcription coupled NER(TC-NER) (b, c, g–j) or if lesion is not in the transcribing strand then the damage is processed via global genomic NER(GG NER) (d–j) (b). In TC-NER, factors such as CSB and UVSSA/USP7 complex travel along with RNAPII. In case of lesion, RNAPII is stalled and CSA–RNAPII interaction is stabilized followed by UVSSA/USP7 assembly. (c) CSB triggers recruitment of CSA and assist the chromosomal remodeling events in aid with p300 and HMGN1. (d) In GG-NER, the lesion is recognized by XPC-RAD23B, which binds to the opposite side of damaged DNA. (e) TFIIH is recruited, which cuts open the DNA, aided by XPB subunit. (f) XPD subunit of TFIIH scans the DNA and stalls at the site of damage, thus allowing recruitment of XPA, RPA, XPG. (g) This is followed by ERCC1-XPF recruitment and its interaction with XPA leads to 5′ incision. (h) This is followed by 3′ end cleavage by XPG to remove the damaged segment, and repair synthesis is done by Pol δ and Pol κ or Pol ε. (i) The synthesis is completed by sealing of the nick by DNA ligase I or III/XRCC. (j) The repaired DNA.

exposure to UV radiation or protein adducts to DNA. NER comprises global genomic NER (GG NER) or transcription coupled (TCR) NER pathway (Fig. 1.7B). The detailed mechanism of this pathway is described earlier (review articles [27]). In short, the basic steps include (1) damage recognition, (2) dual incision bracketing of lesion, (3) release of oligomer, (4) gap filling, (5) ligation. The main enzyme of this pathway is XPC, which on loss of function mutation results in xeroderma pigmentosum, a disorder causing skin cancer and is key player in damage recognition

[27]. The detailed NER process and the factors involved in each step are depicted in Fig. 1.7B.

The other type of NER, i.e., TCR, which mainly repairs the lesions that are on DNA strands of active transcribed genes. Lesions are recognized initially by translocating RNA polymerase (RNAP). Other factors such as *Cockayne syndrome B (CSB), XPG, XPA-binding protein (XAB2), CSA*, high-mobility group nucleosome binding domain contacting protein 1 (*HMGN1*), *TFIIS*, and *p300* in association with RNAPII are vital for TCR. CSB recruits CSA, consisting of an E3 ubiquitin ligase complex of DDB1-CUL4-RBX1 (CRL4), which aids recruitment of XAB2 [28]. Transcription elongation factor TFIIS activates the cryptic nuclease activity RNAPII, which cleaves nascent transcript allowing the 3′ end to become again orientated at the active site starting stalled transcription. HMGN1 (a nucleosome-binding protein) enhances acetylation of the LYS14 residue of histone H3, P300 plays a role in chromatin remodeling, while CAF1 is a prerequisite for reestablishment of the nucleosome structure postrepair mechanism [29]. Defects in NER repair proteins are the basis of many developmental disorders and pathologies [30].

5.2.3 Mismatch Repair

Often during replication, the wrong base pair is incorporated in the strand, and this leads to faulty base pairing among the two strands. This error is corrected by mismatch repair pathway. The detailed mechanism has been described in Ref. [31]. In brief, the mismatch is recognized by MutS, a heterodimer of two subunits, Msh2 and Msh6, which have ATPase activity (Fig. 1.7C). This is followed by recruitment of MutLα, comprising of two subunits, MLH1 and PMS2. Simultaneously, replication factor C (RFC) loads PCNA, the sliding clamp, to DNA and thus activates MutLα and leads to incision of the strand in an ATP-dependent manner. The repair process is then continued by DNA polymerase δ, the new strand is synthesized by RPA and RCF and is finally ligated by Ligase.

5.2.4 Interstrand Cross-Linking Repair

DNA intrastrand cross-link repair is associated with DNA replication and involves factors such as endonucleases, recombinases, DNA polymerases, fanconi anemia (FA), along with other factors [32]. There are at least 15 FA gene products, which participate in this pathway; the core FA protein comprising of FANC A/B/C/E/F/G/L/M, activates FANCD2 and FANCI (Fig. 1.7D). FANCM is known to initiate the pathway, forming a

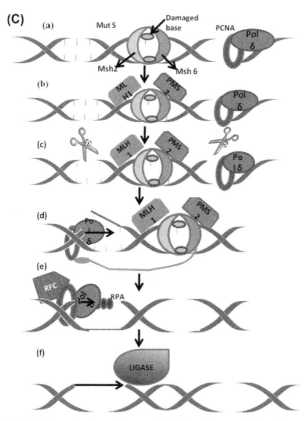

Figure 1.7C Mechanism of mismatch repair. (a) The damaged base is recognized and processed after initiation by MutS which comprises of two subunits Msh2 and Msh6. (b) MutLα binding occurs, which has two subunits MLH1 and PMS2. (c) Proliferating cell nuclear antigen (PCNA) is then recruited to the DNA that activates MutLα allowing it to nick the daughter strand. (d) The ends of DNA are processed, (e) Resynthesis of daughter strand occurs by RPA and RFC. (f) Finally the strand is ligated by ligase enzyme.

complex with FAAP24, which plays pivotal role in activation of the pathway and recruitment of the FA core complex, thus stabilizing the stalled replication fork. This complex also initiates the ATR-mediated checkpoint signaling. Factors such as histone fold protein (MHF1) and MHF2 binds to chromatin and forms a stable association of FANCM with DNA [33]. Monoubiquitinated FANCD2-I (ID) heterodimeric complex is then recruited to the damaged DNA where it controls recruitment of other downstream proteins such as FA proteins (D1/J/N/O) (Fig. 1.7D). This pathway uses multiple nucleases and their recruitment is aided by monoubiquitinated FANCD2 [34]. All these nucleases have a characteristic

(D)

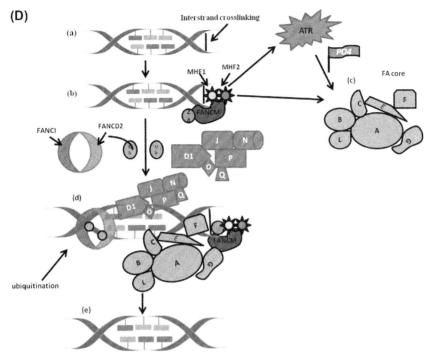

Figure 1.7D Mechanism of interstrand cross-link. (a) On encountering an interstrand cross-link, the two replication forks converge. (b) Factors such as FAAP24, FANCM, and MHF1/MHF2 recognize the stalled replication fork. (c) FANCM is also known to activate ATR/CHK1 dependent checkpoint response, which in turn phosphorylates several FA proteins thus forming the core protein complex. (d) A E3 ubiquitin ligase then monou-biquitinates FANCD2 and FANCI, and this complex is loaded to the damaged DNA. (e) E3 ubiquitin ligase also serves as a platform to recruit other FA proteins (red (gray in print versions)) along with core complex. Finally the damaged is removed and DNA is repaired.

ubiquitin-binding domain called UBZ4 (ubiquitin binding zinc finger 4), which helps recognize the ubiquitin moiety of FANCD2. Nucleases SLX4–MUS81-EME1 and XPF-ERCC1 allow unhooking of the cross-link, thus converting a stalled replication fork to a double-strand break (DSB) [35], and translesion DNA synthesis allows bypassing the unhooked cross-linked oligonucleotide restoring the nascent strand. DSB are repaired via homologous recombination (HR), in association with FA protein D1/J/N/O/P/Q and NER removes the remaining adducts. Modified ID complex is deubiquitinated by the deubiquitinating (DUB) enzyme USP1 (ubiquitin specific peptidase 1), along with UAF1 (USP1-associated factor 1), and thus repair is complete (Fig. 1.7D).

5.3 Double-Strand DNA Breaks Repair

ROS, which are generated by chemicals and ionizing radiation, tend to form free oxygen radicals that react with DNA and causing its damage. DSB repair could be mediated by either NHEJ mechanism (Fig. 1.8A) or HR (Fig. 1.8B). The former joins ends of the DSB while the latter uses an undamaged copy of the parental chromosome [36]. HR allows the retrieval of information of broken DNA from the homologous strand, but it takes place only during S and G2 phase unlike NHEJ, which can function independent of cell cycle stage. Basic steps of NHEJ include (1) end recognition of the broken segment of DNA assembly

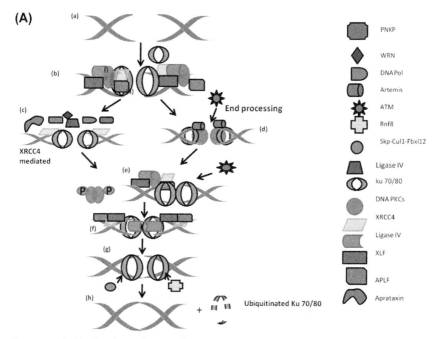

Figure 1.8A Mechanism of DNA double-strand break nonhomologous end joining (NHEJ). (a) A DNA double-strand break, (b) Ku 70/80 is recruited to damaged site of the DNA, followed by recruitment of DNA-PKs, XRCC4, Ligase IV, XLF, and APLF. These factors interact with each other to form a stable complex at DSB. DNA repair is processed either by (c) mediated wherein XRCC4 acts as scaffold in association with Ku70/80 allowing other proteins Aprataxin, ARLF, Ligase IV, PNPK, WRN, and DNA pol to be recruited (d) or DNA repairing could also be processed by Artemis, requiring ATM. These steps ensure ends of DNA to become ligatable. The ends of DNA are ligated and NHEJ complex is dissolved. (e) ATM phosphorylates DNA PKCs. (f) Phosphorylated DNA PKcs breaks open its pincers and releases the DSB. (g) The terminal ligation is aided by Rnf8 and Skp-Cul1-Fbxl12. (h) Ku 70/80 is degraded and DNA is repaired.

and stabilization of NHEJ complex, (2) end stability by bridging DNA ends, (3) processing of DNA ends, (4) broken end ligation and dissolution of the complex [37]. The NHEJ pathway successfully repairs DSBs involving a number of enzymes functioning at tandem as described in Fig. 1.8A. The detailed mechanism and function of each enzyme is explained separately in Refs. [38], [39].

HR is usually completed in three steps: strand invasion, branch migration, and Holliday junction formation. Holliday complex is resolved to duplexes by *endonuclease and resolvase* (Fig. 1.8B). Strand invasion and

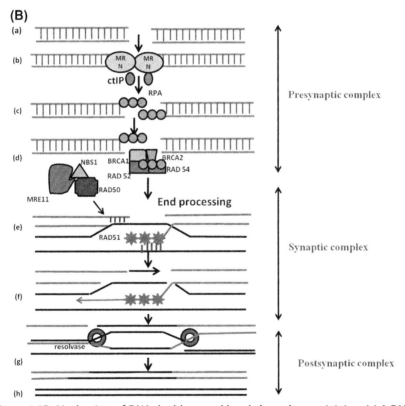

Figure 1.8B Mechanism of DNA double-strand break, homologous joining. (a) A DNA double-strand break. (b) MRN and CtIP are recruited to damaged site. (c) Processing by RPA to remove any secondary structure. (d) Proteins BRCA1-BRCA2, RAD54/RAD52 along with NBS1-MRN11 and NBS1 are recruited, which help in resection generating 3′ ends. (e) RAD51 is loaded that finds suitable homologous DNA and forms D loop, (f) Holliday junctions are formed which are resolved by the MUS81-MMS4 heterodimer. (g) Resolving is done by the enzyme resolvase, thus generating a cross-over product (h) and a normal DNA.

branch migration are mediated by Rad51 [40]. Enzymes such as *Rad52, Rad54, Rad55, Rad57, BRCA1, and BRCA2* are also critical for the HR pathway. Broken DNA ends are processed by the MRN complex (Mre11, Rad50, NBS1) before strand invasion by Rad51. The resulting Holliday junction is then resolved by the MUS81-MMS4 heterodimer [41] (Fig. 1.8B).

6. GENETIC DISORDERS AND DISEASES

The manifestation of any type of unrepaired or incorrectly repaired DNA is the incorporation of mutations, gene deletions, or insertions resulting in DNA sequence distortions ultimately leading to inherited genetic modifications. The various disorders that occur include developmental disorders, neurodegenerative disorders, or predisposition to cancers, which are described in the section below. Some of the common diseases, type of mutation, gene involved, and the symptoms are included in Table 1.1.

Table 1.1 The various common diseases due to mutation, type of mutation, genes involved and the symptoms

Name	Types of mutation and chromosome	Genes involved	Symptoms
Charcot–Marie–Tooth disease	Duplication (19p)	DNM2, YARS MPZ	Genetic disorder of peripheral nervous system, associated with progressive weakness and atropy of peroneal muscle of arms and later distal muscles [83].
Color blindness	Point mutation (X)	OPN1LW, OPN1MW, OPN1SW	Patients have red, blue, or green cones loss and cannot differentiate among these colors. Males are more affected than females [84].
Cri du chat	Deletion of short arm (5p)	SEMAF, CTNND2	High-pitched cat-like cry, typical facial dysmorphism, microcephaly, mental retardation, micrognathia, down slanting palpebral fissures.

Table 1.1 The various common diseases due to mutation, type of mutation, genes involved and the symptoms—cont'd

Name	Types of mutation and chromosome	Genes involved	Symptoms
Cystic fibrosis	Point mutation (7q)		CFTR gene functions as chloride ion channel, effects mucus layer lining in epethelial surface [85].
Down syndrome	Aneuploidy trisomy (21)	GATA1 gene JAK 2	Mental retardation [86], atrioventricular septal defect, duodenal stenosis, and Hirschsprung disease. Rare disease risk increases with increased age of mother.
Duchenne muscular dystrophy	Deletion (X) Point mutation Duplication	DMD gene RAB40AL PLP1	Mutation occurs in dystrophin protein, absence of which makes glycoprotein structure of muscle sarcolemma less stable. Muscle damage, necrosis, affecting skeletal and cardiac muscle primarily.
Klinefelter syndrome	Aneuploidy (X) XXY, XXXY; XXYY; XXXXY		First described in 1942 as a endocrine disorder, characterized by gynecomastia, hypogonadism, small firm testes, abnormally high concentration of FSH, muscle weakness, reading abnormalities etc. [87].
Sickle cell anemia	Point mutation (11p1 (GAG to GTG)	HBB	Disease associated with hemoglobin polymerization, erythrocyte rigidity, vasoocclusion, chronic anemia, hemolysis. There is a variant of hemoglobin present, the HbS, where glutamic acid is substituted by valine, leading to loss of elasticity of RBC [88].

Continued

Table 1.1 The various common diseases due to mutation, type of mutation, genes involved and the symptoms—cont'd

Name	Types of mutation and chromosome	Genes involved	Symptoms
Huntington Disease(HD)	Trinucleotide repeat (CAG)n (4p)	HTT	Autosomal dominant neurodegenerative disorder characterized by chorea, incoordination, cognitive decline, dystonia and behavioral difficulties, selective neuronal loss, atrophy in caudate and putamen [89].
Tay Sachs disease (TSD)	Point mutation (15p)	HEXA	Autosomal recessive neurodegenerative disorder with mutated α subunit of hexosaminidase A gene. Mental retardation from age of 2 years, dementia, paralysis, and ultimately death at an early age. The missing enzyme is vital for breakdown of glycolipids in lysosomes and results in accumulation of protein the brain thus affecting the normal physiological function of brain [90].

6.1 Noonan Syndrome (OMIM 163950)

Noonan syndrome [71] is an autosomal dominant developmental disorder named after Noonan, a pediatric cardiologist. Symptoms include congenital cardiac disorder; muscular, skeletal, and cutaneous aberrations; endocrine anomalies as well as bleeding disorders. Patients with this disorder have distinctive clinical hallmarks such as broad forehead, down slanting palpebral fissures, hypertelorism, ptosis, and high arched palate along with short stature. Patients normally die at an early age. A missense mutation in PTPN11 gene at chromosome 12q24.1 is the primary cause for anomaly in half of the clinically examined patients. PTPN11 encodes a nonreceptor protein tyrosine phosphatase SHP2, which regulates cell division, cell growth, and cell movement, thus playing an important role in Ras-MAPK signaling pathway. SHP2 N308D substitution makes it active for a longer time in response to stimulation [42]. This prolonged activation elevates the Ras-MAPK pathway signaling, thus disrupts the regulation of cell growth and cell division resulting in characteristic Noonan syndrome features. Deregulation of SOS1, RAF1, and KRAS also has been reported to be associated with a small percentage of Noonan syndrome individuals. The detailed mechanism underlying the manifestation of this disease is explained by Romano et al. [43].

6.2 Costello Syndrome (OMIM 218040)

Costello syndrome is a rare congenital disorder with patients associated with increased birth weight, craniofacial dysmorphology, neurocognitive delay, cardiac anomalies, and grim chances of survival [44]. Typical features include prominent macrocephaly, down slanting palpebral fissures, short nose characterized with depressed nasal bridge, and a terminating broad tip. Dermatological features include soft and excessively wrinkled skin, deep plantar and palmer creases. The disease has been explained in detail by Gripp et al. [45]; briefly, a missense mutation in HRAS [46] at G12S accounts for about 80% of this syndrome. Rare mutations are also reported in HRAS at T58, L117, and A146 [47] that hamper the normal functioning of guanine nucleotide binding thereby leading to a reduced GAP-induced GTPase activity and permanent activation of RAS, thus an increased MAPK activity (Fig. 1.9).

6.3 Cardio–Facio–Cutaneous Syndrome (OMIM 115150)

This disease entails an abnormal phenotype with characteristic craniofacial dysmorphy, cardiac defects, and ectodermal abonormalities. Marked features include broad forehead, sparse curly hair with fewer eyebrows and eyelashes, hyperkeratosis, keratosis pilaris, and ichthyosis [48]. Cardiac disorders include pulmonic stenosis, hypertrophic cardiomyopathy and septal defects with poor life expectancy. Neurological anomalies are a typical hallmark, which make a cardio–facio–cutaneous (CFC) patient distinct from Noonan's and include speech, learning disability, motor delay, and hypotonia [49]. Such individuals have altered Ras-MAPK pathway, and mutations in BRAF, MAPK1, MAPK2, and KRAS have been identified [50]. Mutations in BRAF genes account for almost 80% cases of CFC syndrome. Mutations in exons Q257R, E501, and G469E of BRAF gene make the protein overactive that alters tightly regulated RAS-MAPK

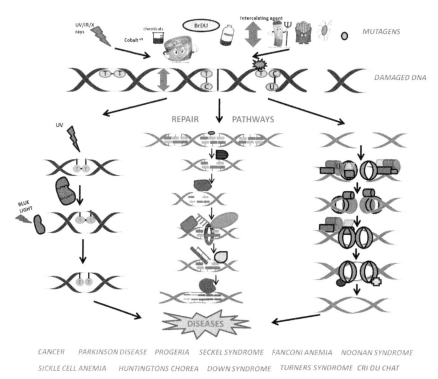

Figure 1.9 Various types of DNA damage, their causative agent and mechanism of repair.

signaling pathway. This transformed signal in turn interferes with cell growth resulting in alteration of organ and tissue development and thus giving rise to CFC syndrome.

6.4 Hirschsprung Disease (OMIM 142623)

Hirschsprung disease (HSCR) is an anomaly of early development that entails the absence of enteric neurons from variable lengths of the distal bowel. Since enteric neurons are imperative for intestinal mobility, babies with this disorder often complain of severe constipation and distended abdomen. The disease is described in detail by Trainor [51]. Mutations in either of RET, EDNRB, or EDN3 gene causes manifestation of the disease. Mutation in RET genes, which are crucial for neural crest development, results in a nonfunctional version of the RET protein that fails to transmit signals within cells causing improper development of enteric nerves which, in turn leads to intestinal problems, a characteristic of HSCR [52].

6.5 Hutchinsons–Gilford Progeria Syndrome (OMIM 176670)

Hutchinsons–Gilford progeria syndrome (HGPS) is attributed to defects in the mature form of Lamin A/C. Lamins are the major architectural protein of the nucleus in animal cells and function to bind chromatin as well as to provide mechanical stability [53]. Lamins are encoded by LMNA gene, primarily Lamin A and C. There is a single base substitution of LMNA at exon 11. Transition of C to T at chromosome 1824 activates the otherwise cryptic splice sites resulting in an in-frame deletion of about 50 amino acids around the C terminus of prelamin A and this unusual protein product formed is called "progerin," which retains a CAAX motif. Progerin causes nuclear morphological abnormalities, cellular decline, mitochondrial dysfunction, chromatin changes, faulty DNA repair, hastened telomeric shortening, and premature senescence. The symptoms of premature aging appear since childhood in a manner where phenotypic expression of a 6-year old kid would resemble that of a 70-year old person. Such children grow slower than normal kids, fail to gain weight, and have poor life expectancy. Typical features include prominent eyes, aged wrinkled skin, small chin, alopecia, arteriosclerosis, and joint abnormalities. However, affected patients have normal intellectual development as well as motor skills, but fail to thrive and exhibit a shorter life span [54]. The disease is explained in detail by Gonzalo, S. and J.C. Eissenberg [55].

6.6 Cancer

Normally, cells of the body are predestined to execute a particular function in the body. They divide in an orderly fashion, and once they are worn out or damaged, cells undergo apoptosis, an exquisite process finetuned by a number of enzymes, thereby ensuring homeostasis in the body. Whenever the DNA is damaged, the cell cycle halts at the checkpoints and give the repair machinery enough time to rectify the damage. If, however, the damage is not repaired, the cell undergoes apoptotic death thus preventing any hazardous effect on human system. But some cells have devised mechanisms to evade such machinery and grow indefinitely. This unregulated growth of cells is often described as **cancer**. Cancers occur due to mutation in two major types of genes: tumor suppressor genes and protooncogenes. Protooncogenes assist cells to grow through gain of function mutation or overexpression. Known protein products of protooncogenes include tyrosine kinases, growth factors, GTP-binding proteins, and DNA-binding proteins [56]. Under physiological conditions, tumor suppressor genes regulate major cell functions involving cell duplication and cell growth. Loss of function mutations or deletion of these genes perpetually free cells of the checkpoints, leading to unchecked growth as is the case of protooncogenes and ultimately stimulated cell proliferation [57]. The first known tumor suppressor gene to be isolated was the retinoblastoma (Rb) gene. Some extensively studied tumor suppressor genes include rb, p53, wt-1, nf-1, and men1 [58]. Till now, there are almost 100 types of cancers known (NCI, 2014), and statistical data submitted by the American Cancer Society revealed that "Cancer accounts for the second most common cause of death in the US" (Cancer facts and figures, 2015). One out of four individuals would probably die of cancer, The World health Organization revealed that worldwide, there were 14 million new cancer cases and 8.2 million cancer-related deaths in 2012 (Cancer facts and figures, 2015). Types of some cancers, genes or proteins affecting them, pathways involved, and other factors are all described in Table 1.2.

Table 1.2 List of various types of cancers, the affected gene, and the mutated pathway

Types of cancer	Affected gene	Mutated pathway
Breast cancer	BRCA1 and BRCA2 ebrB2 (HER2 or neu)	Tyrosine kinase [91]
Bladder cancer	MCM4, ERCC2, ATM, FANCA, STAG2, FGFR3 HRAS or KRAS ERBB2 and ERBB APC, CTNNB1	RAS-MAPK pathway [92] or PI3K pathway Hedgehog and WNT signaling
Cervical cancer	p53 and E2F TF, histone deacetylases components of the AP1 transcription complex and the cyclin-dependent kinase inhibitors p21 and p27	ERBB growth-factor receptors [93] RAS/MAP kinase pathway and the phosphatidylinositol 3 kinase- AKT pathway
Colorectal cancer (CRC)	APC gene, β catenin, and axin	Wnt pathway or TGF beta [94]
Lung cancer	KRAS, EGFR, BRAF, HER2, ALK, ROS1, DDR2, TP53, PTEN, RB1, LKB11, and p16/CDKN2A	RAS/RAF/MEK/MAPK or PI3K/AKT/mTOR JAK/STAT [95]
Pancreatic cancer	K-ras p16, p53	(PI3K)/AKT/mammalian target of Rapamycin (mTOR), Raf/mitogen-activated protein kinase (MAPK) or nuclear factor Kappa B (NF–B) [96]
Prostate cancer	AR, GSTP1, PTEN, TP53, NKX3.1	EGF NK-κB Wnt [97]

6.7 Parkinson Disease (OMIM 168600)

PD is the second most common age-related neurodegenerative disease after Alzheimer disease. The cause of PD is unknown and is sporadic in 95% cases, while genetic mutations account for the remaining 5% of cases. The main anatomical feature of PD is the degeneration of dopaminergic neurons in the SNpc region of the midbrain [59]. Disease symptoms include tremor, bradykinesia, akinesia, rigidity, hypomimia, and micrographia. Loss of nigrostriatal dopaminergic neurons and presence of intraneuronal Lewy bodies (proteinaceous cytoplasmic inclusions, discovered by Frederic Lewy in 1972) are the pathological characteristics of PD. Five percent of PD can be caused by the mutation in SNCA, LRRK2, PINK1, PINK2/PINK7, and VPS35. The SNCA gene encodes α-synuclein protein, which is crucial for dopamine formation [60]. These genes play important roles in autophagy, striatal neurotransmission, mitophagy, and retrograde protein transport. Thus, mutated form of these genes hampers the normal functioning of proteins and leads to manifestation of PD. The disease has been explained in detail by Trinh and Farrer [61].

6.8 Alzheimer Disease (OMIM 104300)

Alzheimer disease is an irreversible and progressive brain disorder, which hampers a person's memory and thinking ability, thereby preventing the patient from performing even the basic day-to-day routine activities. Reports suggest that more than 5 million of the American population are affected by this disorder, and this disease ranks third as the leading cause of death after heart disease and cancer. There is unusual deposition of protein plaques, called amyloid bodies and tau tangles, throughout the brain. The disease is explained in detail by Van Cauwenberghe [62]. In brief, the damage is initially localized to the hippocampus of brain, which is significant for memory, but as the disease progresses, neurons die affecting additional parts of the brain and finally leading to extensive shrinking of brain tissue [63]. The mutated form of genes of Aβ precursor protein (APP), presenilin 1 (PSEN1) [64], and presenilin 2 (PSEN2) is known to set in Alzheimer disease. ADAM10, which encodes α-secretase in brain, is required for cleaving the Aβ domain of APP precluding β amyloid formation. Mutations in Q170H and R181G impair the normal activity of α-secretase in the brain.

6.9 Ataxia-Telangiectasia (OMIM 208900)

Ataxia-telangiectasia (AT) is a rare, autosomal recessive disorder characterized by cerebellar ataxia, immune defects, a predilection for malignancies, and telangiectasia or spider veins (OMIM 607585). The disease is caused by mutation in the ATM gene with progressive cerebellar ataxia followed by conjunctival telangiectasia, neurological degeneration, and malignancies that develop between the age of 3 and 5 years. These symptoms are often preceded by oculomotor apraxia. AT patients have progressive spinal muscular atrophy mostly affecting hands and feet [65] and have a proclivity toward malignancy [66]. Detailed description of the disorder is described by Mark Ambrose and Richard A. Gatti [66].

6.10 Seckel Syndrome (OMIM 210600)

Seckel syndrome is a rare autosomal recessive disorder with stunted growth with intrauterine growth retardation, microcephaly with mental retardation, dwarfism, and characteristic "bird-headed" facial features as pathological hallmarks [67]. Hypomorphic mutations in ATR, FA, ATRIP are major players in Seckel syndrome. The disease is explained in detail by Al-Dosari, M. S [68].

6.11 Cockayne Syndrome (OMIM 216400)

Cockayne syndrome is a rare autosomal recessive disorder characterized by microcephaly, delayed development, progressive neurological dysfunction, short stature, and cataract development [69]. The symptoms become apparent at infancy and worsen over time. The disease could be CSI, CSII, or CSIII type depending on the clinical severity. Of these three, CSII is the severe form wherein prenatal developmental disorders are manifested [70]. Such patients have either CSB/ERCC6 or CSA/ERCC8, both of which are crucial for the TC-NER. Recent studies show that mutated ERCC1 or ERCC4 (XPF), both of which are factors for TC-NER, along with functional ERCC8/ERCC6 could also play a role in this anomaly [70]. The anomaly has been discussed in details by V. Laugel and C. Dalloz [71].

6.12 Fanconi Anemia (OMIM 227650)

Fanconi anemia is a typical anemic condition with bone marrow failure, physical abnormalities, organ defects, and an increased risk of certain cancers [72]. Affected people develop extreme fatigue, frequent infections due to neutropenia, and clotting problem due to thrombocytopenia. They also exhibit an increased risk of developing acute myeloid leukemia as well as tumors of the head, neck, skin, gastrointestinal system, or genital tract [73]. Mutation in 15 genes involved in Fanconi anemia, of which FANCA, FANCC, and FANCG, results in 80%–90% of the disease prevalence. These genes are involved in the FA pathway (a mechanism to repair interstrand cross-links, as discussed in the "DNA Repair Pathways") and constitute components of the FA core complex. Mutation in any of these genes causes the core complex to be nonfunctional and disrupts the entire FA pathway. This leads to inefficient DNA damage repair resulting in an inability to make new DNA or uncontrolled cell growth and eventually cells that divide quickly (such as bone marrow cells or developing fetus cells) are affected. Ultimately, decrease in blood cell levels and physical abnormalities occur, which define the characteristics of Fanconi anemia. The disease is explained in more detail by Toshiyasu Taniguchi and Alan D. D'Andrea [74].

6.13 Bloom Syndrome (OMIM 210900)

Bloom syndrome is a rare autosomal recessive disorder, characterized by short stature; development of telangiectasia after sun exposure across eyes, nose, and cheeks; and skin rashes. Such patients have predisposition towards cancer [75]. Patients have learning disability; high pitched voice; prominent nose, ear, and respiratory dysfunctions.

The disease occurs due to mutated BLM gene, which is known to encode the BLM RecQ helicase homolog protein [76]. This gene is important in maintaining genomic stability. Defective helicase activity of BLM RecQ protein leads to the formation of abnormal protein structures, as the enzyme is crucial for DNA unwinding during replication. Chromosomal breakage is frequent in such individuals due to gaps and breaks introduced into the DNA, causing mutated gene expression.

The BLM gene is also known to repair UV-induced DNA damage and thus Bloom syndrome patients show increased sun sensitivity [75].

Detailed descriptions of this disease have been discussed by Kaneko H and Kondo N [76].

6.14 Werner Syndrome (OMIM 277700)

Werner syndrome is a rare autosomal recessive disorder, which is associated with premature aging and hence is often called "adult progeria." Such patients are normal before the onset of puberty; however, they fail to show the normal growth spurt and develop short stature. Aging starts in the early twenties and patients develop gray hair, thinning and hardening of skin, hair loss, and have "birdlike" facial expressions [77].

Mutated WRN gene, a member of the RecQ helicase family that is essential for genome stability and maintenance, is known to account for manifestation of this disease [78]. WRN gene has 3'-5'helicase and 3'-5'exonuclease activity [79] and plays important roles in DNA replication, DNA repair, transcription, and telomere maintenance. Loss of WRN function leads to arrest of cell cycle and hence such patients have stunted growth. This disease is described in detail by Kudlow et al. [79].

6.15 Xeroderma Pigmentosum (OMIM 278700)

Xeroderma pigmentosum is a rare autosomal recessive disorder, which is characterized by extreme sensitivity to sunlight. Such patients develop sunburn even during minor sun exposure (within few minutes), premature aging, photosensitivity, and a predisposition to lung cancer along with telangiectasia and neurodegeneration. The disease has been explained in detail by Alan R Lehmann and James E. Cleaver [80,82]. In brief, mutated genes of the NER pathway are known to underlie the manifestation of such diseases, they include *XPA, ERCC3, XPC, ERCC2, DDB2, ERCC4, ERCC5, ERCC1* [81]. Neurodegeneration arises due to mutated XPA, XPB, XPD, and XPG gene [82].

7. SUMMARY

Mutation could be regarded both as a boon and bane for mankind. These mutations, as a boon are raw materials for evolution and help in incorporation of new alleles within the population, which are passed to new generation. But often due to various genotoxic insults, that a human body is subjected to, day by day, DNA is damaged. A single change in nucleotide causing sickle cell anemia, when expressed in both the chromosomes (homozygous) results in death of the individual, but this same mutant when present in heterozygous condition proves, a natural resistance against malaria for inmates of South Africa.

The entire mechanism right from DNA replication to transcription and translation is very intricately woven and scrupulously controlled process, but still there remains a chance of incorporation of wrong base pairing. Our body has devised several response mechanisms to repair the damaged DNA and prevent the onset of diseases, but often this meticulous system fails to work and causes various types of developmental disorders and neurodegenerative diseases.

There occurs a very intricate balance between the mankind and the nature and often because of changing lifestyle unknowingly, we incorporate many practices that poses threat to our delicate machinery of replication and repair of DNA. Thus we should try to prevent such deliberate insults for betterment of not only ourselves but also the future generation.

ACKNOWLEDGMENT

The authors thank Ms. Poorwa Awasthi and Mr. Vipin Yadav for their critical reading of the chapter and editorial assistance. Work in A.K. lab is supported by the CSIR-Indepth and—Epigenetics in health and disease network projects and funding from Department of Science and Technology, Govt. of India and Wellcome Trust DBT India Alliance.

REFERENCES

[1] Collins F. Finishing the euchromatic sequence of the human genome. Nature 2004;431(7011):931–45.

[2] Kunkel TA. DNA replication fidelity. J Biol Chem 2004;279(17):16895–8.

[3] Kodym A, Afza R. Physical and chemical mutagenesis. Methods Mol Biol 2003;236:189–204.

[4] Freese EB. Transitions and transversions induced by depurinating agents. Proc Natl Acad Sci USA 1961;47:540–5.

[5] Schuster H. The method of reaction of desoxyribonucleic acid with nitrous acid. Z Naturforsch B 1960;15B:298–304.

[6] Liu X, et al. Pesticide-induced gene mutations and Parkinson disease risk: a meta-analysis. Genet Test Mol Biomarkers 2013;17(11):826–32.

[7] Rous P. A sarcoma of the fowl transmissible by an agent separable from the tumor cells. J Exp Med 1911;13(4):397–411.

[8] Collins DW, Jukes TH. Rates of transition and transversion in coding sequences since the human-rodent divergence. Genomics 1994;20(3):386–96.

[9] Stenson PD, et al. The human gene mutation database: providing a comprehensive central mutation database for molecular diagnostics and personalized genomics. Hum Genomics 2009;4(2):69–72.

[10] Driscoll DA, Gross S. Clinical practice. Prenatal screening for aneuploidy. N Engl J Med 2009;360(24):2556–62.

[11] Ebersberger I, et al. Genomewide comparison of DNA sequences between humans and chimpanzees. Am J Hum Genet 2002;70(6):1490–7.

[12] Graur D. Single base mutation. In: Nature encyclopedia of the human genome. Macmillan Pub Ltd, Nat Pub Grp; 2003. pp. 287–290.

[13] Mort M, et al. A meta-analysis of nonsense mutations causing human genetic disease. Hum Mutat 2008;29(8):1037–47.

[14] Hospach T, et al. Pseudodominant inheritance of the hyperimmunoglobulinemia D with periodic fever syndrome in a mother and her two monozygotic twins. Arthritis Rheum 2005;52(11):3606–10.

[15] Murnane JP. Telomere dysfunction and chromosome instability. Mutat Res 2012;730(1–2):28–36.

[16] Mackie Ogilvie C, Scriven PN. Meiotic outcomes in reciprocal translocation carriers ascertained in 3-day human embryos. Eur J Hum Genet 2002;10(12):801–6.

[17] Griffith B, Scott J, Carpenter JW, Reed C. Translocation as a species conservation tool: status and strategy. Science 1989;245(4917):477–80.

[18] Sullivan AE, et al. Recurrent fetal aneuploidy and recurrent miscarriage. Obstet Gynecol 2004;104(4):784–8.

[19] Jacobs PA, et al. The somatic chromosomes in mongolism. Lancet 1959;1(7075):710.
[20] Lempiainen H, Halazonetis TD. Emerging common themes in regulation of PIKKs and PI3Ks. EMBO J 2009;28(20):3067–73.
[21] Matsuoka S, et al. ATM and ATR substrate analysis reveals extensive protein networks responsive to DNA damage. Science 2007;316(5828):1160–6.
[22] Sancar A, et al. Molecular mechanisms of mammalian DNA repair and the DNA damage checkpoints. Annu Rev Biochem 2004;73:39–85.
[23] Sancar A. Structure and function of DNA photolyase and cryptochrome blue-light photoreceptors. Chem Rev 2003;103(6):2203–37.
[24] Daniels DS, Tainer JA. Conserved structural motifs governing the stoichiometric repair of alkylated DNA by O(6)-alkylguanine-DNA alkyltransferase. Mutat Res 2000;460(3–4):151–63.
[25] Krokan HE, Bjoras M. Base excision repair. Cold Spring Harb Perspect Biol 2013;5(4):a012583.
[26] Quinones JL, Demple B. When DNA repair goes wrong: BER-generated DNA-protein crosslinks to oxidative lesions. DNA Repair (Amst) 2016;44:103–9.
[27] Sugasawa K. Molecular mechanisms of DNA damage recognition for mammalian nucleotide excision repair. DNA Repair (Amst) 2016;44:110–7.
[28] Vermeulen W, Fousteri M. Mammalian transcription-coupled excision repair. Cold Spring Harb Perspect Biol 2013;5(8):a012625.
[29] Hanawalt PC, Spivak G. Transcription-coupled DNA repair: two decades of progress and surprises. Nat Rev Mol Cell Biol 2008;9(12):958–70.
[30] Menck CF, Munford V. DNA repair diseases: what do they tell us about cancer and aging? Genet Mol Biol 2014;37(1 Suppl.):220–33.
[31] Kunkel TA, Erie DA. Eukaryotic mismatch repair in relation to DNA replication. Annu Rev Genet 2015;49:291–313.
[32] Deans AJ, West SC. DNA interstrand crosslink repair and cancer. Nat Rev Cancer 2011;11(7):467–80.
[33] Singh TR, et al. MHF1-MHF2, a histone-fold-containing protein complex, participates in the Fanconi anemia pathway via FANCM. Mol Cell 2010;37(6):879–86.
[34] Crossan GP, Patel KJ. The Fanconi anaemia pathway orchestrates incisions at sites of crosslinked DNA. J Pathol 2012;226(2):326–37.
[35] Yamamoto KN, et al. Involvement of SLX4 in interstrand cross-link repair is regulated by the Fanconi anemia pathway. Proc Natl Acad Sci USA 2011;108(16):6492–6.
[36] Thompson LH. Recognition, signaling, and repair of DNA double-strand breaks produced by ionizing radiation in mammalian cells: the molecular choreography. Mutat Res 2012;751(2):158–246.
[37] Davis AJ, Chen DJ. DNA double strand break repair via non-homologous end-joining. Transl Cancer Res 2013;2(3):130–43.
[38] Grundy GJ, et al. APLF promotes the assembly and activity of non-homologous end joining protein complexes. EMBO J 2013;32(1):112–25.
[39] Mari PO, et al. Dynamic assembly of end-joining complexes requires interaction between Ku70/80 and XRCC4. Proc Natl Acad Sci USA 2006;103(49):18597–602.
[40] Ogrunc M, Sancar A. Identification and characterization of human MUS81-MMS4 structure-specific endonuclease. J Biol Chem 2003;278(24):21715–20.
[41] Tidyman WE, Rauen KA. Noonan, costello and cardio-facio-cutaneous syndromes: dysregulation of the Ras-MAPK pathway. Expert Rev Mol Med 2008;10:e37.
[42] Tartaglia M, et al. Diversity and functional consequences of germline and somatic PTPN11 mutations in human disease. Am J Hum Genet 2006;78(2):279–90.
[43] Romano AA, et al. Noonan syndrome: clinical features, diagnosis, and management guidelines. Pediatrics 2010;126(4):746–59.
[44] Costello JM. A new syndrome: mental subnormality and nasal papillomata. Aust Paediatr J 1977;13(2):114–8.

[45] Gripp KW, et al. A novel HRAS substitution (c.266C>G; p.S89C) resulting in decreased downstream signaling suggests a new dimension of RAS pathway dysregulation in human development. Am J Med Genet A 2012;158A(9):2106–18.

[46] Rauen KA. HRAS and the Costello syndrome. Clin Genet 2007;71(2):101–8.

[47] Denayer E, et al. Mutation analysis in Costello syndrome: functional and structural characterization of the HRAS p.Lys117Arg mutation. Hum Mutat 2008;29(2):232–9.

[48] Weiss G, et al. Cutaneous manifestations in the cardiofaciocutaneous syndrome, a variant of the classical Noonan syndrome. Report of a case and review of the literature. J Eur Acad Dermatol Venereol 2004;18(3):324–7.

[49] Yoon G, et al. Neurological complications of cardio-facio-cutaneous syndrome. Dev Med Child Neurol 2007;49(12):894–9.

[50] Niihori T, et al. Germline KRAS and BRAF mutations in cardio-facio-cutaneous syndrome. Nat Genet 2006;38(3):294–6.

[51] Butler Tjaden NE, Trainor PA. The developmental etiology and pathogenesis of Hirschsprung disease. Transl Res 2013;162(1):1–15.

[52] McKeown SJ, et al. Hirschsprung disease: a developmental disorder of the enteric nervous system. Wiley Interdiscip Rev Dev Biol 2013;2(1):113–29.

[53] Burke B, Stewart CL. Functional architecture of the cell's nucleus in development, aging, and disease. Curr Top Dev Biol 2014;109:1–52.

[54] Gonzalo S, Kreienkamp R, Askjaer P. Hutchinson-Gilford Progeria Syndrome: a premature aging disease caused by LMNA gene mutations. Ageing Res Rev 2016.

[55] Gonzalo S, Eissenberg JC. Tying up loose ends: telomeres, genomic instability and lamins. Curr Opin Genet Dev 2016;37:109–18.

[56] Furth ME, Aldrich TH, Cordon-Cardo C. Expression of ras proto-oncogene proteins in normal human tissues. Oncogene 1987;1(1):47–58.

[57] Kastan MB, Bartek J. Cell-cycle checkpoints and cancer. Nature 2004;432(7015):316–23.

[58] Macleod K. Tumor suppressor genes. Curr Opin Genet Dev 2000;10(1):81–93.

[59] Srivastava G, et al. Proteomics in Parkinson's disease: current trends, translational snags and future possibilities. Expert Rev Proteom 2010;7(1):127–39.

[60] Lesage S, et al. G51D alpha-synuclein mutation causes a novel parkinsonian-pyramidal syndrome. Ann Neurol 2013;73(4):459–71.

[61] Trinh J, Farrer M. Advances in the genetics of Parkinson disease. Nat Rev Neurol 2013;9(8):445–54.

[62] Van Cauwenberghe C, Van Broeckhoven C, Sleegers K. The genetic landscape of Alzheimer disease: clinical implications and perspectives. Genet Med 2016;18(5):421–30.

[63] Sherrington R, et al. Cloning of a gene bearing missense mutations in early-onset familial Alzheimer's disease. Nature 1995;375(6534):754–60.

[64] Levy-Lahad E, et al. Candidate gene for the chromosome 1 familial Alzheimer's disease locus. Science 1995;269(5226):973–7.

[65] Gatti RA, et al. Ataxia-telangiectasia: an interdisciplinary approach to pathogenesis. Medicine (Baltimore) 1991;70(2):99–117.

[66] Ambrose M, Gatti RA. Pathogenesis of ataxia-telangiectasia: the next generation of ATM functions. Blood 2013;121(20):4036–45.

[67] Shanske A, et al. Central nervous system anomalies in Seckel syndrome: report of a new family and review of the literature. Am J Med Genet 1997;70(2):155–8.

[68] Al-Dosari MS, et al. Novel CENPJ mutation causes Seckel syndrome. J Med Genet 2010;47(6):411–4.

[69] Weidenheim KM, Dickson DW, Rapin I. Neuropathology of Cockayne syndrome: evidence for impaired development, premature aging, and neurodegeneration. Mech Ageing Dev 2009;130(9):619–36.

[70] Kashiyama K, et al. Malfunction of nuclease ERCC1-XPF results in diverse clinical manifestations and causes Cockayne syndrome, xeroderma pigmentosum, and Fanconi anemia. Am J Hum Genet 2013;92(5):807–19.

[71] Laugel V, et al. Mutation update for the CSB/ERCC6 and CSA/ERCC8 genes involved in Cockayne syndrome. Hum Mutat 2010;31(2):113–26.

[72] Alter BP. Fanconi's anemia and malignancies. Am J Hematol 1996;53(2):99–110.

[73] Poole SR, et al. Monozygotic twin girls with congenital malformations resembling fanconi anemia. Am J Med Genet 1992;42(6):780–4.

[74] Taniguchi T, D'Andrea AD. Molecular pathogenesis of Fanconi anemia: recent progress. Blood 2006;107(11):4223–33.

[75] Amor-Gueret M. Bloom syndrome, genomic instability and cancer: the SOS-like hypothesis. Cancer Lett 2006;236(1):1–12.

[76] Kaneko H, Fukao T, Kondo N. The function of RecQ helicase gene family (especially BLM) in DNA recombination and joining. Adv Biophys 2004;38:45–64.

[77] Oshima J, et al. Lack of WRN results in extensive deletion at nonhomologous joining ends. Cancer Res 2002;62(2):547–51.

[78] Rossi ML, Ghosh AK, Bohr VA. Roles of Werner syndrome protein in protection of genome integrity. DNA Repair (Amst) 2010;9(3):331–44.

[79] Kudlow BA, Kennedy BK, Monnat Jr RJ. Werner and Hutchinson-Gilford progeria syndromes: mechanistic basis of human progeroid diseases. Nat Rev Mol Cell Biol 2007;8(5):394–404.

[80] Lehmann AR, McGibbon D, Stefanini M. Xeroderma pigmentosum. Orphanet J Rare Dis 2011;6:70.

[81] Kraemer KH, DiGiovanna JJ. Xeroderma Pigmentosum. 1993.

[82] Cleaver JE, Lam ET, Revet I. Disorders of nucleotide excision repair: the genetic and molecular basis of heterogeneity. Nat Rev Genet 2009;10(11):756–68.

[83] Wakerley BR, et al. Charcot-Marie-Tooth disease associated with recurrent optic neuritis. J Clin Neurosci 2011;18(10):1422–3.

[84] Deeb SS, Motulsky AG. Red-green color vision defects. In: Pagon RA, et al., editor. GeneReviews(R). 1993. Seattle (WA).

[85] O'Sullivan BP, Freedman SD. Cystic fibrosis. Lancet 2009;373(9678):1891–904.

[86] Down JL. Observations on an ethnic classification of idiots. 1866. Ment Retard 1995;33(1):54–6.

[87] Lanfranco F, et al. Klinefelter's syndrome. Lancet 2004;364(9430):273–83.

[88] Rees DC, Williams TN, Gladwin MT. Sickle-cell disease. Lancet 2010;376(9757):2018–31.

[89] Walker FO. Huntington's disease. Lancet 2007;369(9557):218–28.

[90] Fernandes Filho JA, Shapiro BE. Tay-Sachs disease. Arch Neurol 2004;61(9):1466–8.

[91] Tan M, Yu D. Molecular mechanisms of erbB2-mediated breast cancer chemoresistance. Adv Exp Med Biol 2007;608:119–29.

[92] di Martino E, et al. Mutant fibroblast growth factor receptor 3 induces intracellular signaling and cellular transformation in a cell type- and mutation-specific manner. Oncogene 2009;28(48):4306–16.

[93] Mathur SP, Mathur RS, Young RC. Cervical epidermal growth factor-receptor (EGF-R) and serum insulinlike growth factor II (IGF-II) levels are potential markers for cervical cancer. Am J Reprod Immunol 2000;44(4):222–30.

[94] Sancho E, Batlle E, Clevers H. Signaling pathways in intestinal development and cancer. Annu Rev Cell Dev Biol 2004;20:695–723.

[95] Cooper WA, et al. Molecular biology of lung cancer. J Thorac Dis 2013;5(Suppl. 5): S479–90.

[96] Xiong HQ. Molecular targeting therapy for pancreatic cancer. Cancer Chemother Pharmacol 2004;54(Suppl. 1):S69–77.

[97] Wang G, Wang J, Sadar MD. Crosstalk between the androgen receptor and beta-catenin in castrateresistant prostate cancer. Cancer Res 2008;68(23):9918–27.

CHAPTER TWO

Detection of Mutation in Prokaryotic Cells

Ashutosh Kumar[1], Rishi Shanker[2], Alok Dhawan[2]

[1]Ahmedabad University, Ahmedabad, India; [2]CSIR-Indian Institute of Toxicology Research, Lucknow, India

1. INTRODUCTION

Mutagenicity testing is the first step to screen the chemicals for their potential to be a pesticide, food additive, or drug. The most widely used mutation test is Ames test, developed by Ames [1], which is performed in different strains of *Salmonella typhimurium* and in *Escherichia coli*. Ames test is the preferred in vitro primary screening test for gene mutation, which detects most of the mutagens and carcinogens [2–5]. The assay is based on the premise that a substance, mutagenic to the bacterium (in presence/absence to exogenous metabolic activation system), is likely to be a carcinogen in laboratory animals, and hence, poses a risk of cancer in humans.

In this test, the bacterial strains already mutated at an easily detectable locus (e.g., histidine) are tested with compounds that produce a second mutation (nullifying the first mutation) and revert to normal. Bacterial strains mutated at histidine locus, do not synthesize histidine and thus die when plated on an agar medium lacking histidine. Different bacterial strains are treated with varying concentrations of the test compound in absence and presence of exogenous metabolic activation system. When a mutation occurs in these cells that reverts the original *his* mutation, the reverse-mutated cells (revertants) grow on a histidine–deficient agar plate and form a visible colony after 48 h of incubation. A reproducible dose-dependant response in at least one tester strain indicates a positive result, which suggests that the substance is a mutagenic and could be a carcinogen as well. If no increase in mutant colonies is seen after testing various strains under different culture conditions, the test chemical is considered to be nonmutagenic.

The sensitivity of Ames assay to mutagens and specific classes of chemicals [6,7] has been enhanced through strain engineering. Several recombinant strains of *S. typhimurium*, e.g., TA1535, TA1537, TA97/TA97a, TA98,

Mutagenicity: Assays and Applications
ISBN 978-0-12-809252-1
http://dx.doi.org/10.1016/B978-0-12-809252-1.00002-X

Table 2.1 Bacterial strains used in the Ames test and their characteristics

Bacterial strain	Target DNA sequence	Target allele	Plasmid and other characteristics	Mutation detected	Primary mutations
Salmonella typhimurium TA97a	CCCCCC GGGGGG	*hisD6610*	pKM101 (Apr), *rfa*, Δ*uvrB*	Frameshift	C or G deletion
S. typhimurium TA98	CGCGCGCG GCGCGCGC	*hisD3052*	pKM101 (Apr), *rfa*, Δ*uvrB*	Frameshift	GC or CG deletions and complex frameshift
S. typhimurium TA100	CCC GGG	*hisG46*	pKM101 (Apr), *rfa*, Δ*uvrB*	Base pair substitution	GC → AT (ts), AT → GC (ts), TA → GC (tv)
S. typhimurium TA102	CAAGTAAGAGC GTTCAT TCTCG	*hisG428*	pKM101 (Apr) and pAQ1 (Tcr), *rfa*	Base pair substitution, oxidative and cross–linking mutagens	AT → CG (tv), AT → TA (tv)
S. typhimurium TA104	TAA GC ATT and CG	*hisG428*	pKM101 (Apr), *rfa*, Δ*uvrB*	Base pair substitution oxidative and cross–linking mutagens	GC → AT (ts), GC → TA (tv), AT → GC (ts), AT → CG (tv), AT → TA (tv)
S. typhimurium TA1535	CC GG	*hisG46*	None, *rfa*, Δ*uvrB*	Base pair substitution	GC → AT (ts), AT → GC (ts), TA → GC (tv)
S. typhimurium TA1537	CCCCC GGGGG	*hisC3076*	None, *rfa*, Δ*uvrB*	Frameshift	C or G deletion
Escherichia coli WP2 *uvrA*	AT TA	*trpE*	None, *uvrA*	Base pair substitution and cross–linking	GC → AT (ts), GC → TA (tv), AT → GC (ts), AT → CG (tv), AT → TA (tv)
E. coli WP2 *uvrA* (pKM101)	AT TA	*trpE*	pKM101, *uvrA*	Base pair substitution and cross–linking	GC → AT (ts), GC → TA (tv), AT → GC (ts), AT → CG (tv), AT → TA (tv)

Apr, ampicillin resistance selection of plasmid; *Tcr*, tetracyclin resistance selection of plasmid; *ts*, transition; *tv*, transversion; *uvrB*, deletes excision repair. plasmid pKM101 encodes *mucAB* gene that participates in SOS repair system; *rfa* increases the permeability of cell wall to large molecules.

TA100, TA102, and TA104 (Refs. [8–11]; Table 2.1), exist for use in the assay. The strains can detect changes at guanine–cytosine (G-C) or adenine–thymine (A-T) base pairs sites within the histidine genes. The use of different strains in the assay helps to assess the specific mutagenic mechanism of the compound.

The tester strains have additionally been modified with deep rough (*rfa*) mutation, which eliminates the polysaccharide side chain of the lipopolysaccharide on the bacterial surface, making the bacteria more permeable to the test chemicals as well as to the new chemical entities. Deletion of excision repair system (Δuvr) confers sensitivity to many mutagens, whereas the plasmid pKM101 carrying the *muc*AB confers sensitivity to compounds, which act via the SOS system. The TA1535 set (TA1535, TA1536, TA1537, TA1538) is most sensitive to mutagenesis and is recommended for general testing for mutagens and carcinogens in vitro [2,3]. The compounds suspected to be oxidative mutagens may be detected by strains TA102 and TA104. Plasmids encoding for mutagenesis and metabolic function (e.g., specific enzymes) provide broad substrate specificity and high mutagenic sensitivity toward chemicals [6,12–14]. Additionally, recombinant *S. typhimurium* strain, YG5185 (carrying a plasmid encoding for *E. coli* DNA polymerase IV), has been specifically used for the sensitive detection of the genotoxicity of benzopyrene and other polyaromatic hydrocarbons in complex mixtures. A new bioluminescent strain of *Salmonella*, developed by Aubrecht et al. [15] employs bioluminescence for identifying histidine-independent revertant cells and provides an economical, high-throughput tool for assessment of mutagenicity.

Along with *Salmonella* strains, the bacterial tryptophan reverse mutation assay with the *E. coli* WP2 strains is also recommended [16,17]. Inclusion of the DNA repair–proficient strain, *E. coli* WP2 (pKM101), and excision repair–deficient strains *E. coli* WP2 *uvr*A for detection of cross-linking agents has been accepted by the international guidelines in place of the *Salmonella* strain TA102 because both carry an AT base sequence at the target site, which leads to the detection of transitions and transversion mutation [17].

A bacterial reverse mutation assay has also been designed utilizing the *E. coli lacZ*⁻ strain. *E. coli* uses lactose as a carbon source, and each strain carries a *lacZ* allele encoding for an inactive β-galactosidase protein, which upon mutation results in *lacZ*⁺ cells that can grow in a medium lacking lactose. This assay can detect base pair substitution and frameshift mutations [6].

Many chemicals are biologically inactive in their native forms and need to be metabolized, usually in the mammalian liver, to an active form. It is this active form that, in many cases, is the causative agent for mutation, cancer, and other effects. Because the bacteria used in the test are not capable of performing this metabolism, Prof. Ames and his colleagues added a rat liver homogenate (called S9 because it is prepared from a $9000 \times g$ supernatant liver fraction) and enzymatic cofactors, to the test. The S9 is prepared from rats that were pretreated with a polychlorinated biphenyl mixture or phenobarbital-plus-β-naphthoflavone to induce the cytochrome P450 enzymes needed for the metabolic activation. The S9 has also been prepared from mice, hamsters, and other animals, but the rat S9 is most widely used. Human S9 has also been used but there is currently no evidence that it is more effective in identifying mutagens or predicting carcinogenicity than induced rodent S9, and may be less effective for many classes of chemicals.

It has been established that there is a high predictivity of a positive mutagenic response in the Ames test for rodent carcinogens [7,13]. International guidelines have been developed for this assay (e.g., OECD 471 [18]; ICH) to ensure the uniformity of testing procedures. The extensive database on the assay justifies its inclusion as the initial screening test for detection of mutagens [19]. However, bacterial assays may not provide sufficient information for the assessment of genotoxicity of some compounds. These include compounds that are highly toxic to bacteria (e.g., some antibiotics), those which would interfere with mammalian cell-specific systems (e.g., topoisomerase inhibitors, nucleoside analogues, or certain inhibitors of DNA metabolism), or those which may be clastogens and do not produce mutations (heavy metal). Hence, Ames test should be used with a second test, which detects both clastogenicity and gene mutations in mammalian cells, e.g., mouse lymphoma assay (MLA; [19]).

2. MATERIALS AND METHODOLOGIES

Chemicals: The chemicals used in this assay are of analytical grade (purity >98%). D-biotin, L-histidine·HCl, $MgSO_4·H_2O$, 2-aminoanthracene, 2-nitroflourene, sodium ammonium phosphate·$4H_2O$, 4-nitroquinoline-N-oxide are obtained from Sigma (St. Louis, MO). Some chemicals such as citric acid monohydrate, sodium phosphate monobasic, sodium phosphate dibasic are purchased from Merck (New Delhi, India). While NADP disodium salt, D-glucose 6-phosphate, dipotassium hydrogen phosphate, and dextrose are procured from SRL (Mumbai) India. Bacto Agar is obtained from BD, Sparks, USA, and Nutrient broth No. 2 is from Oxoid Limited, United Kingdom.

Strains: The *Salmonella* tester strains can be obtained from American Type Culture Collection (ATCC; www.atcc.org).

2.1 Preparation of Mammalian Liver S9 Fraction

2.1.1 Induction of Rat Liver Enzymes

For cytochrome P450 induction, it is recommended that a polychlorinated biphenyl (PCB) mixture (Aroclor 1254) be administered in rats [5]. Dilute the Aroclor 1254 in corn oil to a concentration of 200 mg/mL and administer a single ip injection of 500 mg/kg to each rat (200 g) for 5 days before sacrifice.

2.1.2 Removal of Liver From Rats

The liver has to be excised under sterile conditions. Sacrifice the animal by cervical dislocation and place its back on an autopsy board. Swab the fur of the animal with 70%–90% ethanol. Cut the skin with a sterile pair of scissors and scalpel and excise the liver.

Precaution: Care should be taken that the esophagus and intestine are not damaged as this will result in contamination.

2.1.3 Preparation of Liver S9 Fraction

All steps of the procedure should be carried out at 0–4°C using sterile solution and glassware. Place the excised liver in a beaker containing 1 mL of chilled 0.15 M KCl per gram of liver. Wash the liver properly with prechilled 0.15 M KCl to remove hemoglobin because the presence of hemoglobin inhibits the activity of cytochrome P450. Transfer the washed livers to a beaker containing three volume of 0.15 M KCl (3 mL/g liver), mince with sterile scissor and homogenize properly. Centrifuge the homogenate for 10 min at 9000 g. Discard the pellet and aliquot the supernatant for storage at −80°C. This liver fraction can be used for metabolic activation for 1 year. The protein concentration of the S9 fraction should be 40 mg/mL and sterility should be tested by plating 100 μL of S9 fraction on minimum glucose (GM) plate.

2.2 Preparation of Reagents

2.2.1 Vogel-Bonner (VB Salt) Solution (50×)

1. In 650 mL of warm distilled water (about 50°C) mix 10 g of magnesium sulfate monohydrate, 100 g of citric acid monohydrate, 500 g of potassium phosphate dibasic anhydrous, and 175 g of sodium ammonium phosphate tetrahydrate.

2. Dissolve each salt thoroughly by stirring on magnetic stirrer before adding the next salt.
3. Adjust the volume to 1000 mL.
4. Distribute in aliquots of 20 mL and autoclave for 30 min at 121°C.
5. Store at room temperature in dark.

2.2.2 Glucose Solution (10% W/V)
1. Mix 100 g of dextrose in 700 mL of distilled water.
2. Stir on a magnetic stirrer until the mixture is clear.
3. Bring the final volume up to 1000 mL.
4. Autoclave the solution at 121°C for 20 min or filter sterilize with 0.22 µm polyvinylidene fluoride (PVDF) filter.
5. Store the glucose solution at 4°C.

2.2.3 Minimal Glucose Agar Plates
1. In 900 mL of distilled water mix 15 g agar into the flask.
2. Autoclave the flask for 30 min at 121°C and allow it to cool for 45 min to reach the temperature approximately about 65°C.
3. Add 20 mL of 50× VB salt solution and 50 mL of 10% glucose solution (as prepared above) to the flask.
4. Mix thoroughly to dissolve the precipitate.
5. Allow the flask to cool up to 37°C and then dispense 25–30 mL media to the petri dishes.
6. Leave the petri dishes to solidify the medium and thereafter keep the plates at 4°C.

2.2.4 Histidine–Biotin Solution (0.5 mM)
1. In 1000 mL of distilled water, mix 124 g of D-biotin and 96 mg of L-histidine·HCl.
2. Mix the ingredients properly.
3. Filter sterilize through a 0.22 µm (PVDF) filter or can be autoclave at 121°C for 20 min.
4. Store the bottle at 4°C.

2.2.5 Top Agar Supplemented With Histidine–Biotin Solution
1. In 900 mL of distilled water, mix 6 g of agar and 6 g of sodium chloride in the bottle.
2. Autoclave the bottle for 20 min at 121°C.

3. Add 100 mL of His–Bio solution (0.5 mM) to the top agar maintained at 40–50°C.
4. Store the bottle at room temperature in dark.

2.2.6 Buffers for Metabolic Activation
Two types of buffer system are required for the Ames test.

2.2.6.1 Sodium Phosphate Buffer (0.1 mM, pH 7.4)
1. Sodium phosphate buffer (0.1 mM, pH 7.4) is required for the system, which lacks S9 fraction.
2. Sodium phosphate buffer (0.1 mM, pH 7.4) can be prepared by mixing 120 mL of sodium phosphate monobasic (0.1 M; 13.8 g $NaH_2PO_4 \cdot H_2O$ in 1000 mL d H_2O) with 880 mL of sodium phosphate dibasic (0.1 M; 14.2 g $Na_2HPO_4 \cdot H_2O$ in 1000 mL d H_2O).
3. Mix both and autoclave it at 121°C for 20 min.
4. Store the buffer at room temperature in dark.

2.2.6.2 To Activate the NADP Regenerating System in Presence of Liver S9 Fraction
1. Prepare the individual stock solution of 1 M KCl (by dissolving 3.78 g in 50 mL of distilled water), 0.25 M $MgCl_2 \cdot 6H_2O$ (by dissolving 2.5 g in 50 mL of distilled water), 0.2 M NaH_2PO_4 (by dissolving 1.56 g in 50 mL of distilled water), and store them at 4°C.
2. Prepare the stock solution of 0.2 M Glucose 6-phosphate (by mixing 0.564 g in 10 mL of distilled water) and 0.04 M NADP (by mixing 0.306 g in 10 mL of distilled water), and store at −20°C.
3. For the final working S9 mixture (30%; 5 mL) mix 166 μL of 1 M KCl, 160 μL of 0.25 M $MgCl_2 \cdot 6H_2O$, 2.5 mL of 0.2 M NaH_2PO_4, 126 μL of 0.2 M Glucose 6-phosphate, 500 μL of 0.04 M NADP, 50 μL sterile water, and 1.5 mL of S9 fraction (Fig. 2.1).

2.3 Methodology
2.3.1 Culture Preparation
1. Inoculate the individual culture flask with each strain (use frozen stock cultures).
2. Grow the tester strain culture overnight in oxoid nutrient broth No. 2 containing ampicillin (100 μg/mL) to a density up to approximately 1×10^9 CFU/mL.
3. Incubate the culture at 37°C and 130 rpm.

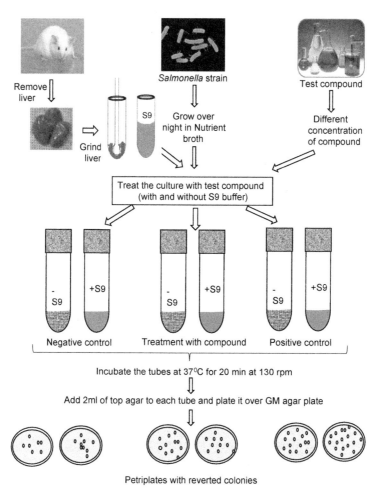

Figure 2.1 Schematic representation of the Ames test.

2.3.2 Genetic Analysis of the Tester Strains

1. The tester strains should be analyzed for their genetic integrity and spontaneous mutation rate when frozen cultures are prepared.

2. An analysis of the genetic integrity of every strain used should also be performed at the time of every experiment, and in parallel with the experiment, to ensure that each strain used contains the uvr and cell wall (rfa) mutations, and plasmid, where appropriate.

3. The strain check is performed with the nutrient broth overnight cultures prior to the start of the experiment, and the results are evaluated before

scoring the experiment. If the tester strain does not behave appropriately in the strain check, all test plates from that strain/culture should be discarded and not scored.

2.3.3 Histidine and Biotin (his, bio) or Tryptophan (trp) Dependence

1. The uvr B deletion in the *Salmonella* strains also deletes the biotin (bio) genes, thus the uvr B deleted strains also require biotin for growth.
2. Streak a loop of the culture across a minimal agar plate supplemented with biotin alone and with an excess of biotin and histidine (*Salmonella* strains) or tryptophan (*E. coli* strains).
3. Growth should be observed with all strains in the presence of histidine or tryptophan, but not on the plates containing only biotin. When using this technique, more than one strain can be streaked on the same plate, taking care not to have the individual streaks touch each other.

2.3.4 rfa (Deep Rough) Mutation

1. Streak a loop-full of the overnight culture across a nutrient agar plate supplemented with an excess of biotin and histidine or tryptophan.
2. Place a sterile filter paper disk in the center of the streak and apply 10 μL of a sterile 0.1% crystal violet solution. The crystal violet disks can be prepared in advance and stored aseptically at room temperature.
3. All strains should show a zone of growth inhibition surrounding the disk.

2.3.5 Treatment of Culture

1. In a 13 × 100 mm sterile glass tube, add the following with mild mixing after each step.
   a. 500 μL of metabolic activation (S9) mix or buffer.
   b. 50 μL of the test compound dilution.
   c. 100 μL of overnight culture of *Salmonella* strain.
   d. Mix all properly.
2. Incubate the culture at 37°C for 20 min at 130 rpm.

2.3.6 Plating of Treated Culture

1. To each tube add 2 mL of molten top agar maintained at 43–48°C. The contents of the tube are then mixed and poured onto the surface of GM agar plate.
2. When the top agar has hardened (2–3 min), place the plates inverted and keep it in an incubator at 37°C for 48–72 h.

Table 2.2 Average number of spontaneous revertants for different strains

Serial number	Strains	Average number of revertants without S9 fraction	Average number of revertants with S9 fraction
1	TA97	75–200	75–200
2	TA98	20–50	20–50
3	TA100	75–200	75–200
4	TA102	100–300	200–400
5	TA1535	5–20	5–20
6	TA1537	5–20	5–20
7	TA1538	5–20	5–20

The values given above may vary, however, they should be near the ranges given.

3. RESULT EVALUATION

1. Count the colonies after 48–72 h and report the results as the number of revertant colonies per plate.
2. Spontaneous revertant control value: The spontaneous mutation can also lead to the conversion of auxotrophic strain into wild type. Table 2.2 shows average number of spontaneous revertants for different strains.
3. A clear positive response in any one of the tester strains is sufficient to demonstrate that a substance is mutagenic.
4. However, a substance should not be considered nonmutagenic unless it has been tested in, at least five strains.
5. The most widely used is the so called modified twofold rule. In this, a test is considered positive if there is a dose-related response reaching at least twofold over the control.

4. RESULT INTERPRETATION

1. A positive response in the assay in any of the tester strains indicates the mutagenic potential of the compound and can be assumed that it will have a high probability of being carcinogenic in laboratory animals (typically rats and mice).
2. The probability of the chemical being a carcinogen is the same regardless of which strain, or the number of tester strains, in which the chemical is mutagenic, or the potency of the mutagenic response (Fig. 2.2).

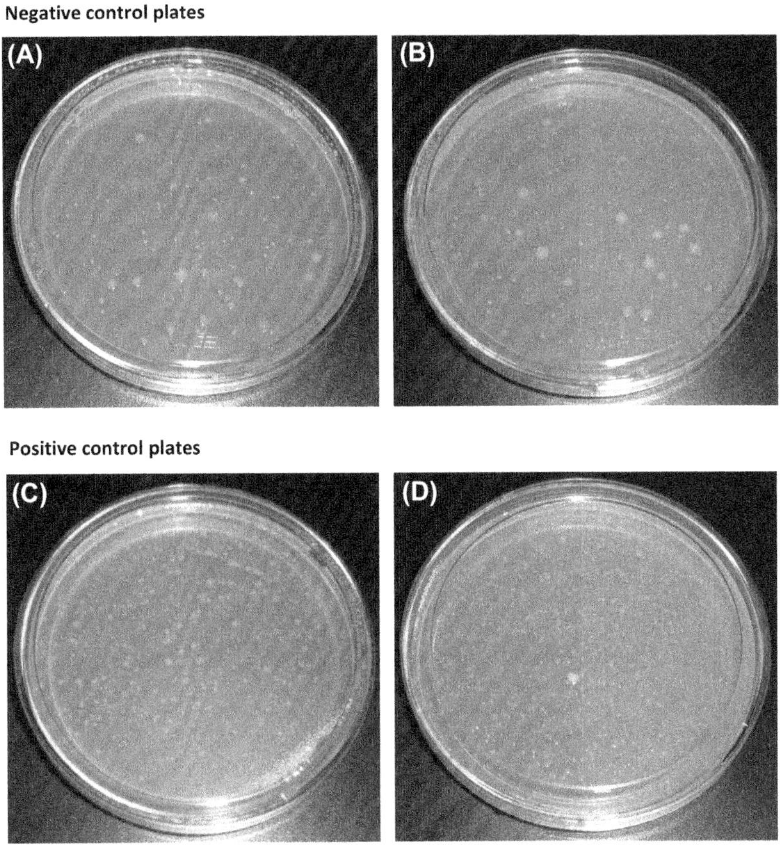

Negative control plates

Positive control plates

Figure 2.2 Representative pictures showing the revertant colonies of *Salmonella typhimurium* TA100. (A) Negative control without S9 fraction, (B) negative control with S9 fraction, (C) positive control (sodium azide) without S9 fraction, (D) positive control (2-aminoanthracin) with S9 fraction.

5. EXPERIMENTAL DESIGN

Given below is a model experiment to test the mutagenicity of a compound.

The tester strains should be seeded in a culture tube a night before the start of the experiment at the dilution of 1:100 with growth media. After 15–16 h of the seed culture incubated at 37°C, the cell number becomes approximately 1×10^9 CFU/mL. Prepare the various concentration of the compound (to be tested). Incubate the test compound with tester strain (with/without S9

Table 2.3 Representative positive control chemicals

	Positive control chemical (µg/plate)	
Strain	**Without activation**	**With activation**
TA97	9-Aminoacridine (50)	2-Aminoanthracene (1–5)
TA98	4-Nitro-*o*-phenylenediamine (2.5)	2-Aminoanthracene (1–5)
TA100	Sodium azide (5)	2-Aminoanthracene (1–5)
TA102	Mitomycin C (0.5)	2-Aminoanthracene (5–10)
TA104	Methyl methane sulfonate (250)	2-Aminoanthracene (5–10)
TA1535	Sodium azide (5)	2-Aminoanthracene (2–10)
TA1537	9-Aminoacridine (50)	2-Aminoanthracene (2–10)
TA1538	4-Nitro-*o*-phenylenediamine (2.5)	2-Aminoanthracene (2–10)

Concentration based on each petri plate containing 20–25 mL GM agar.

fraction) for 20 min at 37°C. After 20 min of exposure, 2 mL top agar is mixed in each tube and plated/poured on GM plates. Each experiment should be done in replicate and must be repeated thrice. The positive and negative controls must be performed at the same time, with the same culture and same buffer. For negative control use only the solvent that has been used to dissolve the test compound. The positive controls for different strains are given in Table 2.3.

6. PRECAUTION

1. Same as other basic bacteriological experiments, there should be a minimum exposure to the *Salmonella* tester strain.
2. Surface area must be disinfected before and after the use of tester strain.
3. Aseptic techniques, being a part of the basic bacteriological laboratory procedure, are also essential to prevent contamination to the overnight culture, GM agar plates, solutions, and the reagent
4. Contaminated working culture of *Salmonella* tester strain is useless and should be discarded.
5. As we consider the chemicals are mutagenic or carcinogenic, so handling of the chemicals should be performed in a chemical safety cabinet and the worker should wear the gloves while using the chemicals.
6. All contaminated materials (e.g., test tubes, pipettes, petri plates) should be properly discarded, autoclave all the things before disposal.

REFERENCES

[1] Ames BN. The detection of chemical mutagens with enteric bacteria. In: Holleander A, editor. Chemical mutagens: principle and methods for their detection. New York: Plenum Press; 1971. p. 267–82.

[2] Ames BN, Lee FD, Dursten WE. An improved bacterial test system for detection and classification of mutagens and carcinogens. Proc Natl Acad Sci USA 1973; 70:782–6.

[3] Ames BN, Dursten WE, Yamasaki E, Lee FD. Carcinogens are mutagens: a simple test system combining liver homogenate for activation and bacteria for detection. Proc Natl Acad Sci USA 1973;70:2281–5.

[4] Ames BN, McCann J, Yamasaki E. Methods for detecting carcinogens and mutagens with the *Salmonella*/mammalian-microsome mutagenicity test. Mutat Res 1975;31: 347–64.

[5] Maron D, Ames BN. Revised methods for the *Salmonella* mutagenicity test. Mutat Res 1983;113:173–215.

[6] Josephy PD, Gruz P, Nohnii T. Recent advances in the construction of bacterial genotoxicity assays. Mutat Res 1997;386:1–23.

[7] McCann J, Choi E, Yamasaki E, Ames BN. Detection of carcinogens as mutagens in the *Salmonella*/microsome test: assay of 300 chemicals. Proc Natl Acad Sci USA 1975;72(12):5135–9.

[8] Bonneau D, Thybaud V, Melcion C, Bouhet F, Cordier A. Optimum associations of tester strains for maximum detection of mutagenic compounds in the Ames test. Mutat Res 1991;252(3):269–79.

[9] Levin DE, Hollstein M, Christman MF, Schwiers EA, Ames BN. A new *Salmonella* tester strain (TA102) with A×T base pairs at the site of mutation detects oxidative mutagens. Proc Natl Acad Sci USA 1982;79(23):7445–9.

[10] Ohe T, Shaughnessy DT, Landi S, Terao Y, Sawanishi H, Nukaya H, Wakabayashi K, DeMarini DM. Mutation spectra in *Salmonella* TA98, TA100 and TA104 of two phenyl-benzotriazole mutagens (PBTA-1 and PBTA-2) detected in the Nishitakase River in Kyoto, Japan. Mutat Res 1999;429(2):189–98.

[11] Venitt S, Crofton-Sleigh C, Forster R. Bacterial mutation assays using reverse mutations. In: Venitt S, Parry JM, editors. Mutagenicity testing. Washington: IRC Press; 1984. p. 45–98.

[12] Carroll CC, Warnakulasuriyarachchi D, Nokhbeh MR, Lambert IB. *Salmonella typhimurium* mutagenicity tester strain that overexpresses oxygen insensitive nitro-reductase nfsA and nfsB. Mutat Res 2002;501:79–98.

[13] Mortelmans K, Zieger E. The Ames *Salmonella*/microsome mutagenicity assay. Mutat Res 2000;455:29–60.

[14] Suzuki A, Kushida H, Iwata H, Watanabe M, Nohnii T, Eujita K, Gonzalez T, Kamatki FJ. Establishment of a *Salmonella* tester strain highly sensitive to mutagenic heterocyclic amines. Cancer Res 1998;58:1833–8.

[15] Aubrecht J, Osowski JJ, Persaud P, Cheung RJ, Ackerman J, Lopes SH, Ku WW. Bioluminescent Salmonella reverse mutation assay: a screen for detecting mutagenicity with high throughput attributes. Mutagenesis 2007;22(5):335–42.

[16] Gatehouse D, Haworth S, Cebula T, Gocke E, Kier L, Matsushima T, Melcion C, Nohmi T, Ohta T, Venitt S, Zeiger E. Recommendations for the performance of bacterial mutation assays. Mutat Res 1994;312:217–33.

[17] Mortelmans K, Riccio ES. The bacterial tryptophan reverse mutation assay with *Escherichia coli* WP2. Mutat Res 2000;455:61–9.

[18] OECD (Organisation for Economic Co-operation, Development) Guideline for Testing of Chemicals. Test guideline 471: bacterial reverse mutation test, OECD, Paris, France. 1997.

[19] Seifried HE. A compilation of two decades of mutagenicity test results with the Ames *Salmonella typhimurium* and L5178Y mouse lymphoma cell mutation assays. Chem Res Toxicol 2008;21(2):554–5.

FURTHER READING

[1] Yamada M, Matsui K, Nohmi T. Development of a bacterial hyper-sensitive tester strain for species detection of the genotoxicity of polycyclic aromatic hydrocarbons. Genes and Environ 2006;28:23–30.

CHAPTER THREE

Detection of Gene Mutation in Cultured Mammalian Cells

Jayant Dewangan, Prabhash Kumar Pandey, Aman Divakar, Sakshi Mishra, Sonal Srivastava, Srikanta Kumar Rath
CSIR–Central Drug Research Institute, Lucknow, India

1. INTRODUCTION

The genetic information encoded in the DNA is transcribed to messenger RNA, which is further translated to proteins. Protein processing, folding, transport, and incorporation also contribute in flow of message encoded in genes. Damaged DNA which has evaded the DNA repair machinery undergoes replication subsequently leading to a mutation. The alteration in the nucleotide sequence in coding region of the DNA may change the amino acid sequence of the protein upon translation. However, a change in the noncoding DNA sequence might influence the gene expression by altering the promoter strength. Such alteration can be either confined to one base pair or multiple base pairs including addition, deletion, amplification, and recombination [1]. Mutations are broadly categorized depending on its occurrence in the DNA or chromosomes. Various mechanisms cause mutation which ranges from single-base alteration to loss, duplication, or chromosomal rearrangement [2]. Generally, mutations are detected in terms of changes in phenotype per unit time indicated as cell generations [3] or days [4]. With the time, new molecular techniques have been developed for precise understanding of the genome and potential gene therapies. Genetic screening of mammalian cultured cells is an effective and powerful tool for identifying unique mechanism or pathway which regulates the biological process [5,6]. This chapter provides a brief overview of mutation detection techniques that are currently available. The purpose of this chapter is to make familiar with the molecular techniques and tools used for detection of gene mutation in mammalian cells.

Mutagenicity: Assays and Applications
ISBN 978-0-12-809252-1
http://dx.doi.org/10.1016/B978-0-12-809252-1.00003-1

2. PCR-BASED MUTATION DETECTION

Since long, many singleplex polymerase chain reaction (PCR)–based techniques have been in use for detection of specific gene mutation. New PCR based methods such as allele-specific priming or amplification refractory mutation system (ARMS), allele-specific oligonucleotide (ASO) hybridization, gap-PCR and restriction endonuclease analysis of PCR products (RE-PCR) are now in use.

1. *Restriction fragment length polymorphism (RFLP).* RFLP is a DNA-based molecular marker used for analysis of polymorphism of different alleles of same gene. Restriction fragment length is a site in genome where the distance between two restriction sites varies among normal and mutated cell lines. These sites are identified by restriction enzyme digest of chromosomal DNA. Basically, RFLP is used to find mutations within restriction sites [7]. Variation in fragment length is observed in the DNA of mutated cell lines upon digestion with the same restriction enzymes if a gene polymorphism is present at the restriction site. The Southern blotting is used for identification of specific fragments and genetic analysis which requires a radioactive probe. The variants of RFLP are codominant and measure the variation at level of DNA sequence and not the protein sequence. The only drawback is that it requires large amount of DNA. After gel electrophoresis, the unique band profile allows to distinguish the variation at the DNA level without costly DNA sequencing. This technique has been used widely in the detection of c-Ki-Ras mutations in pancreatic adenocarcinomas [8,9], 12K-Ras mutation in human lung adenocarcinoma as well as in human colon cancer cell lines viz HT29 and SW480 [10]. The PIK3CA exon-20 (H1047R) point mutation is a potential marker for EGFR-targeted therapies. In a recent study, use of PCR-RFLP method in the detection of the PIK3CA[H1047R] mutation in colorectal cancer cell lines LoVo, SW620, LS174T, HT29, HCT-8, and Colo205 has been suggested as a simple, effective, and sensitive tool [11]. Further, RFLP has been found to be useful for screening EGFR mutations in non–small cell lung cancers [12]. PCR-RFLP has a limitation as defective restriction enzyme activity could be confused with the loss of the restriction site; therefore, to overcome this problem double RFLP technique was proposed where two modified nested PCRs are performed at the same time. First set of PCR primers is used to introduce a site-specific restriction for the wild-type allele, whereas the other one is designed to introduce a restriction site specific for the mutant allele [13].

2. *Amplification refractory mutation system (ARMS) PCR.* Tetra primer ARMS is an application of PCR in which DNA is amplified by specific allele primers for detecting SNPs or mutation. It is an efficient molecular technique for identification of point mutations and polymorphism. It is also able to identify whether DNA mutation is heterozygous or homozygous. The heterozygotic and homozygotic conditions are differentiated by using primers for mutant/polymorphic and wild-type alleles. Generally, the reactions for both the mutant and normal alleles are carried out in separate tubes, but by using the primers labeled with different fluorescent dyes the reaction can be done in the same tube. A comparative study conducted by Machniki et al. concluded that simple ARMS–PCR is a cost-effective rapid screening method for BRAF p.V600E mutation in A375 melanoma cell line harboring homozygous *BRAF* c.T1799A (p.V600E) mutation, pancreatic cancer cell line MIA PaCa-2 harboring homozygous *KRAS* c. G34T (p. G12C) mutation as well as in gastric cancer samples obtained from patients [14].

3. *Multiplex PCR.* PCR is the most widely used technique in both biomedical research and clinical diagnosis to amplify DNA of interest [15,16]. In a multiplex PCR, more than one target regions are amplified simultaneously using different pairs of primers in single reaction. Multiplex PCR is used in genetic screening such as gene deletion analysis [17], mutation and polymorphism analysis [18,19], quantitative analysis [20], reverse-transcription (RT)-PCR [21], and other diagnostic purposes for amplifying several targets at the same time. Multiplex PCR has contributed important role for identification of infectious viruses [21–25] as well as bacterial [26] and parasitic [27] diseases. It has more significant advantages over conventional PCR in the clinical diagnosis [28–32]. Apart from its high specificity, this method can save time and is economical in detection of pathogens in patients or the sick animals. Milbury et al. have demonstrated an efficient and alternative multiplex PCR approach on genomic DNA and cell lines viz T47D, MDA-MB-435, SNU-182, HCC2157, and HCC2218 with TP53 mutations to overcome the limitations for limited available DNA sample such as samples of biopsies, circulating DNA or tumor cells, and histology using improved mutation detection sensitivity via COLD-PCR amplification. The conventional method for the detection point mutations uses nested PCR method; however, recently Chen and coworkers developed a TaqMan-based real-time duplex PCR assay for detection of the A2058G, A2059G mutation, associated with macrolide resistance in *Treponema pallidum* [33,34].

4. *Nested PCR (nPCR).* It consists of two consecutive PCRs where the primary PCR product is used as a template for the secondary PCR. It enhances the sensitivity and specificity of DNA and RNA (nRT-PCR) amplification and confirms the integrity of a specific amplicon. It involves the use of two pairs of primers for single locus. The external or first pair of primers amplify the target of interest alike any PCR reaction. The internal or second pair of so-called nested primers binds within the primary PCR product and produces a secondary product that is shorter than the previous one. nPCR is used to ensure the low probability of amplification of wrong locus by the specific internal primers. Studies carried out by Fukayama et al. [35] detected K-ras codon 12 mutations by nPCR in clinical samples of colon carcinoma and pancreatic carcinoma. Further studies performed by Banerjee et al. revealed that use of two-step enriched nPCR/RFLP technique was more sensitive in detecting 100% of K-ras codon 12 mutations of pancreatic adeno-carcinomas in comparison to single-step PCR/RFLP method, where only 9 out of 15 mutations could be detected [36]. As an advancement to nPCR, Nested Patch PCR technique is also used, which enables identification of highly multiplexed gene mutations [37]. 5382insC, 185delAG, and C61G gene mutations in BRCA1 are responsible for ovarian cancer. Reports suggest that results of nPCR assay were quite promising as compared with whole genome amplification in case of BRAC1 gene mutation detection for unfertilized oocytes [38].

3. DETECTION OF MUTATIONS BY DENATURING GRADIENT GEL ELECTROPHORESIS

For the screening of unknown point mutations, denaturing gradient gel electrophoresis (DGGE) serves to be a crucial technique [2]. DGGE is an approach for the detection of single-base mutations in a particular gene. Sequence-based melting behavior of DNA is the principle of this method, and melting behavior of the DNA differs in mutated and nonmutated DNA samples. In a polyacrylamide gel against a gradient of chemical denaturant, separation of DNA strand at a domain in a DNA fragment is pointed out by the reduction in the mobility of the fragment. In a denaturing gradient gel, same sized but differing in sequence, two double-stranded DNA fragments melt at discrete points. In mutation detection, these gels play a crucial role in the screening of a number of individuals. Due to the whole strand dissociation, this method was unable to separate DNA fragments differing by nucleotide changes in the highest melting domain. By incorporating a GC clamp tail, this issue has been now sorted out, and by doing this it became more efficient in the detection of mutation. DGGE with PCR allows easier way of mutation detection. For further analysis such as sequencing, mutant molecules can also be isolated from gels, and by doing this it can be used in the study of PCR amplification errors [39]. By DGGE method we can screen a newly characterized disease gene for known and unknown mutations, e.g., rhodopsin mutations, mutations in porphyria carriers and α-1-antitrypsin deficiency. This is a relatively low-throughput method and needs extensive optimization for each analysis. Designing of complex primers to incorporate the GC clamps in the optimum position as well as maintaining the fragment to be scanned as a single melting domain is also a limitation of this approach. In spite of the above drawbacks its high sensitivity makes it relatively much popular within the diagnostic setting. A fully automated high-throughput temperature-gradient capillary electrophoresis technique that works on the same principle as DGGE with no prior labeling of the sample required has recently been described. Tan et al. identified the first androgen receptor missense mutation in human prostate cancer in the LNCaP cell line by DGGE method. Analysis of androgen receptor exon H by DGGE revealed extra doublet bands, suggesting a mismatch in the heteroduplex of wild-type and CWR22 DNA [40]. Guldberg et al. developed a method based on DGGE to seek for mutations in the MMACJIPTEN gene. The cell lines were SK-MEL-3, SK-MEL-24, and SK-MEL-31 [41].

4. SINGLE-STRAND CONFORMATION POLYMORPHISM

For the detection of unknown mutations such as small deletions, single-base substitutions, small insertions, single-strand conformation polymorphism (SSCP) is one of the simplest screening methods. Within gel electrophoresis during migration, a DNA variation causes alterations in the conformation of denatured DNA fragments.

In SSCP under special conditions holding the formation of dsDNA and permitting the formation of conformational structures in single-stranded fragments, DNA fragments are denatured and renatured. Unique conformation is revealed from the primary nucleotide sequence, and based on these unique conformation structures variations are detected. In nondenaturing polyacrylamide gels, mobility of these fragments is different. For SSCP the optimal fragment size should be around 150–200 bp; therefore the PCR used to amplify the fragments is called PCR-SSCP. The upper limit of the fragment size in SSCP is 300 bp.

SSCP is quite sensitive, and approximately 80%–90% of potential point mutations are detected by it. It can detect even single nucleotide difference that occupies a different conformation. This method is not 100% successful in the detection of mutation, although it is very popular due to its simplicity. For the detection of polymorphism in a family, in disease gene linked studies, or finding mutations in a particular gene to imply it in a particular disease this method can be used [39]. By RNA-SSCP, this method has been improved in the proportion of mutations detection but not close to 100%. Use of capillary electrophoresis and fluorescence enhances the sensitivities in blinded trials and in the detection of larger fragments.

Vidal et al. detected the mutations in metastatic melanoma cell lines by SSCP analysis [42]. Danks et al. detected mutations by the method of SSCP analysis of the M(r) 170,000 isozyme of DNA topoisomerase II in human tumor cells [43].

5. HETERODUPLEX ANALYSIS

The principle behind the heteroduplex analysis assay of mutation detection technique is that, upon denaturation DNA strands can be separated and in similar fashion it can also be renatured to form a homoduplex. In the process of reannealing, if there is a mutation in any one of the strands, then the DNA forms a heteroduplex [44]. This heteroduplex state formation property of DNA can be used for the point mutations detection. In comparison with the homoduplex DNA the heteroduplex DNA move slower on a polyacrylamide gel.

Two types of heteroduplex are present: the first one is "bubble"-type heteroduplex, it is formed when the sequence difference between the two DNA fragments is one or more point mutations and the "bulge"-type heteroduplex is formed when the sequence difference between the two fragments is a small insertion or deletion. Bulges are large structural perturbations that arise from the ds homoduplexes. These can be resolved on polyacrylamide gels but the bubble (change in overall structure) heteroduplexes cannot be resolved from the homoduplexes on neither agarose nor polyacrylamide gels under standard conditions [45]. Heteroduplex analysis on mutation detection enhancement gels has made this mutation detection technique more viable. Another type of gel matrix named conformation-sensitive gel electrophoresis (CSGE), formed by mixing a novel cross-linker bis-acrolylpiperazine in polyacrylamide matrix, is not sufficient for the complete separation of DNA strands, but it is helpful in destabilizing the conformational disruptions in the double-stranded DNA structure introduced by the presence of mismatched bases. CSGE enhances the reproducibility of the patterns obtained with specific mutations. CSGE is a very useful tool for the repetitive analyses of those genes that have disease-causing mutations or multiple benign polymorphisms. It is a very simple technique and has been successfully applied to fragments of >1 kb in size, but the sensitivity is reduced in larger fragments. Baker et al. performed heteroduplex analysis with human cervical carcinoma cell lines SiHa and CaSki; however, in the study no deletions, insertions, or rearrangements were found [46].

6. MICROARRAY

Microarray technology has been employed for several applications such as gene expression analysis, gene discovery, gene mapping, and genotyping studies [47]. Microarray is one of the alternative approaches for genotyping and wide variety of mutation detection in genome. Microarray is an array of different probes spotted on surface matrix. Specific DNA fragments viz cDNA or PCR amplicon also known as probe are ligated or synthesized silicon chips in form of microspotting. Spotting is performed through robotic machinery, which is highly flexible, time saving, and relatively low cost [48,49]. Microarray principle is based on the complimentary nucleic acid sequence hybridization. For identification of hybridization, probes can be labeled with fluorescent moiety [50].

DNA microarray system has been used for a long time in mutation detection studies. Samples of multiple diseases have been screened for detecting mutations in genes such as BRCA1, CFTR, ATM, and RET [51–54]. Berber et al. designed an oligonucleotide microarray platform for resequencing and mutational analysis by taking human hemophilia A mutation as a model. There are more than 900 known mutations that cause hemophilia A. Out of the 720 probes, 25 bp each from randomly picked 6 exons of Factor VIII gene were designed and ligated on microarray system. Wild-type and hemophilic A patient DNA samples were labeled with different fluorescent dyes and cohybridized to the array. Any kind of mutation, which is changing in the wild-type sequence, led to loss of hybridization and fluorescence intensity. Microarray provides efficient and reliable platform to detection mutations. However, it is not as cost-effective as DNA sequencing approach for mutation detection [55].

A different strategy has been employed for single nucleotide polymor-phism detection by modification in conventional microarray technology. To increase hybridization specificity, probe having a stem loop structure is designed. It consists of two target-specific GC-rich palindromic sequences at flanking end. Once it is hybridized with its complimentary sequence, stem loop or hair loop structure is changed into double-stranded linear duplex form. Else, unhybridized probes again reform into stem loop–like structures and remain stable. Probes are ligated on microarray platform and give fluorescence on binding of labeled target DNA. Fluorescence intensity indicates degree of hybridization [56].

7. ARRAYED PRIMER EXTENSION TECHNOLOGY

High-throughput screening for detection of known mutation on microarray platform can be performed by arrayed primer extension (APEX) technology. Here 20–25 bp oligonucleotides are attached from its 5′ region on surface, and 3′ region is free for enzymatic reaction. This technique is performed on two-dimensional microarray platform. APEX works on the principle of designing a probe, which terminates amplifica-tion of the sequence just before mutation site. Hence, wild-type alleles can be differentiated from mutated alleles. Nucleotide extension is carried out by incorporation of fluorescent labeled dideoxynucleotide (ddNTP), which is complimentary to the nucleotide base of target DNA. To detect a particular mutation, two different types of probes are ligated, i.e., for sense and antisense DNA [50,57]. In the Jewish population, comprehen-sive APEX array was performed for the detection of 59 sequence variants in Tay–Sachs disease, Bloom syndrome, Canavan disease, Niemann–Pick A, familial dysautonomia, torsion dystonia, mucolipidosis type IV, Fanconi anemia, Gaucher disease, factor XI deficiency, glycogen storage disease type 1a, maple syrup urine disease, nonsyndromic sensorineural hearing loss, familial Mediterranean fever, and glycogen storage [58].

8. SINGLE-BASE EXTENSION-TAGS TECHNOLOGY

Single-base extension (SBE) tagged array system is one of the flex-ible and alternative approaches for SNP detection. Similar to APEX, in SBE tagged array there are some changes in minisequencing reaction. Bifunctional primers having a specific sequence attached to specific locus are used in primer extension reaction, which is carried out next to mutation site by the addition of fluorescent labeled ddNTP which are detected in further rounds of amplification, extension, and termination. Wild-type and mutated alleles are detected by different colored dyes. In SBE microarray allele, specific tagged probes are used, which are independent to the particular mutation being analyzed. Thus it is a simple, inexpensive, and highly accurate technique for detection of SNPs [50,59].

9. ATOMIC FORCE MICROSCOPY

The atomic force microscopy (AFM) is a technique used for the measurement of intermolecular or interatomic force between molecules in air or aqueous solution by nanometer-sized mechanical probe, which is mounted on a microcantilever [60]. Ionizing radiations induce unrepairable double-strand break in DNA leading to cell death [61]. AFM is a very useful tool for detecting damaged or short fragmented DNA of a few tens to a few hundred nanometer length. DNA fragments of different sizes are compared with the standards of predefined size, which is used to calculate the average number of double-strand break per unit of DNA by application of mathematical formula [62].

MutS is an important protein that plays a crucial role in DNA mismatch repair system where it binds the DNA at the base mismatch site [63]. Tanigawa et al. performed AFM to study the structure of DNA protein complex and also reported it to be helpful in determination of transcription factor binding region in DNA. AFM is utilized to detect base substitution in DNA by detecting MutS binding at mutated region [64]. Later, Zhang et al. used this approach to find out heteroduplex in DNA. They found that MutS protein had more affinity to bind with heteroduplex DNA rather than homoduplex. Binding ratio of DNA to MutS is reflected between homoduplex and heteroduplex. Free DNA is easily distinguished from DNA–protein complex, therefore this microscopy technique is used in gene mutation detection.

10. FLUORESCENCE IN SITU HYBRIDIZATION

Fluorescence in situ hybridization (FISH) is used to detect chromosomal abnormalities or mutations using fluorescent labeled probes. Abnormalities such as aneuploidy, gene fusion, or loss of chromosomal region can be monitored for diagnostic and research purposes. Gene mapping and identification of oncogene by FISH are helpful in cancer diagnostics. From single-gene mutation identification to multiple aberrations can be identified by this method [65].

A fluorescent labeled probe recognizes and binds the unique complementary sequence on single-stranded DNA. It is carried out in interphase or metaphase stage of cell division. Further, DNA is fixed and followed by labeling with specifically designed tagged probes, which can be monitored under fluorescence microscope for detection of any chromosomal alterations [66]. Two types of labeling strategies are followed in fluorescent tagging of a probe. First is indirect labeling, in which the nucleotide probe is slightly modified and attached with a Hapten. This probe and target DNA are denatured to get a linearized structure and are allowed to hybridize with complementary sequence. In indirect labeling, additional fluorophore has to be provided, which binds to Hapten and generates fluorescence. The second approach is direct labeling where the probe is directly tagged to the fluorophore. The probe binds with its complementary sequence in target DNA and can be easily detected under fluorescence microscope [67].

Fluorophore-tagged probe specifically detect its complementary target. However, more than one target can be visualized using multicolor labeling approach, which is useful in parallel detection of multiple chromosomal aberrations. Probe selection is very important in fluorescent in situ hybridization. Depending on various applications, three types of probes are available. First one is whole-chromosome painting probes. This helps in labeling and imaging of separate chromosome in interphase or metaphase stage of cell cycle. Chromosome-specific fluorescent detection is helpful in pathological conditions and identification of chromosomal anomalies. Chromosome-specific probe is designed by amplification and labeling of chromosome-specific region by PCR. Using this type of probe chromosomal abnormalities can be visualized only in metaphase and not in interphase as the signal gets diffused. Chromosomal painting probes were used to diagnose the genetic translocation t(12; 21) (p12; q22) in acute lymphoblastic leukemia. There are 24 different color chromosomal painting probes available to label human chromosomes. Second type of probe is repetitive sequence probe, which detects the repetitive sequence of a specific chromosomal region. One of the examples of using this probe is α and β satellite targeting Pan-telomeric probes. The third type of probe is locus-specific probe. This probe comes with variable size depending on application and is helpful in chromosomal rearrangement studies such as inversion, deletion, or translocation. BCR/ABL gene translocation can be detected in leukemia by locus-specific probes [65].

11. DNA SEQUENCING

1. *Maxam–Gilbert and Sanger sequencing.* Maxam–Gilbert sequencing is a chemical sequencing method that requires radioactive labeling of the DNA at the 5′ end. Chemical treatments such as depurination, hydrolysis, and methylation generate breakage of terminally labeled nucleotides, which are analyzed on the polyacrylamide gel [68]. Whereas, Sanger approach of DNA sequencing is dideoxynucleotide chain termination reaction method, in which 2′,3′-dideoxynucleotide triphosphates (ddNTPs) are used for termination of DNA elongation [69]. Small alterations in genome such as, deletion, insertion, or point mutation can be easily identified by sequencing [70]. In a study conducted by Furitsu et al., mutations in the coding region of the Proto-oncogene c-kit were identified in a human mast cell leukemia cell line by following the Sanger sequencing procedure [71].

2. *Pyrosequencing.* Pyrosequencing is a real-time approach of DNA sequencing that works on the principle of release of pyrophosphate during DNA polymerase reaction. Conversion of these pyrophosphates to ATP is done by sulfurylase reaction and can be quantified through visible light production by firefly luciferase. After each cycle of amplification, addition of enzyme is required. DNA to be sequenced is allowed to hybridize with primer and incubated with mixture of polymerase, ATP sulfurylase, luciferase, and apyrase. Deoxynucleotide is incorporated in the growing chain, and pyrophosphate is subsequently released. Sulfurylase converts pyrophosphate into ATP, which in turn is converted into light by firefly luciferase and detected through CCD camera [72]. Ogino et al. assessed mixtures of DNA from mutant *KRAS* cell lines as well as wild-type *KRAS* cell line and reported that the mutation detection rates for pyrosequencing were superior to dideoxy sequencing [73].

3. *Next generation sequencing.* Next generation sequencing is a non–Sanger-based sequencing approach having immense role in many scientific achievements. It is known for its speed, sensitivity, and reduced cost for sequencing [74]. It is utilized to identify mutation in cells. For example, in case of cancer there is a need to identify genetic alterations in many genes, which can be done rapidly and reliably by next generation sequencing. Identification of chromosomal aberrations in fetal DNA, somatic or inherited mutations, or disorders associated with Mendelian

genetics can also be detected [75]. Tumor heterogeneity is also detected by identification of mutant allele frequency. In clinical samples, colorectal cancer mutations were detected in PIK3CA, NRAS, KRAS, and BRAF genes by next generation sequencing [76].

12. KARYOTYPING

Karyotyping is the process of pairing all the chromosomes of an individual and detecting any alterations on the basis of morphological features, thereby providing entire genome information. Karyotyping is one of the cytogenetic approaches used for detection of structural and numerical changes in chromosomes. Chromosomal aberrations such as deletion, duplication, translocation, inversion can be identified by karyotyping [2]. Aberrant chromosomes and cytogenetic abnormalities were detected using karyotyping in chronic lymphocytic leukemia patients having 17p deletion in TP53 gene [77].

13. HYPOXANTHINE PHOSPHORIBOSYL TRANSFERASE, THYMIDINE KINASE, AND XANTHINE–GUANINE PHOSPHORIBOSYL TRANSFERASE GENE MUTATION DETECTION METHODS

There might be a possibility that a small proportion of mammalian-specific mutations may not be detected by using both chromosomal aberrations and bacterial gene mutation assays in mammalian cells. Gene mutations induced by chemicals can be detected by the in vitro mammalian cell gene mutation tests with great efficiency. In mutagenicity, testing with cultured mammalian cells has some benefits over microbial tests. Firstly in bacteria the cell division apparatus, i.e., genomic organization is absent and secondly the cell metabolism specific to mammals cannot be replicated in bacteria [78]. The two common assays for the detection of gene mutations in mammalian cells are (1) hypoxanthine phosphoribosyl transferase (HPRT) and (2) thymidine kinase (TK) assay. Forward mutations are the main basis that confers resistance to a toxic chemical [79]. In both gene mutation detection techniques, forward mutations inactivate sex linked loci (e.g., HPRT$^+$) or a wild-type gene at heterozygous autosomal loci (e.g., TK$^{+/-}$). For the analysis of mutations in mammalian cell lines, HPRT gene is used as a model gene, which is located on the X chromosome of mammalian cells. HPRT gene–based gene mutation assay is used to detect any mutation caused by DNA damage by a wide range of chemicals, which alters the gene expression/function of an HPRT⁻ mutant.

The methodology of both the TK and HPRT test follows a very similar pattern. In HPRT assay, those mutations which ruins the HPRT gene's functionality and or/protein are positively selected by using a toxic analogue, and viable colonies of HPRT⁻ mutants are formed. TK6, AHH-1, MCL-5 human lymphoblastoid cells, AS52, CHO, and V79 cell lines of Chinese hamster cells and L5178Y mouse lymphoma cells are suitable for the detection of gene mutation in cultured mammalian cells [80,81].

The most commonly used genetic endpoints in these cell lines measure mutations at HPRT, TK, and a transgene of xanthine–guanine phosphoribosyl transferase (XPRT). All the abovementioned tests detect different spectra of genetic events. Some genetic events such as large deletions not noticed at the HPRT locus on X-chromosome can be overcome by the autosomal location of TK and XPRT.

Due to the mutation, thymidine kinase–deficient cells are resistant to the cytotoxic effects of the pyrimidine analogue trifluorothymidine (TFT). Thymidine kinase rich cells are sensitive to TFT, which halts further cell division by inhibiting the cellular metabolism. By this way only mutant cells can proliferate in the presence of TFT. In a similar fashion the cells lacking HPRT or XPRT are chosen by resistance to 6-thioguanine (TG) or 8-azaguanine (AG).

The crucial attribute of mammalian HPRT gene mutation assay that makes this technique widely accepted is that the mutations in the same gene can be compared among experimental animals, cell lines, and humans [78]. Small changes such as point mutations and exon deletions can also be detected in the HPRT gene [82].

14. CONCLUSION

Identification of gene mutation is the primary step for the treatment of genetic abnormalities. In this chapter, various techniques available for the detection of known and unknown mutations have been described. Each technique has some advantages and limitations over the other. In summary, PCR is a powerful and cost-effective tool in gene mutation detection, which people are using since long time. Recently, numerous techniques are available for identification of gene mutations, which are highly sensitive and also produce high-throughput results. Screening methods such as SSCP, DGGE, and FISH are very useful and sensitive, but sequencing methodologies remain the gold standard technique to validate and identity a specific mutation. The choice of the technique for the identification of gene mutation largely depends on user's requirement.

ACKNOWLEDGMENT

Jayant Dewangan is thankful to UGC for the award of Senior Research Fellowship. Prabhash Kumar Pandey (PDF/2015/000033) is thankful to Department of Science and Technology (DST) for providing financial assistance. Aman Divakar is thankful to DBT for the award of Junior Research Fellowship. Sonal Srivastava is thankful to Indian Council of Medical Research for the award of Senior Research Fellowship. Sakshi Mishra is thankful to Department of Science and Technology (DST) for providing financial assistance (SR/WOS-A/LS-1290/2015). This work was supported by Council of Scientific and Industrial Research Network project BSC0103. The CSIR-CDRI communication No. for this manuscript is 9532.

REFERENCES

[1] Balin SJ, Cascalho M. The rate of mutation of a single gene. Nucleic Acids Res March 2010;38(5):1575–82. PubMed PMID: 20007603.

[2] Mahdieh N, Rabbani B. An overview of mutation detection methods in genetic disorders. Iran J Pediatr August 2013;23(4):375–88. PubMed PMID: 24427490.

[3] Wabl M, Burrows PD, von Gabain A, Steinberg C. Hypermutation at the immunoglobulin heavy chain locus in a pre-B-cell line. Proc Natl Acad Sci USA January 1985;82(2):479–82. PubMed PMID: 3918310.

[4] Wang CL, Wabl M. Hypermutation rate normalized by chronological time. J Immunol May 01, 2005;174(9):5650–4. PubMed PMID: 15843565. Epub 2005/04/22. eng.

[5] Ashworth A, Bernards R. Using functional genetics to understand breast cancer biology. Cold Spring Harbor Perspect Biol July 2010;2(7):a003327. PubMed PMID: 20519343. Pubmed Central PMCID: PMC2890198. Epub 2010/06/04. eng.

[6] Schlabach MR, Luo J, Solimini NL, Hu G, Xu Q, Li MZ, et al. Cancer proliferation gene discovery through functional genomics. Science February 01, 2008;319(5863):620–4. PubMed PMID: 18239126. Pubmed Central PMCID: PMC2981870. Epub 2008/02/02. eng.

[7] Botstein D, White RL, Skolnick M, Davis RW. Construction of a genetic linkage map in man using restriction fragment length polymorphisms. Am J Hum Genet May 1980;32(3):314–31. PubMed PMID: 6247908. Pubmed Central PMCID: PMC1686077. Epub 1980/05/01. eng.

[8] Urban T, Ricci S, Grange JD, Lacave R, Boudghene F, Breittmayer F, et al. Detection of c-Ki-ras mutation by PCR/RFLP analysis and diagnosis of pancreatic adenocarcinomas. J Natl Cancer Inst December 15, 1993;85(24):2008–12. PubMed PMID: 7902444. Epub 1993/12/15. eng.

[9] Molina-Vila MA, Bertran-Alamillo J, Reguart N, Taron M, Castella E, Llatjos M, et al. A sensitive method for detecting EGFR mutations in non-small cell lung cancer samples with few tumor cells. J Thorac Oncol November 2008;3(11):1224–35. PubMed PMID: 18978556.

[10] Urban T, Ricci S, Danel C, Antoine M, Kambouchner M, Godard V, et al. Detection of codon 12 K- ras mutations in non-neoplastic mucosa from bronchial carina in patients with lung adenocarcinomas. Br J Cancer January 2000;82(2):412–7. PubMed PMID: 10646897.

[11] Li WM, Hu TT, Zhou LL, Feng YM, Wang YY, Fang J. Highly sensitive detection of the PIK3CA(H1047R) mutation in colorectal cancer using a novel PCR-RFLP method. BMC Cancer 2016;16. PubMed PMID: 27405731.

[12] Kawada I, Soejima K, Watanabe H, Nakachi I, Yasuda H, Naoki K, et al. An alternative method for screening EGFR mutation using RFLP in non-small cell lung cancer patients. J Thorac Oncol October 2008;3(10):1096–103. PubMed PMID: 18827604. Epub 2008/10/02. eng.

[13] Grau O, Griffais R. Diagnosis of mutations by the PCR double RFLP method (PCR-dRFLP). Nucleic Acids Res December 25, 1994;22(25):5773–4. PubMed PMID: 7838739.

[14] Machnicki MM, Glodkowska-Mrowka E, Lewandowski T, Ploski R, Wlodarski P, Stoklosa T. ARMS-PCR for detection of BRAF V600E hotspot mutation in comparison with Real-Time PCR-based techniques. Acta Biochim Pol 2013;60(1):57–64. PubMed PMID: 23460942. Epub 2013/03/06. eng.

[15] Carl W. Dieffenbach GSD. PCR primer: a laboratory manual. 2nd ed. Cold Spring Harbor Laboratory Press; 2003.

[16] Ye J, Coulouris G, Zaretskaya I, Cutcutache I, Rozen S, Madden TL. Primer-BLAST: a tool to design target-specific primers for polymerase chain reaction. BMC Bioinformatics June 18, 2012;13:134. PubMed PMID: 22708584. Pubmed Central PMCID: PMC3412702. Epub 2012/06/20. eng.

[17] Chamberlain JS, Gibbs RA, Ranier JE, Nguyen PN, Caskey CT. Deletion screening of the Duchenne muscular dystrophy locus via multiplex DNA amplification. Nucleic Acids Res December 9, 1988;16(23):11141–56. PubMed PMID: 3205741.

[18] Rithidech KN, Dunn JJ, Gordon CR. Combining multiplex and touchdown PCR to screen murine microsatellite polymorphisms. BioTechniques July 1997;23(1). 36, 40, 2, 4. PubMed PMID: 9232223. Epub 1997/07/01. eng.

[19] Shuber AP, Skoletsky J, Stern R, Handelin BL. Efficient 12-mutation testing in the CFTR gene: a general model for complex mutation analysis. Hum Mol Genet February 1993;2(2):153–8. PubMed PMID: 7684636. Epub 1993/02/01. eng.

[20] Zimmermann K, Schogl D, Plaimauer B, Mannhalter JW. Quantitative multiple competitive PCR of HIV-1 DNA in a single reaction tube. BioTechniques September 1996;21(3):480–4. PubMed PMID: 8879588. Epub 1996/09/01. eng.

[21] Casas I, Pozo F, Trallero G, Echevarria JM, Tenorio A. Viral diagnosis of neurological infection by RT multiplex PCR: a search for entero- and herpesviruses in a prospective study. J Med Virol February 1999;57(2):145–51. PubMed PMID: 9892399. Epub 1999/01/19. eng.

[22] Heredia A, Soriano V, Weiss SH, Bravo R, Vallejo A, Denny TN, et al. Development of a multiplex PCR assay for the simultaneous detection and discrimination of HIV-1, HIV-2, HTLV-I and HTLV-II. Clin Diagn Virol November 1996;7(2):85–92. PubMed PMID: 9137864. Epub 1996/11/01. eng.

[23] Markoulatos P, Samara V, Siafakas N, Plakokefalos E, Spyrou N, Moncany ML. Development of a quadriplex polymerase chain reaction for human cytomegalovirus detection. J Clin Lab Anal 1999;13(3):99–105. PubMed PMID: 10323473. Epub 1999/05/14. eng.

[24] Markoulatos P, Mangana-Vougiouka O, Koptopoulos G, Nomikou K, Papadopoulos O. Detection of sheep poxvirus in skin biopsy samples by a multiplex polymerase chain reaction. J Virol Methods February 2000;84(2):161–7. PubMed PMID: 10680965. Epub 2000/02/19. eng.

[25] Markoulatos P, Georgopoulou A, Kotsovassilis C, Karabogia-Karaphillides P, Spyrou N. Detection and typing of HSV-1, HSV-2, and VZV by a multiplex polymerase chain reaction. J Clin Lab Anal 2000;14(5):214–9. PubMed PMID: 11018799. Epub 2000/10/06. eng.

[26] Hendolin PH, Markkanen A,Ylikoski J,Wahlfors JJ. Use of multiplex PCR for simultaneous detection of four bacterial species in middle ear effusions. J Clin Microbiol November 1997;35(11):2854–8. PubMed PMID: 9350746.

[27] Harris E, Kropp G, Belli A, Rodriguez B, Agabian N. Single-step multiplex PCR assay for characterization of New World Leishmania complexes. J Clin Microbiol July 1998;36(7):1989–95. PubMed PMID: 9650950. Pubmed Central PMCID: PMC104966. Epub 1998/07/03. eng.

[28] Henegariu O, Heerema NA, Dlouhy SR,Vance GH,Vogt PH. Multiplex PCR: critical parameters and step-by-step protocol. BioTechniques September 1997;23(3):504–11. PubMed PMID: 9298224. Epub 1997/09/23. eng.

[29] Exner MM. Multiplex molecular reactions: design and troubleshooting. Clin Microbiol Newslett 2012;34(8):59–65.

[30] Fernandez S, Pagotto AH, Furtado MM, Katsuyama AM, Madeira AM, Gruber A. A multiplex PCR assay for the simultaneous detection and discrimination of the seven Eimeria species that infect domestic fowl. Parasitology October 2003;127(Pt 4):317–25. PubMed PMID: 14636018. Epub 2003/11/26. eng.

[31] Yan W,Wang W,Wang T, Suo X, Qian W,Wang S, et al. Simultaneous identification of three highly pathogenic Eimeria species in rabbits using a multiplex PCR diagnostic assay based on ITS1-5.8S rRNA-ITS2 fragments.Vet Parasitol March 31, 2013;193 (1–3):284–8. PubMed PMID: 23246036. Epub 2012/12/19. eng.

[32] Mishra B, Sharma M, Pujhari SK, Ratho RK, Gopal DS, Kumar CN, et al. Utility of multiplex reverse transcriptase-polymerase chain reaction for diagnosis and serotypic characterization of dengue and chikungunya viruses in clinical samples. Diagn Microbiol Infect Dis October 2011;71(2):118–25. PubMed PMID: 21865001. Epub 2011/08/26. eng.

[33] Milbury CA, Chen CC, Mamon H, Liu P, Santagata S, Makrigiorgos GM. Multiplex amplification coupled with COLD-PCR and high resolution melting enables identification of low-abundance mutations in cancer samples with low DNA content. J Mol Diagn March 2011;13(2):220–32. PubMed PMID: 21354058.

[34] Chen CY, Chi KH, Pillay A, Nachamkin E, Su JR, Ballard RC. Detection of the A2058G and A2059G 23S rRNA gene point mutations associated with azithromycin resistance in Treponema pallidum by use of a TaqMan real-time multiplex PCR assay. J Clin Microbiol March 2013;51(3):908–13. PubMed PMID: 23284026. Pubmed Central PMCID: PMC3592075. Epub 2013/01/04. eng.

[35] Fukayama N, Sugano K, Kyogoku A, Nose H, Kondou H, Ohkura H. Sensitive detection of K-ras oncogene codon 12 mutations by nested PCR using mismatched primers and selective digestion of non-mutated PCR fragments with restriction enzyme. Rinsho Byori September 1993;41(9):1017–23. PubMed PMID: 8254964. Epub 1993/09/01. jpn.

[36] Banerjee SK, Makdisi WF, Weston AP, Campbell DR. A two-step enriched-nested PCR technique enhances sensitivity for detection of codon 12 K-ras mutations in pancreatic adenocarcinoma. Pancreas July 1997;15(1):16–24. PubMed PMID: 9211488. Epub 1997/07/01. eng.

[37] Varley KE, Mitra RD. Nested Patch PCR enables highly multiplexed mutation discovery in candidate genes. Genome Res November 2008;18(11):1844–50. PubMed PMID: 18849522. Pubmed Central PMCID: PMC2577855. Epub 2008/10/14. eng.

[38] Michalska D, Jaguszewska K, Liss J, Kitowska K, Mirecka A, Łukaszuk K. Comparison of whole genome amplification and nested-PCR methods for preimplantation genetic diagnosis for BRCA1 gene mutation on unfertilized oocytes–a pilot study. Hered Cancer Clin Pract 2013;11(1):10. PubMed PMID: 23941236.

[39] Cotton RG. Current methods of mutation detection. Mutation Res January 1993;285(1):125–44. PubMed PMID: 7678126. Epub 1993/01/01. eng.

[40] Tan J, Sharief Y, Hamil KG, Gregory CW, Zang DY, Sar M, et al. Dehydroepiandrosterone activates mutant androgen receptors expressed in the androgen-dependent human prostate cancer xenograft CWR22 and LNCaP cells. Mol Endocrinol April 1997;11(4):450–9. PubMed PMID: 9092797. Epub 1997/04/01. eng.

[41] Guldberg P, Thor Straten P, Birck A, Ahrenkiel V, Kirkin AF, Zeuthen J. Disruption of the MMAC1/PTEN gene by deletion or mutation is a frequent event in malignant melanoma. Cancer Res September 01, 1997;57(17):3660–3. PubMed PMID: 9288767. Epub 1997/09/01. eng.

[42] Vidal MJ, Loganzo Jr F, de Oliveira AR, Hayward NK, Albino AP. Mutations and defective expression of the WAF1 p21 tumour-suppressor gene in malignant melanomas. Melanoma Res August 1995;5(4):243–50. PubMed PMID: 7496159. Epub 1995/08/01. eng.

[43] Danks MK, Warmoth MR, Friche E, Granzen B, Bugg BY, Harker WG, et al. Single-strand conformational polymorphism analysis of the M(r) 170,000 isozyme of DNA topoisomerase II in human tumor cells. Cancer Res March 15, 1993;53(6):1373–9. PubMed PMID: 8383009. Epub 1993/03/15. eng.

[44] Tian H, Brody LC, Landers JP. Rapid detection of deletion, insertion, and substitution mutations via heteroduplex analysis using capillary- and microchip-based electrophoresis. Genome Res September 2000;10(9):1403–13. PubMed PMID: 10984458. Pubmed Central PMCID: PMC310899. Epub 2000/09/14. eng.

[45] Zielenski J, Aznarez I, Onay T, Tzounzouris J, Markiewicz D, Tsui L-C. CFTR mutation detection by multiplex heteroduplex (mHET) analysis on MDE gel. In: Skach WR, editor. Cystic fibrosis methods and protocols, vol. 70. Humana Press; 2002. p. 3–19.

[46] Baker CC, Phelps WC, Lindgren V, Braun MJ, Gonda MA, Howley PM. Structural and transcriptional analysis of human papillomavirus type 16 sequences in cervical carcinoma cell lines. J Virol April 1987;61(4):962–71. PubMed PMID: 3029430.

[47] Fan JB, Chee MS, Gunderson KL. Highly parallel genomic assays. Nat Rev Genet August 2006;7(8):632–44. PubMed PMID: 16847463. Epub 2006/07/19. eng.

[48] Heller MJ. DNA microarray technology: devices, systems, and applications. Ann Rev Biomed Eng 2002;4:129–53. PubMed PMID: 12117754. Epub 2002/07/16. eng.

[49] Barrett JC, Kawasaki ES. Microarrays: the use of oligonucleotides and cDNA for the analysis of gene expression. Drug Discov Today February 1, 2003;8(3):134–41. PubMed PMID: 12568783. Epub 2003/02/06. eng.

[50] Cremonesi L, Ferrari M, Giordano PC, Harteveld CL, Kleanthous M, Papasavva T, et al. An overview of current microarray-based human globin gene mutation detection methods. Hemoglobin 2007;31(3):289–311. PubMed PMID: 17654067. Epub 2007/07/27. eng.

[51] Salvado CS, Trounson AO, Cram DS. Towards preimplantation diagnosis of cystic fibrosis using microarrays. Reprod Biomed Online January 2004;8(1):107–14. PubMed PMID: 14759297. Epub 2004/02/05. eng.

[52] Hacia JG, Brody LC, Chee MS, Fodor SP, Collins FS. Detection of heterozygous mutations in BRCA1 using high density oligonucleotide arrays and two-colour fluorescence analysis. Nat Genet December 1996;14(4):441–7. PubMed PMID: 8944024. Epub 1996/12/01. eng.

[53] Hacia JG, Sun B, Hunt N, Edgemon K, Mosbrook D, Robbins C, et al. Strategies for mutational analysis of the large multiexon ATM gene using high-density oligonucleotide arrays. Genome Res December 1998;8(12):1245–58. PubMed PMID: 9872980. Epub 1999/01/05. eng.

[54] Kim IJ, Kang HC, Park JH, Ku JL, Lee JS, Kwon HJ, et al. RET oligonucleotide microarray for the detection of RET mutations in multiple endocrine neoplasia type 2 syndromes. Clin Cancer Res February 2002;8(2):457–63. PubMed PMID: 11839664. Epub 2002/02/13. eng.

[55] Berber E, Leggo J, Brown C, Berber E, Gallo N, Feilotter H, et al. DNA microarray analysis for the detection of mutations in hemophilia A. J Thromb Haemost August 2006;4(8):1756–62. PubMed PMID: 16879218. Epub 2006/08/02. eng.

[56] Baaj Y, Magdelaine C, Ubertelli V, Valat C, Talini L, Soussaline F, et al. A highly specific microarray method for point mutation detection. BioTechniques January 2008;44(1):119–26. PubMed PMID: 18254389. Epub 2008/02/08. eng.

[57] Gemignani F, Perra C, Landi S, Canzian F, Kurg A, Tonisson N, et al. Reliable detection of beta-thalassemia and G6PD mutations by a DNA microarray. Clin Chem November 2002;48(11):2051–4. PubMed PMID: 12406995. Epub 2002/10/31. eng.

[58] Schrijver I, Külm M, Gardner PI, Pergament EP, Fiddler MB. Comprehensive arrayed primer extension array for the detection of 59 sequence variants in 15 conditions prevalent among the (Ashkenazi) Jewish population. J Mol Diagnost 2007;9(2):228–36.

[59] Hirschhorn JN, Sklar P, Lindblad-Toh K, Lim YM, Ruiz-Gutierrez M, Bolk S, et al. SBE-TAGS: an array-based method for efficient single-nucleotide polymorphism genotyping. Proc Natl Acad Sci USA October 24, 2000;97(22):12164–9. PubMed PMID: 11035790. Pubmed Central PMCID: PMC17312. Epub 2000/10/18. eng.

[60] Binnig G, Quate CF, Gerber C. Atomic force microscope. Phys Rev Lett March 3, 1986;56(9):930–3. PubMed PMID: 10033323. Epub 1986/03/03. Eng.

[61] Mladenov E, Magin S, Soni A, Iliakis G. DNA double-strand break repair as determinant of cellular radiosensitivity to killing and target in radiation therapy. Front Oncol 2013;3:113. PubMed PMID: 23675572. Pubmed Central PMCID: PMC3650303. Epub 2013/05/16. eng.

[62] Pang D, Thierry AR, Dritschilo A. DNA studies using atomic force microscopy: capabilities for measurement of short DNA fragments. Front Mol Biosci 2015;2:1. PubMed PMID: 25988169. Pubmed Central PMCID: 4429637.

[63] Li GM. Mechanisms and functions of DNA mismatch repair. Cell Res January 2008;18(1):85–98. PubMed PMID: 18157157. Epub 2007/12/25. eng.

[64] Tanigawa M, Gotoh M, Machida M, Okada T, Oishi M. Detection and mapping of mismatched base pairs in DNA molecules by atomic force microscopy. Nucleic Acids Res May 1, 2000;28(9):E38. PubMed PMID: 10756205. Pubmed Central PMCID: PMC103311. Epub 2000/04/11. eng.

[65] Bishop R. Applications of fluorescence in situ hybridization (FISH) in detecting genetic aberrations of medical significance. Biosci Horizons 2010;3(1):85–95.

[66] Gozzetti A, Le Beau MM. Fluorescence in situ hybridization: uses and limitations. Semin Hematol October 2000;37(4):320–33. PubMed PMID: 11071355. Epub 2000/11/09. eng.

[67] Speicher MR, Carter NP. The new cytogenetics: blurring the boundaries with molecular biology. Nat Rev Genet October 2005;6(10):782–92. PubMed PMID: 16145555. Epub 2005/09/08. eng.

[68] Maxam AM, Gilbert W. A new method for sequencing DNA. Proc Natl Acad Sci USA February 1977;74(2):560–4. PubMed PMID: 265521.

[69] Sanger F, Nicklen S, Coulson AR. DNA sequencing with chain-terminating inhibitors. Proc Natl Acad Sci USA December 1977;74(12):5463–7. PubMed PMID: 271968.

[70] Montgomery KT, Iartchouck O, Li L, Perera A, Yassin Y, Tamburino A, et al. Mutation detection using automated fluorescence-based sequencing. Curr Protoc Hum Genet 2008. Chapter 7. 2008/04/23, Unit 7 9.

[71] Furitsu T, Tsujimura T, Tono T, Ikeda H, Kitayama H, Koshimizu U, et al. Identification of mutations in the coding sequence of the proto-oncogene c-kit in a human mast cell leukemia cell line causing ligand-independent activation of c-kit product. J Clin Invest 1993;92(4):1736.

[72] Ronaghi M, Uhlen M, Nyren P. A sequencing method based on real-time pyrophosphate. Science July 17, 1998;281(5375). 363, 5. PubMed PMID: 9705713. Epub 1998/08/15. eng.

[73] Ogino S, Kawasaki T, Brahmandam M, Yan L, Cantor M, Namgyal C, et al. Sensitive sequencing method for KRAS mutation detection by pyrosequencing. J Mol Diagnost 2005;7(3):413–21.

[74] Schuster SC. Next-generation sequencing transforms today's biology. Nat Methods January 2008;5(1):16–8. PubMed PMID: 18165802. Epub 2008/01/01. Eng.

[75] Meldrum C, Doyle MA, Tothill RW. Next-generation sequencing for cancer diagnostics: a practical perspective. Clin Biochem Rev November 2011;32(4):177–95. PubMed PMID: 22147957.

[76] Haley L, Tseng LH, Zheng G, Dudley J, Anderson DA, Azad NS, et al. Performance characteristics of next-generation sequencing in clinical mutation detection of colorectal cancers. Modern Pathol October 2015;28(10):1390–9. PubMed PMID: 26226847. Pubmed Central PMCID: 4618462.

[77] Dicker F, Herholz H, Schnittger S, Nakao A, Patten N, Wu L, et al. The detection of TP53 mutations in chronic lymphocytic leukemia independently predicts rapid disease progression and is highly correlated with a complex aberrant karyotype. Leukemia January 2009;23(1):117–24. PubMed PMID: 18843282. Epub 2008/10/10. eng.

[78] Johnson GE. Mammalian cell HPRT gene mutation assay: test methods. Methods Mol Biol 2012;817:55–67. PubMed PMID: 22147568. Epub 2011/12/08. eng.

[79] DeMarini DM, Brockman HE, de Serres FJ, Evans HH, Stankowski Jr LF, Hsie AW. Specific-locus mutations induced in eukaryotes (especially mammalian cells) by radiation and chemicals: a perspective. Mutation Res January 1989;220(1):11–29. PubMed PMID: 2643030. Epub 1989/01/01. eng.

[80] Parry JM, Parry EM, Johnson G, Quick E, Waters EM. The detection of genotoxic activity and the quantitative and qualitative assessment of the consequences of exposures. Exp Toxicol Pathol July 2005;57(Suppl. 1):205–12. PubMed PMID: 16092728. Epub 2005/08/12. eng.

[81] Elsa Nielsen GO, Larsen JC. Toxicological risk assessment of chemicals: a practical guide. Taylor & Francis Group; 2008.

[82] Zoulikha M, Zaïr GEJ. The applicable use of the HPRT gene mutation assay as a practical tool in mutagenesis and DNA repair studies. In: María Sierra L, Gaivão I, editors. Genotoxicity and DNA repair. New York: Springer; 2014. p. 185–97.

CHAPTER FOUR

Chromosomal Aberrations

Abhishek K. Jain, Divya Singh, Kavita Dubey, Renuka Maurya, Alok K. Pandey
CSIR-Indian Institute of Toxicology Research, Lucknow, India

1. INTRODUCTION

Chromosomes are heritable material inside the cell having unique properties that discriminate them from other cellular organelles. The number of chromosomes varies between different species but remains constant from cell to cell within same organism. The word *chromosome* comes from the Greek word, *Chroma* (color) and *soma* (body) as they possess the characteristic of strong staining with certain dyes used in scientific research [1]. Although chromosomes have been noticed as cell structure in the 19th century, it was only after the discovery of Mendel's principle that their fundamental role in heredity was established. At the time of cell division, a chromosome looks like a worm-shaped structure, which is easy to monitor under a microscope [2]. A chromosome keeps DNA tightly bound around the histone protein (a protein present in the nuclei of eukaryotic cell that packages DNA into structural units). Chromosomes play an important role during cell division, which ensures that the DNA replication is accurate and distributed between newly formed cells after cell division. Each chromosome in a mammalian cell nucleus, as well as in nuclei of other eukaryotes, occupies a limited part of the nuclear volume, forming a chromosome territory [3]. Chromosomes can vary largely in their size between different organisms as the difference can be over 100 folds. As a matter of fact, within a single genome, the length of the chromosome can vary considerably as in human, difference between chromosome 1 (the biggest) and chromosome 21 (the smallest).

Chromosomal alterations lead to genetic instability, which is a major cause of various genetic disorders such as Down syndrome, Triple X

Mutagenicity: Assays and Applications
ISBN 978-0-12-809252-1
http://dx.doi.org/10.1016/B978-0-12-809252-1.00004-3

syndrome, Chronic myeloid leukemia, Burkitt lymphoma, and many more. These chromosomal alterations (loss or gain) in complete set of chromosome may be lethal, i.e., aneuploidy, as many give rise to genetic disorders. Consequently, chromosomal abnormality is also responsible for failure of gametogenesis, fertilization, birth defects, deformities in live-born infants and mental retardations [4]. Depending on the size, location, and timing, structural changes could lead to birth defects, syndromes or even cancer. Families that carry a chromosome rearrangement are generally offered genetic counseling.

2. CHARACTERISTICS OF HUMAN CHROMOSOMES

Most human cells have a complete diploid set of 46 chromosomes divided into 23 pairs. Mature gametes, oocytes and spermatozoa, have normally haploid sets of chromosomes. Chorionic villi in liver cells and amniotic fluid cells comprise of four sets of chromosomes. Mature red blood cells have no nuclei and, subsequently, have no chromosomes. The haploid number between most of the organism is between 6 and 25. Humans have 23 pairs of chromosomes, out of which, 1 is the sex chromosome and other 22 are the autosomes, which are homologous in nature as they are morphologically similar [5]. In each homologous pair of autosomes, the lengths of both chromosomes are equal and the positions of the centromeres are also same. Sex chromosome is present as either XX (in females) or XY (in males). Genes for the same feature appear on the same locus of each matching pair of chromosome in every human cell.

A typical metaphase chromosome consists of two arms separated by primary constriction and may contain a secondary constriction too that is often near the end. This constriction marks the location of centromere, which is essential for the normal movement of the chromosomes with respect to the spindles. A chromosome without the centromere is termed as an acentric fragment. A typical metaphase chromosome is often recognized by the length of chromosomes and the position of centromeres. A chromosome that contains centromere in the middle is termed as metacentric. A chromosome in which centromere is at the end is termed as telocentric. In an acrocentric chromosome, both the arms are unequal and the position of the centromere is located quite near to one end of the chromosome [6,7].

3. SOURCES OF SPECIMEN FOR CYTOGENETIC ANALYSIS

Chromosome analysis can be performed using different cells and tissues. Sampling of the cells can be done before birth too for the collection of chorionic villi, amniotic fluid cells. The first step while performing chromosome analysis is the proper collection of samples. The blood samples should be kept by mixing anticoagulant to prevent the clotting of blood. The time required for chromosome analysis depends on the type of tissue or sample. As in the case of prenatal diagnosis, the cells that are rapidly dividing, i.e., chorionic villi of cytotrophoblast layer, permit direct and immediate chromosome analysis, whereas amniotic fluid of mesenchymal core requires approximately 6–14 days for chromosomal analysis. Amniocentesis is a process by which amniotic fluid is collected from developing fetus usually around the 16th week of pregnancy and is done to diagnose fetal genetic and developmental disorders, whereas chorionic villus sampling is done in the eighth and ninth week of pregnancy to diagnose biochemical and cytogenetic defect in the fetus. To conduct genetic testing, a number of methods are being developed to isolate fetal genetic material present in maternal blood as a target for non invasive prenatal diagnosis.

4. CYTOGENETIC ANALYSIS

This field was initiated in the year 1956. The analysis of the structure and number of chromosome within human and animal species is termed as cytogenetics. Cytogenetic analysis is generally performed during pregnancy to determine if fetus is at the risk of common aneuploidy (syndrome causing a change in the number of chromosomes) or syndromes causing structural changes (translocation or inversion) or missing genetic material through cytogenetic microarray. With the information provided by such analysis, more accurate diagnosis can be made, allowing for proper treatment of the child. The pairing and ordering of chromosomes of an organism is termed as karyotyping, which helps the clinical cytogeneticists to detect gross genetic changes and changes in the chromosome number related with aneuploidy condition as in the case of trisomy 21 (Down syndrome). A routine cytogenetic analysis involves the evaluation of at least 15–20 cells. The karyotype is generated on the basis of size, centromeric position, and banding pattern of the chromosome. An important part of any routine cytogenetic analysis is that the chromosome must exhibit 350 to 400 bands in the complement. The presence of abnormal chromosome complement in all the cells paves the way forward to analyze a small number of cells for diagnostic purposes [8,9].

Modern cytogenetic approaches allow a researcher to do many things such as the accurate labeling of the chromosomal location of a gene, identifying a cell that has got a translocation, losing or gaining a specific chromosome or set of genes on the chromosome. However, cytogenetic approaches have evolved over time with the development of recent techniques such as flow cytometry to sort chromosomes. Spectral karyotyping (SKY) or multiplex FISH (M-FISH) is another technique that permits the simultaneous tracking of all human chromosomes. There is yet another technique that is able to execute genome wide scans to detect chromosomal regions linked with gain or loss of genetic events known as comparative genome hybridization (CGH). This single microarray can provide information that is equal to that derived from 1000 fluorescence in situ hybridization (FISH) experiments.

5. TYPES OF CHROMOSOME ABERRATION

Chromosomal abnormalities are generated in chromosome due to alteration in genetic materials through loss, gain or rearrangement of particular segments. It can be divided into two main categories: (1) numeric chromosomal aberration (CA) and (2) structural CA.

5.1 Numeric Chromosomal Aberration

In this type of CA, increase or decrease in the number of chromosomes is seen. Numeric CA exists in two conditions: (1) euploidy and (2) aneuploidy.

5.1.1 Euploidy

Euploidy is a condition of a cell, tissue, or organism in which changes in chromosome number can occur by addition of one or more complete sets of chromosomes. Euploidy occurs frequently in plants but are rarely found in animals.

Euploidy can be divided into three categories:
- **Monoploidy**—(One set of chromosome present)
- **Haploidy**—(Half the number of chromosome present)
- **Polyploidy**—(Number of chromosomes present in multiple copies)

5.1.2 Aneuploidy

Aneuploidy is the presence of an abnormal number of chromosome in cells either in the autosomes or the sex chromosomes. Certain agents capable of causing aneuploidy are called "aneuploidogens." Mutagens, carcinogens, X-rays, and also chemicals such as colchicines can produce aneuploidy by affecting microtubules polymerization.

Aneuploidy exists in four categories:
- **Nullysomy**—(Lack of one pair of chromosome than the normal)
- **Monosomy**—(There is one less chromosome than the normal)
- **Trisomy**—(Presence of one extra chromosome than the normal)
- **Tetrasomy**—(Presence of one pair of chromosomes than the normal)

5.2 Structural Chromosomal Aberration

Structural chromosomal abnormalities occur when there is a change in the parts of a chromosome. It involves the rearrangement through gain, loss, and reallocation of chromosomal segment. This left one segment with too much or other segment with too little genetic material. These chromosome abnormalities lead to some birth defects.

There are four major types of aberration: deletion, duplication, inversion, and translocation.

5.2.1 Inversion

In general, inversion occurs when a segment of a chromosome is clipped off, inverted around 180 degrees and reallocated into the same chromosome. Detection of inversion in humans in quite difficult, and it does not change the phenotype of the individual unless the clipped region of the inversion is within the regulatory or structural region of the gene. Inversion is classified into various types on the basis of the location of the centromere corresponding with the inverted segment:
- **Paracentric**—A type of chromosomal rearrangement in which the segment does not include the centromere, i.e., it has been snipped out of the chromosome, turned through 180 degrees, and inserted again in the original position of chromosome is termed as paracentric inversion.
- **Pericentric inversions**—The pericentric inversion region mostly includes the centromere and a break point in each arm.
- **Acentric inversions**—Crossing over within the inversion loop of a paracentric inversion connects homologous centromere in a dicentric bridge while also producing an acentric fragment, which is a fragment without a centromere. However, the crossing over produces chromatids that contain duplication and a deficiency for different parts of the chromosome. Consequently, pericentric inversion may change the length of the two arms of the chromosomes, whereas a paracentric inversion has no such effect.

5.2.2 Translocation

When two nonhomologous chromosomes exchange their parts, the resulting chromosomal rearrangements are translocation. Translocation may be reciprocal or nonreciprocal.

- **Nonreciprocal translocation**—It involves the transfer of a segment in one direction from one chromosome to another.
- **Reciprocal translocation**— This is generally the exchange of materials between nonhomologous chromosomes. These translocations are harmless and can be detected by prenatal diagnosis. The exchange of chromosome parts between nonhomologous chromosomes establish new linkage relation. A special pattern of translocation involving two acrocentric chromosomes is called centric fusion type or Robertsonian translocation. Typically, the break occurs close to the centromere, affecting the short arms of both chromosomes. Transfer of the segments leads to one very large chromosome and one extremely small one (Fig. 4.1).

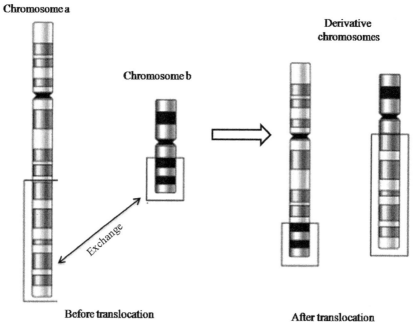

Figure 4.1 Representative image of reciprocal translocation in chromosomes a and b.

5.2.3 Deletion

A deletion is the loss of a segment of a chromosome. Deletion of a large segment of chromosome makes part of the genome hyperploids, which are associated with a phenotypic effect. Deletion of chromosome segment leads to the various syndromes, which are as given below:

- **Cri-du-chat syndrome**

  Cri-du-chat syndrome occurs when the variable portion of the short arm of chromosome 5 is deleted or missing. The symptoms can vary from person to person depending on the size and location of deleted genetic material. The individual suffering from these syndromes are severely impaired, mentally as well as physically, cat like crying gives the syndrome its name [10].

- **Wolf–Hirschhorn syndrome (WHS)**

  Wolf–Hirschhorn syndrome is an extremely rare congenital chromosomal defect, which results due to partial deletion of the short arm (p) of chromosome 4. The typical features of this disorder are heart disease, ophthalmic defect such as crossed eyes, delayed growth and skeletal development retardation, and seizures [11].

- **Jacobsen syndrome**

  Jacobsen syndrome is caused by deletion that occurs on long (q) arm of chromosome 11. People affected with this disease have delayed development of speech and motor skills, cognitive impairment, learning difficulties and also have distinctive facial features such as small and low-set ears, wide eyes with droopy eyelids, and also have a large head.

- **Angelman syndrome**

  It is a rare neurogenetic disorder that occur 1 in 15,000 live births. Angelman syndrome is caused by genetic mutation on chromosome 15 or by paternal disomy. Characteristics feature of this disease include developmental delay, walking and balance disorder, lack of speech, frequent laughter or smiling, and happy comportment [12].

5.2.4 Duplication

Duplication is the occurrence of a segment of chromosomes in two or more copies per genome. Duplication can be tandem or reverse. The duplicate region can be located adjacent to each other, or dispersed on the same chromosome. Duplication of certain genetic regions may produce specific phenotypes.

Examples of duplication aberration include the following:

- **The bar eye phenotype in Drosophila**—The wild-type fruit fly (Drosophila) normally consists of 800 facets in each eye. Bar eye phenotype occurs as a result of duplication of 16A locus of the X chromosome of Drosophila. The bar eye heterozygote has about 350 facets, whereas bar eye homozygote has about 70.

- **Charcot–Marie–Tooth disease**—One of the most common congenital neurological disorders that is caused by the duplication of the gene on chromosome 17. A typical symptom of this disease includes weakness in legs, ankle and feet, loss of muscle bulk, curved toes; and high foot arches, which may result in foot drop. It normally occurs when there are mutations in the genes, which affect the nerves of our feet, ankle and legs.

5.2.5 Isochromosome

An isochromosome shows loss of one arm with duplication of the other. The most probable explanation for the formation of an isochromosome is that the centromere has divided transversely rather than longitudinally [13].

5.2.6 Ring Chromosome

A ring chromosome is formed when the arms are fused together to form a ring following genetic damage by mutagens or spontaneous mutation.

6. MECHANISM OF THE FORMATION OF CHROMOSOME ABERRATION

It was earlier reported that during each cell division, around 5000 DNA single strand breaks (SSBs) are generated per nucleus by oxidants, in which 1% of total SSBs are converted into double strand breaks (DSBs) [14,15].

DNA DSB is a serious threat for the cell when unrepaired or mis-repaired, as they can result in genomic instability and may lead to chromosomal alterations or even cell death. CA formation is one of the major alterations formed during DSBs. The molecular mechanism of CA formation is still unclear but recent theories try to explain how CA formation takes place.

1. The "breakage and reunion" theory [16].
2. The "exchange theory" [17].

The breakage and reunion theory explains that breaks in the chromosome may (1) rejoin and form the original structure (restitution), (2) lead to exchange type aberrations by rejoining another breaks. The exchange type aberration may result from a single DNA DSB by recombinational repair, which may be characterized as one DSB on exchange [18]. Ultraviolet radiation–induced and chemical–induced mutagens lead to lesions in chromosomal DNA that indirectly progresses toward DSB or ultimately to CA [19].

In eukaryotic cells, three pathways are involved in DNA DSBs repair pathways induced by ionizing radiation (IR).

- **Homologous recombination repairing (HRR)**—(Restore original sequence)
- **Nonhomologous DNA end joining (NHEJ)**—(Usually generate small alterations such as base pair substitution, insertions, deletions at the breaks site)
- **Single strand annealing (SSA)**—(Lead to formation of interstitial deletions) (Fig. 4.2).

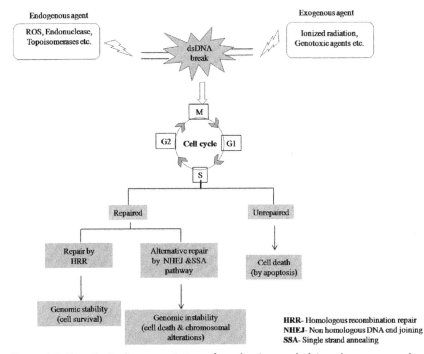

Figure 4.2 Hypothetical representation of mechanism underlying chromosome aberration. Different causative agents induce double strand break, which ultimately leads to genomic instability.

Out of the three pathways, HRR and NHEJ are considered important pathways in eukaryotic cells. A single DSB is sufficient for HRR while at least two initial DSBs are required for NHEJ and SSA. In mammals, NHEJ is the most predominant pathway in cells. HRR seems to be the important pathway in yeast cells [20,21] (Fig. 4.3).

7. CAUSES OF CHROMOSOME ABERRATION

CA is the microscopically visible part of wide spectrum of DNA changes generated during repair mechanism of DNA DSBs. Previous study revealed that dsDNA breaks is the most important lesion for radiation-induced CA [18].

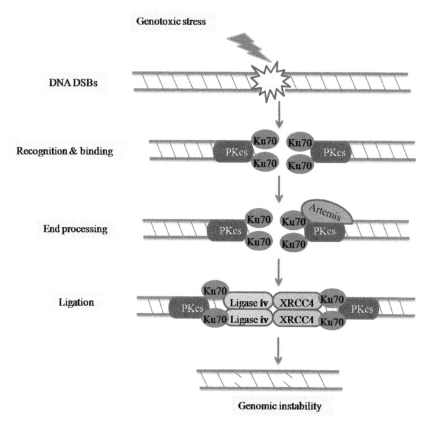

Figure 4.3 Representative image of genomic instability in response to genotoxic stress through nonhomologous DNA end joining (NHEJ) pathway.

7.1 Ionizing Radiation

IRs (X-rays, gamma rays, ultraviolet rays) have excess energy, and therefore they ionize the atoms and molecules. These IRs can generate damaging intermediates through the interaction with water and generate reactive oxygen species (ROS). Excessive ROS may induce oxidative stress and facilitates DNA damage in inducing dsDNA break. dsDNA break generated by IR are challenging to repair because it is converted into intermediate containing 5′ phosphate and 3′ hydroxyl groups, which works as a substrate for cellular enzymes [18].

7.2 Spontaneous dsDNA Break

Double strand DNA break can also be induced spontaneously and occurs when there is an error during different phases of cell cycle following mitosis and meiosis.

7.2.1 Endogenous Reactive Oxygen Species

Endogenous ROS are by-products of normal cellular metabolism of cells and are responsible for cell signaling. Amount of ROS (superoxide radical, hydroxyl radical, hydrogen peroxide) inside cells depend on cell and tissue types. These molecules are extensively reactive and remove electrons from any molecule in its path; resultant molecules form free radical and further lead to DNA damage.

7.2.2 Topoisomerases

These enzymes solve the topological problems associated with DNA. Basically these are of two types: (1) Type I topoisomerases (cleaves single strand in DNA) and (2) Type II topoisomerases (cleaves double strands of DNA). During the cell division, topoisomerase-mediated DNA cleavage takes place. This enzyme is responsible for maintaining genome stability and can promote inappropriate recombination, which ultimately lead to the formation of chromosomal abnormalities.

7.2.3 Replication Error

Chromosomal abnormalities usually occurs during S-phase (synthesis phase) of cell cycle in which DNA synthesis takes place. In human, normal cellular activities as well as environmental factors can cause DNA damage. Human cells are estimated to suffer approximately 10 dsDNA break per cell cycle.

The rate of DNA damage depend on the cell type, life span of cell, as well as the environmental factors.

7.3 Chemicals

Genotoxicity is generally induced by chemicals when DNA damage interferes with DNA replication. Alkylating agent is more efficient in inducing CA in plant and mammalian cells particularly in heterochromatin region of chromosome. These alkylating agents induced CA in lymphocyte cells [22] and rodent cells [23]. Nanoparticles and pesticides are also known to cause CA in different cell types (Table 4.1).

8. FREQUENCY OF CHROMOSOME ABERRATION

The frequency of CA in the general population is estimated to be 0.5% of live births, but the frequency of CA within the high risk population was higher than that of the general population [28]. As earlier reported, increased frequencies of CA lead to elevated risk of cancer [29]. Types and frequency of chromosomal anomalies detected in prenatal diagnosis [30] (Table 4.2).

Table 4.1 Major toxicants/chemicals inducing chromosomal aberration in various cell types

Toxicants	Cells	References
Nanoparticles		
Titanium dioxide (TiO$_2$)	In human lymphocyte	[24]
Single-walled carbon nanotubes	In human lymphocyte	[24]
Double-walled carbon nanotubes	In mouse macrophages	[25]
Pesticides		
Thiodan insecticide Folithion insecticide Lebaycid insecticide Kitazin fungicide	In meiotic cells of barley	[26]
Chemical mutagens		
Diethyl sulfate (DES) Ethyl methane sulfonate (EMS) Ethyleneimine	In *Allium cepa*	[27]

Table 4.2 Frequency of chromosomal aberration in general population having different chromosomal anomalies

Chromosomal anomalies	Number	Abnormal frequency (%)
Numerical anomalies	**87**	**58.0**
Autosomal anomalies		
Trisomy 21	36	24.0
Trisomy 18	22	14.7
Trisomy 13	5	3.3
Trisomy 9	2	1.3
Trisomy 8	1	0.7
Sex chromosome anomalies		
45,X	9	6.0
47,XXX	3	2.0
47,XXY	4	2.7
47,XYY	2	1.3
48,XXXX	1	0.7
Triploidy (69,XXX)	2	1.3
Structural anomalies	**63**	**42.0**
Translocation		
Reciprocal	26	17.3
Robertsonian	21	14.0
Deletion	6	4.0
Inversion	5	3.3
Addition	4	2.7
Marker(47,XY + mar)	1	0.7

9. METHODS FOR DETECTION OF CHROMOSOMAL ABERRATION

Different types of methods have been developed for CA detection, which are given below:

9.1 Cytogenetic Testing

9.1.1 Chromosomal Aberration Test

CA test can be used in vitro using cultured mammalian cells, i.e., in vitro mammalian CA test, OECD-473 [31,32]; as well as in vivo, i.e., mammalian bone marrow CA test, OECD-475 [32]. The cells are treated with metaphase-arresting substances (example: colchicine or colcemid) that appear in metaphase stage of cell cycle. The metaphase cells are harvested, stained, and analyzed microscopically for detection of CA.

9.1.2 Micronucleus Assay

The micronucleus (MN) formation is an alternative test to CA assay. This test is gaining importance because of some advantages as compared with traditionally used CA test for genotoxicity assessment, performed in both in vitro and in vivo studies. Micronuclei formation can be assessed by two different methods either by conventional methods such as Giemsa staining, cytokinesis-blocked micronucleus (CBMN) assay or automatically by flow cytometry analysis (Flow MN assay).

9.1.2.1 Cytokinesis-Blocked Micronucleus Assay

The conventional method for MN detection is CBMN assay, which is used widely for assessing chromosome damage. In this assay, cytochalasin-B, inhibitor of spindle assembly, was used to prevent cytokinesis that occurs after nuclear division. Cells that have completed one nuclear division are blocked during cytoplasmic division to produce binucleate and multinucleate cells after cell division. For the scoring of micronuclei, only binucleated cells are selected [33]. This assay is inexpensive and rapid for detection of chromosomal damage. It has recently been proposed that MN assay can be used for genotoxicity assessment of new chemicals instead of metaphase analysis; the standard criteria for selecting for MN was reported earlier [33].

9.1.2.2 Micronucleus Assay Using Flow Cytometry

Flow cytometry–based method of MN detection is extensively used because microscopic slide preparation produces artifacts. The flow cytometry–based assay is rapid and, it analyzes large number of samples in a short time and produces large number of data than others such as CBMN assay. Flow cytometry–based assay helps us to study the multiple state of aneuploidy [34]. Flow cytometry method analysis of MN formation is based on differences in DNA content between micronuclei and nuclei, usually calculated through automated data analysis system, hence it is a high-throughput method. A working protocol that can be used for MN formation is taken from an earlier study [35]. Physical separation of micronuclei from nuclei is also possible based of its size by flow cytometry if the flow cytometer is having the facility of sorting also.

9.1.3 Karyotyping

Within an organism or species, the number or appearance of chromosome in the cell nucleus is termed as karyotype. The chromosome of higher organism is mostly studied in mitotic metaphase stage because the chromosomes in this stage reach at the stage of highest condensation [36]. During mitotic metaphase, the chromosomes of the karyotype species being studied appear in identifiable shapes and characteristics. Different stains and dyes produce different banding pattern specific to each chromosome. Chromosome banding and other techniques provide information about changes in the chromosome number like aneuploid condition, for example, trisomy 21. It is also used to analyze very minute chromosomal structural changes such as duplication, deletion, inversion and translocation. There are various sets of information that can be obtained for human chromosome by the construction of karyotypes such as number and sex chromosome content, presence and absence of specific chromosome, and nature and extent of large structural abnormalities.

For karyotyping, tissue samples are obtained from various sources such as peripheral blood system, skin biopsy, and for prenatal diagnosis samples are obtained from amniotic fluid and chorionic villus. Karyotyping generation started with culture derived from specimens. After cell growth, dividing cells are arrested by using colchicines and then a hypotonic solution is used that causes nucleus to swell and cell to burst. Nucleus is treated with a chemical fixative and various stains that reveal features of chromosome. To make CA more effective researchers have developed various stains that bind with DNA and generate characteristics banding pattern for different chromosomes. Various banding patterns are evolved to identify the integrity, the site of chromosomes break, alterations and the location of specific genes [37]. Among various banding patterns, G banding karyograme is routinely used to identify various chromosomal abnormalities in individuals.

9.2 Molecular Cytogenetic Testing

9.2.1 Fluorescence In Situ Hybridization

A number of other chromosome banding techniques also do exist that employ molecular cytogenetic techniques such as FISH. Most in situ hybridization procedures use fluorescent probes to detect DNA sequences, and the procedure is referred to as FISH. A variety of FISH procedures are there to detect different types of aberrations. The prime components involved in FISH are a DNA probe and a target sequence. The labeling of the probe can be performed after hybridization by various means such as nick translation, random primed labeling, or PCR. The labeled probe and the target DNA are denatured and combined for annealing of complementary DNA sequence. Previously, FISH and other in situ hybridization results were important in mapping genes on human chromosomes [36]. FISH provides a powerful tool for identifying the location of cloned sequence on metaphase chromosome. FISH and other in situ hybridization procedures are important in the clinical diagnosis of chromosomal abnormalities such as deletions, duplications, and translocations. FISH can be very efficiently used to characterize and diagnose various chromosomal abnormalities such as Angleman and Prader–Willi syndromes [38]. According to the researchers, Angleman syndrome patients receive maternal copy of deleted chromosome 15, whereas Prader–Willi syndrome patients receive paternal copy of deleted chromosome number 15. Various modified versions of FISH are used to find structural and numerical abnormalities of chromosome. For instance, multicolor FISH (spectral karyotyping) is used to monitor multiple sites by hybridizing probes, labeled with different fluorophores. Another modified version of FISH use different fluorophores probe to obtain information about organization of chromosome within nucleus, i.e., interphase chromosome.

9.3 Microarray Comparative Genomic Hybridization Testing

Comparative genomic hybridization (CGH) techniques have been developed to identify specific chromosomal regions that accompany with lose or gain of genetic materials. Standard CGH requires metaphase chromosome of control group and experimental group. The control DNA fragment is generally labeled using green fluorescence whereas the experimental one is labeled using red fluorescence and is allowed to compete for hybridization site on metaphase chromosome. The green and the red probe results in yellow/orange color, as they bind equally in control group whereas in experimental group in the case of amplified chromosomal region, increase in red color appears under microscope. Standard CGH methods are extensively laborious. Nowadays, microarray-based methods consist of microchips containing thousands of base pair fragments of human genome. Microarray-based CGH technique is used to exactly determine the chromosomal regions that are amplified or missing [39]. The information gathered from single-array CGH experiments is equal to thousands of FISH experiment.

9.4 Prenatal Screening to Detect Fetal Abnormalities

Chorionic biopsy provides another effective way of detecting prenatal chromosomal abnormalities in developing fetus. Early detection of fetal condition can give parents, the opportunity to terminate the pregnancy in an early state. The first step in prenatal testing is obtaining DNA sample from unborn child. A hollow needle is passed through the pregnant woman's abdomen with the guidance of ultrasound and a small amount of amniotic fluid is drawn out and then cultured for few weeks before the cells are ready to be used for cytogenetic analysis [40]. There is a need of strong ethical regulation in prenatal testing to ensure that it is not misused.

10. CLINICAL MANIFESTATION OF CHROMOSOMAL ABNORMALITIES

Over 100 chromosomal syndromes that have been reported together play a major role in human morbidity and mortality, but on individual basis these events are considered to be rare [41]. 50%–60% of fetal wastage results due to chromosomal abnormalities as the effect of CAs are greatest during fetus development [42]. Chromosomal abnormalities occur in 6% of all recognized congenital malformations. It also accounts for 30%–40% of severe mental retardation and 10% of the mild mental retardation [43,44].

Patients having CA possess distinct kind of clinical features, which can be broadly characterized as postnatal and intrauterine retardation, patterns of dysmorphic signs, multiple malformation, and impaired mental development. Although it should always be kept in mind that there is no possible cure for the adverse impacts of chromosomal alterations on the phenotype of the species, early diagnosis is required for avoidance of repetition of CA.

10.1 Chromosomal Aberration and Spontaneous Abortions

Prenatal ultrasonic finding in autosomal chromosome aberration reveals some interesting findings, which suggest that there could be different alterations in different stages of development such as intrauterine growth retardation that correlates with survival and postnatal growth, double neck contour, abnormal amount of amniotic fluid, and in some cases could lead to the formation of small placenta. In later stages, CA could also lead to congenital malformations such as omphalocele (rare abdominal wall defect), cleft lips and palate, heart and renal malformations.

10.2 Chromosomal Aberration and Cancer

Cancer is uncontrolled and undifferentiated cell division as well as a multistage progressive disease. These changes in cell interfere with the normal cellular mechanism and disrupt the proto-oncogene and tumor suppressor genes that allow additional changes to occur in the genome. Chromosomal rearrangements could result in cancer either by the formation of chimeric fusion gene with new or altered activity or dysregulated expression of functionally normal gene that is unusually short because of the reciprocal translocation and possesses a hybrid gene known as BCR-ABL1, which is associated with chronic myeloid leukemia. Wilms tumor, retinoblastoma and osteosarcoma are some cancer types in which gene deletions and duplications are considered to be responsible for cancer progression [45]. It can also lead to various types of cancers as in these cases the region of the chromosome containing the tumor suppressor gene may be deleted or mutated like deletion in the case of 9p21 segment, which encodes a tumor suppressor gene that is observed in many type of human cancers such as gliomas, non–small cell lung cancers, leukemias, and melanomas [46,47].

10.3 Behavior Peculiarities Associated With Chromosomal Aberration

There are different behavioral changes and distinct personality traits with respect to different cases of special chromosome aberration. Some general behavioral observations with respect to persons with either normal or near to normal intelligence are like having low confidence and non-aggressive. Under unfavorable condition and stress, they will develop the tendency of psychosis and will have higher SQ (social quotient) than the IQ (intelligence quotient). As in the case of Turner syndrome, individual suffering from it will show impairment in visual, spatial, executive and social cognitive tasks such as poor perceptual reasoning, poor processing speed, and poor visual memory whereas women are often reported with psychosocial difficulties and adaptive functions [48]. The individual having Klinefelter syndrome may suffer from language and learning difficulties such as difficulty in saying words clearly, trouble in expressing thoughts, needs and in processing what they hear [49]. They will be quieter, more assertive, at times more restless and anxious.

10.4 Changes in Course of Adolescence and Fertility

The overall growth rate is generally higher in the individuals having CAs than normal adolescents. Premature aging is generally very common in autosomal chromosome aberration and is observed at the age group from 18 to 25 years. Some very common symptoms are graying and loss of hair, osteoporosis, and atrophic changes in skin. Attainment of puberty in adolescent stage depends on the type of aberration. Puberty is either diminished, delayed, or completely absent in individuals. Normal puberty is attained in few aberrations as in the case of ring chromosome and Cat eye syndrome (inversion duplication), whereas puberty is not achieved in many cases and is more common in males having mental retardation. In females, there is frequent occurrence of primary amenorrhea and secondary amenorrhea. It is well known that most of the individuals having CAs generally do not have offspring, which is due to mental deficiency and gonadal hypoplasia that results in the progressive loss of germ cells on the developing gonads of an embryo. Fertility in males is rare whereas in females it is common as compared with males as in trisomy 21 and is never observed in ring chromosome.

10.5 Pattern of Dysmorphic Signs in Chromosomal Aberration

It is said that one in 40 infants is born with dysmorphic signs. Minor malformations can be easily detected by observing certain dysmorphic features, which keep on changing with the type of aberration as in some cases major malformations can have a masking effect. Dysmorphic features are generally observed in infants as these changes are quite impressive in neonates and in childhood, whereas in some aberrations there are marked changes with the age of the individual. Different chromosomal conditions can have different presenting dysmorphic features; as in the case of Down syndrome the infant can have brachycephaly, hypotonia, single palmer crease, sandal gap, and Hirschsprung disease. Cri du chat syndrome possesses presenting features such as mewing cry, microcephaly, cleft palate, ear anomalies, and round face. Holoprosencephaly, cleft, heart defect, renal abnormalities, and microphthalmia are the common presenting features of Patau syndrome that can be diagnosed either by karyotyping or by quantitative fluorescence polymerase chain reaction [50].

10.6 Congenital Malformations and Chromosomal Aberration

These are more variable than dysmorphisms as they involve a number of characteristics than just a single malformation. Rare and early determined malformations are very common in spontaneous abortions. Malformations are the consequence of CAs as they are caused by the retardation of organ development. Some common malformations in autosomal chromosome aberration pertaining to different organs are as follows: the central nervous system can have spina bifida, holoprosencephaly, Dandy–Walker malformation; gastrointestinal malformations may be like malrotation, common mesentery, omphalocele; and limb malformations can be duplication of thumb, hypoplasia of radius, and postaxial hexadactyly. There are some uncommon malformations that result due to autosomal chromosome aberration. In the central nervous system, they can be anencephalus, otocephalus, acecephalus, and exencephalus; gastrointestinal malformations can be atresia of ileum and jejunum, total situs inversus, gastroschisis; and limb malformations can be peromelia, amelia, phocomelia, and arthrogryposis [51].

11. APPLICATIONS OF CHROMOSOMAL ABERRATION ANALYSES

CA tells about the chromosome loss as well as chromosomal breakage, therefore this test can be used routinely for testing of different types of toxicants, which induced chromosomal abnormalities. CA detection can also be useful for other purposes as explained below.

11.1 Radiation- and Chemical-Induced Cancer Risk Assessment

CA study of DNA damage provide information at the chromosome level which is an essential part of cancer risk assessment [33]. It is well known that X-rays are potent mutagenic agents. These mutations play major role in cancer progression, as evident from earlier study [33]. A recent study tells about the relationship between the elevated frequency of CA and increased risk of cancer [52]. Increased frequency of CA in lymphocyte is an early biomarker for the assessment of cancer risk that comes from earlier study [53]. Such studies strongly indicate the urgent need of CA assay for assessment of chemical carcinogens and radiation-induced cancer risk.

11.2 Genotoxicity Assessment of Environmental Chemicals

Release of several substances such as household cleaners, solvents, detergents, cosmetics, volatile organic compounds and unburned hydrocarbons in environment cause serious problems. Increased concentrations of these substances cause severe reactions in many people. These chemicals induced alteration at chromosome level, which further exaggerate cellular dysfunction. CA test can be used routinely for testing of these environmental substances.

11.3 Testing of New Pharmaceuticals and Chemical Substances

Increased industrialization has significantly increased the number of new chemical substances released in the environment such as pesticides, industrial waste, solvents, drugs, etc. Release of these toxins from industries poses serious health risks. Therefore, it becomes mandatory to check the toxic potential of newly formed chemicals and industrial waste [54]. Thus, this test is useful for the assessment of toxic potential of these chemicals before they are released for commercial purpose.

12. FUTURE PERSPECTIVE

CA has significant contribution in infertility, pregnancy loss at different stages, and developmental disorders in newborns. Extensive study needs to be done to understand the connecting link between genetics to the origin of cancer, physical gene mapping, and prenatal testing for avoidance of genetic diseases. Generally, it is presumed that molecular techniques involving DNA "chips" will completely replace the conventional chromosome analysis. Future research should be focused on understanding the mechanism associated with CA and in the development of new techniques so as to improve the resolution, as the available techniques are inadequate to observe small changes in the chromosomes such as exchanges close to telomeres.

13. CONCLUSION

High incidence of unbalanced chromosome aberrations is a characteristic feature of humans, especially trisomy and triploidy. Nature has developed various pathways or mechanisms to correct these unbalanced chromosomes. Chromosomal abnormalities are the important cause of congenital malformations and therefore arises the need of cytogenetic evaluation.

REFERENCES

[1] Connor J, Ferguson-Smith M. Essential medical genetics. Oxford: Blackwell Scientific Publications; 1987.

[2] Wolffe A. Chromatin: structure and function. Academic press; 1998.

[3] Cremer C, Münkel C, Granzow M, et al. Nuclear architecture and the induction of chromosomal aberrations. Mutat Res 1996;366(2):97–116.

[4] Plachot M, Veiga A, Montagut J, et al. Are clinical and biological IVF parameters correlated with chromosomal disorders in early life: a multicentric study. Hum Reprod 1988;3(5):627–35.

[5] Therman E, Susman M. Human chromosomes: structure, behavior, and effects. Springer Science & Business Media; 2012.

[6] Carey G. Human genetics for the social sciences, vol. 4. Sage publications; 2002.

[7] Kreth G, Münkel C, Langowski J, Cremer T, Cremer C. Chromatin structure and chromosome aberrations: modeling of damage induced by isotropic and localized irradiation. Mutat Res 1998;404(1):77–88.

[8] Sanchez O, Yunis JJ. New chromosome techniques and their medical applications. New York: New chromosomal syndromes Academic; 1977. p. 1–54.

[9] Schwartz S, Palmer CG. High-resolution chromosome analysis: I. Applications and limitations. Am J Med Genet 1984;19(2):291–9.

[10] Zhang X, Snijders A, Segraves R, et al. High-resolution mapping of genotype-phenotype relationships in cri du chat syndrome using array comparative genomic hybridization. Am J Hum Genet 2005;76(2):312–26.

[11] Paradowska-Stolarz AM. Wolf-Hirschhorn syndrome (WHS)–literature review on the features of the syndrome. Adv Clin Exp Med 2014;23:485–9. Off Organ Wroclaw Medical University.

[12] Gos M. Epigenetic mechanisms of gene expression regulation in neurological diseases. Acta Neurobiol Exp 2013;73:19–37.

[13] Snustad DP, Simmons MJ. Principles of genetics, binder ready version. John Wiley & Sons; 2015.

[14] Vilenchik MM, Knudson AG. Endogenous DNA double-strand breaks: production, fidelity of repair, and induction of cancer. Proc Nat Acad Sci 2003;100(22):12871–6.

[15] Zhao H, Tanaka T, Mitlitski V, Heeter J, Balazs EA, Darzynkiewicz Z. Protective effect of hyaluronate on oxidative DNA damage in WI-38 and A549 cells. Int J Oncol 2008;32(6):1159–68.

[16] Sax K. Types and frequencies of chromosomal aberrations induced by X-rays. In: Cold Spring Harbor Symposia on quantitative Biology, vol. 9. Cold Spring Harbor Laboratory Press; 1941. p. 93–103.

[17] Revell S. The Breakoge-ond-Reunion theory and the exchange theory for chromosomal aberrations induced by ionizing radiations: a short history. Adv Radiat Biol 1974;4:367–416.

[18] Pfeiffer P, Goedecke W, Obe G. Mechanisms of DNA double-strand break repair and their potential to induce chromosomal aberrations. Mutagenesis 2000;15(4):289–302.

[19] Obe G, Pfeiffer P, Savage J, et al. Chromosomal aberrations: formation, identification and distribution. Mutat Res 2002;504(1):17–36.

[20] Natarajan AT. Chromosome aberrations: past, present and future. Mutat Res 2002;504(1):3–16.

[21] Pastwa E, Błasiak J. Non-homologous DNA end joining. Acta Biochimica Polonica 2003;50(4):891–908.

[22] Shaw MW, Cohen MM. Chromosome exchanges in human leukocytes induced by mitomycin C. Genetics 1965;51(2):181.

[23] Natarajan A, Raposa T. Heterochromatin and chromosome aberrations. A comparative study of three mouse cell lines with different karyotype and heterochromatin distribution. Hereditas 1975;80(1):83–9.

[24] Catalán J, Järventaus H, Vippola M, Savolainen K, Norppa H. Induction of chromosomal aberrations by carbon nanotubes and titanium dioxide nanoparticles in human lymphocytes in vitro. Nanotoxicology 2012;6(8):825–36.

[25] Di Giorgio ML, Di Bucchianico S, Ragnelli AM, Aimola P, Santucci S, Poma A. Effects of single and multi walled carbon nanotubes on macrophages: cyto and genotoxicity and electron microscopy. Mutat Res 2011;722(1):20–31.

[26] Grover I, Tyagi P. Chromosomal aberrations induced by pesticides in meiotic cells of barley. Caryologia 1980;33(2):251–9.

[27] Hassan L, Ahmad S. Chromosomal aberrations induced by chemical mutagens in allium. Pak J Biol Sci 2000;3:1187–9.

[28] Mathews T, Navsaria D, Verma R. Prenatal cytogenetic diagnosis of 1,400 consecutive amniocenteses. Gynecol Obst Invest 1992;34(2):122–3.

[29] Hagmar L, Bonassi S, Strömberg U, et al. Chromosomal aberrations in lymphocytes predict human cancer: a report from the European Study Group on Cytogenetic Biomarkers and Health (ESCH). Cancer Res 1998;58(18):4117–21.

[30] Park SY, Kim JW, Kim YM, et al. Frequencies of fetal chromosomal abnormalities at prenatal diagnosis: 10 years experiences in a single institution. J Korean Med Sci 2001;16(3):290–3.

[31] Galloway SM. Cytotoxicity and chromosome aberrations in vitro: experience in industry and the case for an upper limit on toxicity in the aberration assay. Environ Mol Mutagen 2000;35(3):191–201.

[32] OECD. OECD guidelines for the testing of chemicals. Organization for Economic co-operation and development; 1994.

[33] Fenech M. The in vitro micronucleus technique. Mutat Res 2000;455(1):81–95.

[34] Muehlbauer P, Schuler M. Detection of numerical chromosomal aberrations by flow cytometry: a novel process for identifying aneugenic agents. Mutat Res 2005;585(1):156–69.

[35] Wagner ED, Anderson D, Dhawan A, Rayburn AL, Plewa MJ. Evaluation of EMS-induced DNA damage in the single cell gel electrophoresis (Comet) assay and with flow cytometric analysis of micronuclei. Teratog Carcinog Mutagen 2003;23(S2):1–11.

[36] Walling JG, Zhang W, Jiang J. Fluorescence in situ hybridization techniques for cytogenetic and genomic analyses. Rice Protoc 2013:13–27.

[37] Schreck RR, Distèche CM. Chromosome banding techniques. Curr Protoc Hum Genet 1994.

[38] Knoll J, Nicholls R, Magenis R, et al. Angelman and Prader-Willi syndromes share a common chromosome 15 deletion but differ in parental origin of the deletion. Am J Med Genet 1989;32(2):285–90.

[39] Pinkel D, Segraves R, Sudar D, et al. High resolution analysis of DNA copy number variation using comparative genomic hybridization to microarrays. Nat Genet 1998;20(2):207–11.

[40] Handyside AH, Kontogianni EH, Hardy K, Winston R. Pregnancies from biopsied human preimplantation embryos sexed by Y-specific DNA amplification. Nature 1990;344(6268):768–70.

[41] Dave U, Shetty D. Chromosomal abnormalities in mental retardation: Indian experience. Int J Hum Genet 2010;10(1–3):21–32.

[42] Seashore MR, Wappner RS. Genetics in primary care & clinical medicine: a lange medical book. McGraw-Hill/Appleton & Lange; 1996.

[43] Ahuja A, Thapar A, Owen M. Genetics of mental retardation. Indian J Med Sci 2005;59(9):407.

[44] Raynham H, Gibbons R, Flint J, Higgs D. The genetic basis for mental retardation. Qjm 1996;89(3):169–76.

[45] Mitelman F. Cancer cytogenetics update 2005. Atlas Genet Cytogenet Oncol Haematol 2005.

[46] Cairns P, Polascik TJ, Eby Y, et al. Frequency of homozygous deletion at p16/CDKN2 in primary human tumours. Nat Genet 1995;11(2):210–2.

[47] Sasaki S, Kitagawa Y, Sekido Y, et al. Molecular processes of chromosome 9p21 deletions in human cancers. Oncogene 2003;22(24):3792–8.

[48] Hong D, Scaletta Kent J, Kesler S. Cognitive profile of Turner syndrome. Dev Disabil Res Rev 2009;15(4):270–8.

[49] Visootsak J, Graham JM. Social function in multiple X and Y chromosome disorders: XXY, XYY, XXYY, XXXY. Dev Disabil Res Rev 2009;15(4):328–32.

[50] Clayton-Smith J. Assessment of the dysmorphic infant. Congenit Cond 2008;4(6):206–10.

[51] Schinzel A. Clinical findings in chromosome aberrations. Atlas Genet Cytogenet Oncol Haematol 2005.

[52] Bonassi S, Znaor A, Norppa H, Hagmar L. Chromosomal aberrations and risk of cancer in humans: an epidemiologic perspective. Cytogenet Genome Res 2004;104(1–4):376–82.

[53] Rossner P, Boffetta P, Ceppi M, et al. Chromosomal aberrations in lymphocytes of healthy subjects and risk of cancer. Environ Health Perspec 2005:517–20.

[54] Bajpayee M, Pandey AK, Parmar D, Dhawan A. Current status of short-term tests for evaluation of genotoxicity, mutagenicity, and carcinogenicity of environmental chemicals and NCEs. Toxicol Mech Methods 2005;15(3):155–80.

CHAPTER FIVE

In Vivo Cytogenetic Assays

Pasquale Mosesso[1], Serena Cinelli[2]
[1]Università degli Studi della Tuscia, Viterbo, Italy; [2]Research Toxicology Centre, Roma, Italy

1. INTRODUCTION

The presence of genotoxic agents in the environment may cause chromosomal mutations through different mechanisms, which are associated with serious health effects induced even at low exposure levels. In the germ cells chromosomal mutations contribute to fetal mortality up to 50% of spontaneous abortions (which account only for 15%–20% of identified pregnancies). The total frequency of chromosomal mutations at birth has been estimated to be approximately 6‰: about 2‰ is caused by anomalies in the number of sex chromosome, 1.4‰ by autosomal chromosomes, and 2.5‰ by anomalies in the structure of chromosomes. In somatic cells chromosomal mutations play a key role in processes leading to malignancies if mutations occur in protooncogenes, tumor suppressor genes, and/or DNA damage response genes and are responsible for a variety of genetic diseases [1]. Accumulation of DNA damage in somatic cells has also been proposed to play a role in degenerative conditions, such as accelerated aging, immune dysfunction, cardiovascular, and neurodegenerative diseases [2–5].

Chromosomal mutations are distinguished under two major categories related to changes in the chromosome structure (chromosomal aberrations) and changes in the number of chromosomes (numerical aberrations). Structural chromosome aberrations are induced by physical and chemical genotoxic agents through different repair mechanisms of DNA double-strand breaks (DSBs). These are considered to be the ultimate DNA lesion in the process of formation of chromosomal aberrations [6–8]. DSBs can occur spontaneously (e.g., during DNA replication or after DNA excision repair processes and accumulation of DNA single-strand breaks) and can be directly induced by ionizing radiation, some antibiotics, or endonucleases [8–12]. However, the majority of chemical agents and UV radiation are not capable to induce DSBs directly but induce other DNA lesions, such as base damage, hydrolysis of bases, pyrimidine dimers, and DNA cross-links,

Mutagenicity: Assays and Applications
ISBN 978-0-12-809252-1
http://dx.doi.org/10.1016/B978-0-12-809252-1.00005-5

which, during DNA repair processes or DNA synthesis, may give rise to DSBs and eventually to chromosomal aberrations. Numerical chromosome aberrations are usually induced by a failure of chromosome division, generated by events that involve cell cycle perturbation processes and/or interference with the cell division spindle.

Cytogenetic effects, such as structural and numerical chromosomal aberrations, can be evaluated using in vitro and/or in vivo assays by estimating different classes of chromosome changes scored in either metaphase or interphase (micronuclei) using the light microscope. Usually, in vivo assays are performed in a stepwise approach for hazard identification as follow-up of in vitro positive results to evaluate the level of concern. However, it is important to note that in vivo genotoxicity test(s) provide more relevant data for the evaluation of induced DNA damage because they take into account dynamic whole-animal physiological processes such as absorption and systemic distribution, phase I and phase II metabolism, and physiological excretory system that cannot be reproduced in in vitro systems. Evidences exist for the presence of a small but significant number of genotoxic carcinogens (e.g., procarbazine, urethane, hydroquinone, benzene), which are clearly detected in vivo but are negative, weakly positive, or equivocal in in vitro assays [13].

An in vivo test for cytogenetic effects in rodents fulfills these requirements in the great majority of cases, either by an analysis of micronuclei in erythrocytes in bone marrow or blood, or of chromosomal aberrations at metaphase in bone marrow cells. Nevertheless, the relevance of the in vivo assays is strictly correlated with the target organ exposure. Depending on the specific scenario, cytogenetic analyses can be performed using alternative dividing tissues, mainly using evaluation of micronuclei. In case of substances that are not absorbed and are directly acting genotoxins (e.g., no need of metabolic conversion), the gastrointestinal tract is recommended, whereas, for promutagenic substances that are absorbed but the active metabolites are too short-lived to reach the target organ, the liver of young adult rodents should be employed. However, these last two methods need to be developed further. Research data on the micronucleus assay in G_0 resting splenocytes during subchronic exposure suggest that spleen is also a relevant tissue in addition to hematopoietic cells in bone marrow or peripheral blood.

This chapter will focus on cytogenetic effects in somatic cells illustrating methods to analyze metaphases in bone marrow cells for detection of chromosomal aberrations and micronuclei in immature erythrocytes (e.g., polychromatic erythrocytes (PCEs) in bone marrow or reticulocytes (RETs) in peripheral blood).

2. CYTOGENETIC END POINTS

For chromosomal aberrations in metaphase, two major classes of chromosomal structural changes are identified. This is based on whether one or both chromatids (at the same locus) in a metaphase chromosome are involved in an aberration. The type of aberration observed in metaphase depends on the phase of cell cycle in which the treatment has been administered and the type of mutagenic agent used. Only few clastogenic agents are able to induce structural chromosomal aberrations in any phase of cell cycle: chromosome-type aberration in G_1, chromatid-type aberration in G_2, and a mixture of chromosome-type and chromatid-type aberrations if the treatment is made in S phase [14]. They include ionizing radiations and very few chemical agents, such as bleomycin, cytosine arabinoside, and streptonigrin, which are capable of either inducing DNA strand breaks directly or—as it appears to be the case for cytosine arabinoside—to specifically inhibit the repair of spontaneously induced DNA strand breaks. The majority of chemical agents induce essentially chromatid-type aberrations even when cells are treated in the G_1 phase of cell cycle. On this basis, two classes of chromosome-breaking agents are usually identified: S-dependent and S-independent agents. Chromosome aberrations induced by S-independent agents are generated immediately after primary DNA damage irrespective of the cell stage at which the treatment is performed, and these aberrations are assumed to arise as a consequence of "nonrepair" or "misrepair" of the primary DNA lesions. In contrast, aberrations induced by S-dependent agents are assumed to be formed by a so-called delayed effect due to "misreplication" processes [15,16].

Three types of aberrations are generally scored in metaphase cells: gaps, breaks, and exchanges according to the description of Savage [17].

For the analysis of chromosomal aberrations in interphase, micronuclei are scored. Micronuclei represent a proportion of fragments or whole chromosomes, which lag in anaphase and are not included in the main nucleus. In the subsequent interphase, they condense to form small nuclei, which are easily identified for scoring using microscopic analysis or automated systems such as flow cytometers [18–20], image analysis platforms [21,22], and laser scanning cytometers [23]. Therefore, the test has the potential to detect both clastogenic and aneugenic substances. Micronuclei that have been originated as a consequence of chromosome breakage can be generally distinguished from those by whole chromosomes by means of their size. However, this is not a fully reliable method and it is preferable to classify them on the presence or absence of centromeres. Available methods include the immunochemical labeling of kinetochores [24–26]—often called CREST method because this antibody is derived from the serum of patients with the autoimmune disease scleroderma (CREST syndrome)—and the use of DNA pericentromeric probes in combination with in situ hybridization (FISH) techniques to mark centromeres [27,28].

3. TREATMENT PROTOCOLS

3.1 Selection of Animal Species, Number, and Sex of Animals

Rodent species (mouse and rat) are generally employed as experimental test systems, although other species, such as dog, pig, or nonhuman primates, can be used if scientifically justified as in the case of integration in routine toxicology studies. For the analysis of micronuclei, until recently, the measurement of micronucleated reticulocytes (MN-RETs) in peripheral blood was restricted to the mouse because of presumed splenic elimination of MN-erythrocytes by the rat spleen [29,30]. However, based on accumulated evidence, analysis of micronuclei in peripheral blood RETs is now considered to be acceptable in rats as well as in mice [31]. Furthermore, preliminary data suggest that peripheral blood RETs may be a suitable cell population also in other species, including human [32–36]. Accumulated evidences indicate that, generally, the response of genotoxicity assays is similar between male and female animals. However, in case of relevant differences observed in the preliminary dose range–finding experiments, such as differences in systemic toxicity, absorption, metabolism, and target tissue toxicity, the use of both sexes is then recommended.

3.2 Dose Selection

As not limited by toxicity or solubility, the maximum dose level should be 2000 mg/kg body weight per day for a treatment period shorter than 14 days and 1000 mg/kg body weight per day for longer periods. These doses are defined as limit doses and are considered adequate for a balance between the need to limit false-negative findings and ethical issues, included in the principles of the 3Rs (Replacement, Reduction and Refinement), generally accepted as a fundamental framework for humane animal research. In case of toxicity, it is necessary to identify the maximum tolerated dose (MTD) in a preliminary dose range–finding experiment, as the highest dose level to be used in the main assay. MTD is defined as the highest dose that induces toxicity such as body weight loss or adequate cell toxicity in the target tissue, but not death or evidence of excessive pain, suffering, or distress.

When the cytogenetic assays are integrated into repeated dose toxicity studies, selection of the top dose should be tailored taking into account saturation in exposure, substance accumulation, or decrease in exposure with the time, related to detoxification processes.

To evaluate a possible dose–effect relationship, at least three dose levels are used, which include the highest dose selected as described above and two lower dose levels covering a range from the maximum to a dose producing little or no toxicity.

3.3 Route of Administration

For predicting effects in humans, it is appropriate to utilize a route of exposure that most resembles the one anticipated or known to be the route of human exposure (e.g., oral, subcutaneous, intravenous, topical, inhalation, intratracheal, or implantation). However, in case of compounds that are not systemically absorbed and therefore are not available to the target tissue, it is advisable to modify the route of administration. The administration of test substances in food or drinking water should be only used when scientifically justified because of the uncertainty of the final dose taken up, while the intraperitoneal route is anyway not recommended because it is not considered a physiologically relevant route of human exposure.

3.4 Proof of Exposure

The reliability of in vivo results is strictly dependent on the demonstration of adequate exposure of the relevant target tissue to the test substance. A direct evidence of target tissue exposure may be obtained evaluating specific cytotoxicity effects in the treatment animal groups compared with the concurrent vehicle control group. These effects can consist of significant changes in the proportion of immature erythrocytes among total erythrocytes in bone marrow or blood for micronucleus assays, or a significant reduction in mitotic indices for the chromosomal aberration assay. In the absence of cytotoxicity, measurement of the plasma or blood levels of the test substance or its metabolites is deemed necessary. However, the demonstration of target tissue exposure to the test substance or its metabolites, particularly short-lived metabolites, is often difficult because of technical problems not only in the analytical methods but also in the definition of an adequate plasma level concentration "sufficient" to elicit a genotoxic activity in vivo.

Otherwise, data on absorption, distribution, metabolism, and excretion obtained in an independent study, using the same species and strain and the same route of administration, can be used to demonstrate target tissue exposure. If adequate exposure cannot be achieved, e.g., with compounds showing very poor target tissue availability, conventional in vivo cytogenetic tests should be considered of limited relevance.

3.5 Duration of Treatment and Sampling Time

3.5.1 Micronucleus Test

For the micronucleus assay, based on accumulated evidence, two or more administration of test substance at 24-h intervals are recommended and in this respect the micronucleus test is considered to be the preferred in vivo test for chromosomal aberrations when integration with other toxicity studies is advisable. Single administration of test substance is possible, particularly when information on potential interference with cell cycle progression is envisaged. In this case sampling of bone marrow should be performed from independent groups of animals, in a time frame comprised between 24 and 48 h after treatment, with appropriate interval(s) between samplings. Similarly, when peripheral blood is used as target tissue, samples should be taken at least twice from the same group of animals starting not earlier than 36 h after treatment up to 72 h with appropriate interval(s) between sampling. In case of two treatments at 24-h intervals, bone marrow and peripheral blood should be collected once between 18 and 24 h and 36 and 48 h, respectively, following the final treatment. This is because the consequence of the kinetics of appearance and disappearance of the micronuclei in these two tissues under this treatment regime, being the micronucleated RETs more persistent in the peripheral blood than micronucleated PCEs in the bone marrow [37].

If three or more treatments at 24-h intervals are employed, bone marrow and peripheral blood samples should be collected no later than 24 and 40 h after the last treatment, respectively. This sampling scheme is useful to combine in an acute genotoxicity study, the micronucleus test, and the in vivo Comet assay to evaluate different genotoxic mechanisms and target tissues/organs. The test design comprises administration of the test substance three times to each animal at 48, 24, and 3–6 h prior to sacrifice [38].

The abovementioned treatment and sampling times are also adequate for integration of the micronucleus test with repeated-dose toxicity studies. Under these conditions, experimental evidences using different clastogenic and aneugenic agents [39] demonstrated that the frequencies of micronucleated PCEs in bone marrow and micronucleated RETs in peripheral blood reach a steady-state condition during "continuous" exposure within 2–3 days and remain at steady state for prolonged periods. However, with some classes of test compounds, such as mitomycin-C or dimethylhydrazine, repeated treatments markedly enhance the lethality with the consequence that evaluation of micronucleus induction may be difficult because significantly lower doses need to be applied. In addition, it has been demonstrated that some compounds such as monocrotaline, urethane, or 6-mercaptopurine, following repeated treatments induce lower frequencies of micronuclei with time, thus suggesting a greater sensitivity of young animals because the hematopoietic function in rats change significantly with age [40]. As a consequence, in an integrated micronucleus study, it is advisable to include an early sampling of peripheral blood taken 3–4 days after the start of dosing.

3.5.2 Chromosome Aberration Assay

For the chromosome aberration assay, test substances are usually administered as single treatment because limited data are available on the suitability of a repeated-dose exposure for this assay. For this reason, integration of this test with a repeated-dose toxicity test is only possible when the treatment regime does not produce observable cytotoxic effects such as depression of mitotic index or any other evidence of target tissue cytotoxicity.

In principle the first mitotic division following treatment should be analyzed to detect unstable aberrations and acentric fragments, which are lost during cell division, as well as asymmetrical chromatid and chromosome exchanges, which can affect mechanical segregation of chromosomes inducing cell death, thus escaping detection in later cell divisions. On these bases, bone marrow samples should be collected on two separate occasions, following single treatment, being the first, normally, at 12–18 h (corresponding for both mice and rats to 1.5 normal cell cycle length). The additional sampling of bone marrow should be performed 24 h after the first one to take into account the fact that absorption, metabolism, and/or potential induction of delayed effects on cell cycle kinetics can alter the ideal time for detection of chromosomal aberrations. For dose-repeated treatments, when scientifically justified, a single sampling time at approximately 1.5 cell cycle from the last treatment is recommended.

Prior to sacrifice, animals are injected intraperitoneally with an appropriate dose of the spindle poisons Colcemid or colchicine to accumulate cells in c-metaphases for 3–5 h in mice and for approximately 4–5 h in Chinese hamsters and rats.

4. ANALYSIS

4.1 Target Tissues Processing Staining and Scoring

4.1.1 Micronucleus Test

At the appropriate sampling time, bone marrow from femurs and/or small volumes of peripheral blood from the tail vein are collected and processed according to the method of analysis. Smear preparations from bone marrow are usually stained using conventional methods such as May-Gruenwald and Giemsa, Giemsa single staining, or hematoxylin–eosin to allow identification of mature and immature erythrocytes, which stain in pink and bluish, respectively. Micronuclei are then scored in the immature erythrocyte cell population (PCEs). Figs. 5.1 and 5.2, respectively, show examples of mouse and rat bone marrow smears for analysis of micronucleated erythrocytes. However, staining procedures employing Giemsa are unable to discriminate between DNA-containing micronuclei and other cell inclusions (particularly RNA or other acidic materials that are stained dark blue by Giemsa). The bone marrow of some rodent species, especially the rat, often include large numbers of mast cells, which can be burst during preparation of slides, spreading out large numbers of granules over the cells onto slides. These granules are of similar shape and size of micronuclei and also stain blue with Giemsa. Using hematoxylin–eosin, these bodies are also stained

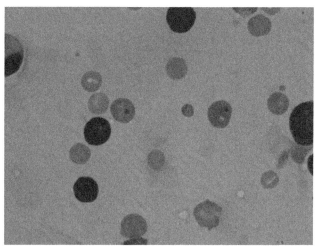

Figure 5.1 Mouse bone marrow smear for micronucleus analysis stained with May-Gruenwald and Giemsa (×1000: high magnification).

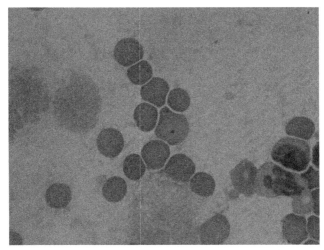

Figure 5.2 Rat bone marrow smear for micronucleus analysis stained with hematoxylin–eosin (×1000: high magnification).

but less intensely than micronuclei, allowing an efficient discrimination. In addition, DNA-specific stains, such as acridine orange or Hoechst 33,258 plus pyronin-Y, can eliminate these artifacts associated with the use of non-DNA specific stain. However, neither of these procedures, being fluorescent staining methods, produces permanent preparations and therefore the slides should be scored within a short time after staining [41].

When supravital staining with acridine orange is used, the abundant presence of nucleated cells generally produces an intense fluorescence, which can interfere with identification of micronuclei. To reduce this side effect, filtration through cellulose columns can be applied to remove the nucleated cells. The validity of this approach has been demonstrated for both clastogens and aneugens at various dose levels [42,43].

For peripheral blood, identification of positive RNA RETs, which represent the cell population to be scored for micronuclei, is generally performed using acridine orange. In case of rat peripheral blood, current recommendations indicate that analyses should be restricted to the most immature RETs [44] to maintain a high sensitivity, despite the splenic filtration function. For this purpose, micronucleated cells should be measured by restricting analysis to type I and type II RETs (i.e., those with reticulum covering at least half the cytoplasm).

The automated scoring using flow cytometry is increasingly used because the method is very fast and allows the analysis of a large number of cells.

Immature RETs can be differentially stained because they are still rich in certain surface proteins such as transferrin receptor (CD71). Using CD71-based fluorescence as an index of RET age, the system has the potential to focus on the analysis of micronuclei in the youngest RETs population. A critical aspect of this method is to ensure the standardization of the flow cytometer configuration before the analysis of micronucleated cells because they are relatively rare and exhibit a heterogeneous DNA content. This is usually performed using fixed blood cells from animals infected with *Plasmodium berghei*, which lends to an adequate number of cells with a homogenous DNA content. Accumulated evidence indicate that, when calibration of the flow cytometer is performed with this biological standard, reproducibility of results both intra- and interlaboratory is much greater than that generated from microscope scoring [31,45].

4.1.2 Chromosomal Aberration Test

Immediately after sacrifice, the bone marrow cells are exposed to an appropriate hypotonic solution treatment and fixed. The cell suspension is then spread onto slides and stained with appropriate dyes, most commonly Giemsa. Examples of rat and mouse normal metaphases are given in Figs. 5.3 and 5.4, respectively. Metaphases to be scored should contain a number of centromeres equal to the diploid number of chromosomes for that

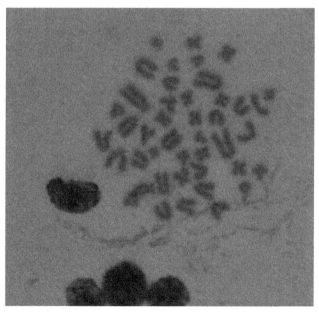

Figure 5.3 Rat bone marrow normal metaphase (×1000: high magnification).

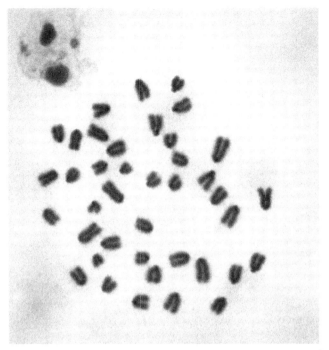

Figure 5.4 Mouse bone marrow normal metaphase (×1000: high magnification).

species ±2. This is because slide preparation procedures may produce bursting of metaphases and chromosome scattering. As a consequence, this assay is not suitable for the analysis of aneuploidy.

Aberrations should be distinguished into chromatid and chromosome type depending on the involvement of both or only one chromatid, with a separate recording of breaks and exchanges. Gaps are also recorded but not included in the subsequent evaluation. A valuable description of the classification of chromosomal aberrations was published by Savage [17]. An example of an aberrant bone marrow metaphase from mouse is shown in Fig. 5.5. The majority of chemical chromosome damaging, which include alkylating agents, nitroso compounds, some antibiotics, and DNA base analogues, act as S-dependent agents inducing exclusively chromatid-type aberrations compared with a limited number of chemicals (radiomimetic agents) and ionizing radiation, which induce chromosome-type aberrations.

Occasionally, in chromosomal aberration test in vivo polyploid and endoreduplicated cells are also observed. Although this could envisage an

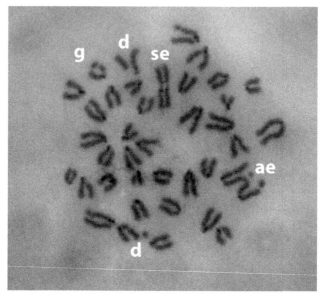

Figure 5.5 Mouse bone marrow aberrant metaphase; *ae*, asymmetrical chromatid exchange; *d*, chromatid deletion; *g*, chromatid gap; *se*, symmetrical chromatid exchange.

indication for numerical chromosomal abnormalities, an increase in poly-ploidy per se does not indicate an aneugenic potential but is rather associated to cell cycle interference or cytotoxicity [46].

Alternatively detection of chromosomal aberration can be performed using fluorescent in situ hybridization (FISH) techniques labeling specific chromosomes with whole chromosome painting probes [47–50]. This method is appropriate when additional information on the nature of chromosomal aberrations is requested. This includes the analysis of stable chromosomal aberrations from symmetrical exchange events, such as translocations, and multichromosome complex aberration, which cannot be detected using conventional staining. However, this technique is not required for general hazard assessment.

4.2 Size of Samples and Statistical Power

4.2.1 Micronucleus Test

Historically, the minimal number of immature erythrocytes to be scored per animal has been 1000 both for mice and rats, including 10 animals per group [51]. This value was later increased up to 2000 PCEs per animal to reduce the group size without loss of sensitivity [52]. Mitchell et al. [53]

analyzed the variability of data in the mouse micronucleus assay, highlighting the importance of outliers in control and treated groups and applied statistical analysis to the historical data to clarify statistically significant but not biologically relevant results. All these studies used data from mice, while Hayes et al. [54], evaluating data from rats, showed that, in case of equivocal results obtained after scoring 2000 PCEs, it is appropriate to increase the sample size up to 6000. However, they demonstrated that no meaningful increase in statistical power is gained by scoring more than 6000 PCEs.

Kissling et al. [55], using flow cytometry as an alternative to microscopy for scoring micronucleated erythrocytes in the peripheral blood of mice, rats, and dogs and in the bone marrow of rats, concluded that scoring of at least 4000 immature erythrocytes per animal (five animals per group) has the power to detect two- to threefold effects with 80% power when the background incidences are relatively high (0.1% or higher). However, in case of lower background values, the Health Canada discussion paper [56] indicated that scoring of 8000 immature erythrocytes per animal is needed to have a sufficient power for detecting a twofold increase in micronucleated cells. Alternatively, using the flow cytometer, the much greater number of RETs that can be scored reduces the counting error to less than the variability between animals. This is normally achieved by analyzing up to 20,000 RETs [22,55].

4.2.2 Chromosome Aberration Assay

The minimum number of 100 metaphases per animal, using five animals per group, has been generally considered adequate. However, Adler et al. [57], based on the evaluation of historical control data, showed that this sample size is insufficient to detect at least one aberrant cell per animal, thus generating scorings with too many zero counts. On this basis, they suggested to score a minimum of 200 cells per animal with the capability to detect a twofold effect with 80% power.

4.3 Cytotoxicity Evaluations

In addition to the analyses of the main cytogenetic end points, cytotoxic effects in target tissues should be determined to ensure that adequate dose levels have been selected. This is because the induction of marked effects on erythropoiesis could significantly affect the sample of the relevant cell population for the analyses of the specific end points. In this respect, the mitotic index, which is an indirect method to evaluate cytotoxicity, is determined for metaphase analysis, while the proportion of immature erythrocytes (PCEs or RETs) among the total erythrocyte population is determined for the micronucleus test. These parameters should not be less than 20% of the concurrent negative control values. However, when the analysis of micronuclei is performed by flow cytometer and RETs CD71$^+$ are the selected target cell population, a reduction up to 5% of the negative control value is considered as acceptable for a valid assay. This is because the CD71$^+$ fraction of RETs is extremely depressed compared with the immature erythrocytes identified by their RNA content at early time points such as those used in acute treatment study designs [58]. This can be observed either using flow cytometer or acridine orange staining in peripheral blood or conventional staining in bone marrow.

4.4 Relevance of Historical Control Data

Historical control data have been generally used to assess the assay acceptance criteria, following comparison with the concurrent control data, and for providing evidence of the technical competence of laboratories for the specific assay [59]. Refinement and further development of both in vitro and in vivo mutagenic assays has led to the identification of a greater number of genotoxic agents but concomitantly has also increased the number of equivocal compounds. In this respect, the biological relevance of genetic toxicity assay results is becoming increasingly important and historical control data are used to aid in this interpretation.

Historical negative and positive control data have to be compiled separately for each species, strain, tissue, cell type, vehicle, treatment, and sampling time because they contribute to total variability and the minimum data set should comprise at least 10 separate experiments as reported by Adler et al. [57]. In addition, they showed that also the year of execution of the assay contributes to the total variability and, hence, the analysis of data set should privilege the more recent data to ensure that they carry greater weight than older data [59].

An example of vehicle control chart for individual values is displayed in Fig. 5.6.

4.5 Data Interpretation and Criteria for a Positive/Negative Result

Evaluation and interpretation of results obtained in the in vivo cytogenetic assays are defined in the two relevant, official, and internationally accepted guidelines on mammalian erythrocyte micronucleus test (TG 474) and mammalian bone marrow chromosomal aberration test (TG 475) issued by the Organisation for Economic Co-operation and Development [60,61].

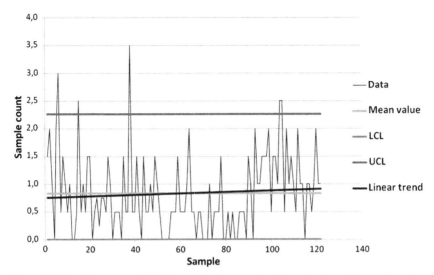

Figure 5.6 Quality control "C" chart of micronucleated polychromatic erythrocytes (sample counts) across animals (sample) from vehicle control groups, *LLC*, lower confidence limit; *ULC*, upper confidence limit.

For a final judgment it is therefore necessary to ascertain first the validity of the study demonstrating that:

1. the concurrent negative control falls within the 95% control limits of the laboratory historical negative control database;
2. the results of concurrent positive control are statistically significantly higher than the concurrent negative control value and compatible with the historical positive control data;
3. the appropriate number of dose levels and cells has been analyzed;
4. the dose levels selected and cytotoxicity level observed in target tissues fulfill the requirements described in paragraphs 3.2 and 4.3, respectively.

If all acceptability criteria are fulfilled, a response is considered clearly positive if all following conditions are achieved:

1. at least one of the treatment group shows a statistically significant increase compared to the concurrent negative control;
2. the observed increases should be dose related as proved by an appropriate trend test analysis;
3. any of these results are outside the distribution of the historical negative control data (95% confidence limits).

On the contrary, the results are considered clearly negative when all the abovementioned conditions are not met and target tissue exposure to the test substance is demonstrated.

When the results are not clearly positive or negative, a scientific expert judgment is recommended to evaluate the biological relevance of the results and the need to score more cells, thus increasing the sample size or to perform further experimental investigations.

4.6 False-Positive Outcome

There is growing evidence that in vivo genotoxicity assays have the potential to give misleading positive results that are not related to a true genotoxic effect of the test substance. Increases in micronucleated PCEs are observed after treatment with test substances that cause hyperthermia, hypothermia, increases in erythropoiesis in the bone marrow, or inhibition of protein synthesis. While hyperthermia disassociate metaphasic chromosome from kinetochores, thus generating chromosome loss and micronuclei, hypothermia enhances erythropoiesis as a consequence of reduction of oxygen tension of the blood. This enhanced cell proliferation causes more erythroblasts to undergo enucleation, resulting in an increase of "spurious" micronuclei. Additional false-positive results can also be generated by compounds that induce marked hematotoxicity followed by a sharp recovery and an associated rebound in erythropoiesis [62]. In such cases it is recommended to perform a follow-up study for induction of chromosome aberrations by bone marrow metaphase analysis.

REFERENCES

[1] Erickson RP. Somatic gene mutation and human disease other than cancer: an update. Mutat Res 2010;705(2):96–106.
[2] De Flora S, Izzotti A. Mutagenesis and cardiovascular diseases molecular mechanisms, risk factors, and protective factors. Mutat Res 2007;621(1–2):5–17.
[3] Frank SA. Evolution in health and medicine Sackler colloquium: somatic evolutionary genomics: mutations during development cause highly variable genetic mosaicism with risk of cancer and neurodegeneration. Proc Natl Acad Sci USA 2010;107(Suppl. 1):1725–30.
[4] Hoeijmakers JH. DNA damage, aging, and cancer. N Engl J Med 2009;361(15):1475–85.
[5] Slatter MA, Gennery AR. Primary immunodeficiencies associated with DNA-repair disorders. Expert Rev Mol Med 2010;12:e9.
[6] Bryant PE. The signal model: a possible explanation for the conversion of DNA double-strand breaks into chromatid breaks. Int J Radiat Biol 1998;73(3):243–51.
[7] Natarajan AT, Obe G. Molecular mechanisms involved in the production of chromosomal aberrations. I. Utilization of Neurospora endonuclease for the study of aberration production in G_2 stage of the cell cycle. Mutat Res 1978;52(1):137–49.
[8] Obe G, Johannes C, Schulte-Frohlinde D. DNA double-strand breaks induced by sparsely ionizing radiation and endonucleases as critical lesions for cell death, chromosomal aberrations, mutations and oncogenic transformation. Mutagenesis 1992;7(1):3–12.

[9] Caldecott KW. Mammalian DNA single-strand break repair: an X-ra(y)ted affair. BioEssays 2001;23(5):447–55.

[10] Dianov GL, O'Neill P, Goodhead DT. Securing genome stability by orchestrating DNA repair: removal of radiation-induced clustered lesions in DNA. BioEssays 2001;23(8):745–9.

[11] Kooistra R, Pastink A, Zonneveld JB, Lohman PH, Eeken JC. The Drosophila melanogaster DmRAD54 gene plays a crucial role in double-strand break repair after P-element excision and acts synergistically with Ku70 in the repair of X-ray damage. Mol Cell Biol 1999;19(9):6269–75.

[12] Pfeiffer P, Goedecke W, Obe G. Mechanisms of DNA double-strand break repair and their potential to induce chromosomal aberrations. Mutagenesis 2000;15(4):289–302.

[13] Tweats DJ, Blakey D, Heflich RH, Jacobs A, Jacobsen SD, Morita T, et al. Report of the IWGT working group on strategy/interpretation for regulatory in vivo tests II. Identification of in vivo-only positive compounds in the bone marrow micronucleus test. Mutat Res 2007;627(1):92–105.

[14] Evans HJ. Chromatid aberrations induced by gamma irradiation. I. The structure and frequency of chromatid interchanges in diploid and tetraploid cells of Vicia faba. Genetics 1961;46:257–75.

[15] Evans HJ, Scott D. The induction of chromosome aberrations by nitrogen mustard and its dependence on DNA synthesis. Proc R Soc Lond B Biol Sci 1969;173(1033):491–512.

[16] Kihlman BA. 1,3,7,9-tetramethyluric acid–a chromosome-damaging agent occurring as a natural metabolite in certain caffeine-producing plants. Mutat Res 1977;39(3–4):297–315.

[17] Savage JR. Classification and relationships of induced chromosomal structual changes. J Med Genet 1976;13(2):103–22.

[18] Torous DK, Dertinger SD, Hall NE, Tometsko CR. Enumeration of micronucleated reticulocytes in rat peripheral blood: a flow cytometric study. Mutat Res 2000;465(1–2):91–9.

[19] De Boeck M, van der Leede BJ, Van Goethem F, De Smedt A, Steemans M, Lampo A, et al. Flow cytometric analysis of micronucleated reticulocytes: time- and dose-dependent response of known mutagens in mice, using multiple blood sampling. Environ Mol Mutagen 2005;46(1):30–42.

[20] Dertinger SD, Torous DK, Hayashi M, MacGregor JT. Flow cytometric scoring of micronucleated erythrocytes: an efficient platform for assessing in vivo cytogenetic damage. Mutagenesis 2011;26(1):139–45.

[21] Parton JW, Hoffman WP, Garriott ML. Validation of an automated image analysis micronucleus scoring system. Mutat Res 1996;370(1):65–73.

[22] Asano N, Torous DK, Tometsko CR, Dertinger SD, Morita T, Hayashi M. Practical threshold for micronucleated reticulocyte induction observed for low doses of mitomycin C, Ara-C and colchicine. Mutagenesis 2006;21(1):15–20.

[23] Styles JA, Clark H, Festing MF, Rew DA. Automation of mouse micronucleus genotoxicity assay by laser scanning cytometry. Cytometry 2001;44(2):153–5.

[24] Degrassi F, Tanzarella C. Immunofluorescent staining of kinetochores in micronuclei: a new assay for the detection of aneuploidy. Mutat Res 1988;203(5):339–45.

[25] Thomson EJ, Perry PE. The identification of micronucleated chromosomes: a possible assay for aneuploidy. Mutagenesis 1988;3(5):415–8.

[26] Eastmond DA, Tucker JD. Identification of aneuploidy-inducing agents using cytokinesis-blocked human lymphocytes and an antikinetochore antibody. Environ Mol Mutagen 1989;13(1):34–43.

[27] Eastmond DA, Pinkel D. Detection of aneuploidy and aneuploidy-inducing agents in human lymphocytes using fluorescence in situ hybridization with chromosome-specific DNA probes. Mutat Res 1990;234(5):303–18.

[28] Marshall RR, Murphy M, Kirkland DJ, Bentley KS. Fluorescence in situ hybridisation with chromosome-specific centromeric probes: a sensitive method to detect aneuploidy. Mutat Res 1996;372(2):233–45.

[29] Schlegel R, MacGregor JT. The persistence of micronucleated erythrocytes in the peripheral circulation of normal and splenectomized Fischer 344 rats: implications for cytogenetic screening. Mutat Res 1984;127(2):169–74.

[30] Hayashi M, Kodama Y, Awogi T, Suzuki T, Asita AO, Sofuni T. The micronucleus assay using peripheral blood reticulocytes from mitomycin C- and cyclophosphamide-treated rats. Mutat Res 1992;278(2–3):209–13.

[31] MacGregor JT, Bishop ME, McNamee JP, Hayashi M, Asano N, Wakata A, et al. Flow cytometric analysis of micronuclei in peripheral blood reticulocytes: II. An efficient method of monitoring chromosomal damage in the rat. Toxicol Sci 2006;94(1):92–107.

[32] Dertinger SD, Chen Y, Miller RK, Brewer KJ, Smudzin T, Torous DK, et al. Micronucleated CD71-positive reticulocytes: a blood-based endpoint of cytogenetic damage in humans. Mutat Res 2003;542(1–2):77–87.

[33] Abramsson-Zetterberg L, Zetterberg G, Bergqvist M, Grawe J. Human cytogenetic biomonitoring using flow-cytometric analysis of micronuclei in transferrin-positive immature peripheral blood reticulocytes. Environ Mol Mutagen 2000;36(1):22–31.

[34] Dertinger SD, Camphausen K, Macgregor JT, Bishop ME, Torous DK, Avlasevich S, et al. Three-color labeling method for flow cytometric measurement of cytogenetic damage in rodent and human blood. Environ Mol Mutagen 2004;44(5):427–35.

[35] Stopper H, Hempel K, Reiners C, Vershenya S, Lorenz R, Vukicevic V, et al. Pilot study for comparison of reticulocyte-micronulei with lymphocyte-micronuclei in human biomonitoring. Toxicol Lett 2005;156(3):351–60.

[36] Grawe J, Biko J, Lorenz R, Reiners C, Stopper H, Vershenya S, et al. Evaluation of the reticulocyte micronucleus assay in patients treated with radioiodine for thyroid cancer. Mutat Res 2005;583(1):12–25.

[37] Higashikuni N, Sutou S. An optimal, generalized sampling time of 30 ± 6 h after double dosing in the mouse peripheral blood micronucleus test. Mutagenesis 1995;10(4):313–9.

[38] Pfuhler S, Kirkland D, Kasper P, Hayashi M, Vanparys P, Carmichael P, et al. Reduction of use of animals in regulatory genotoxicity testing: identification and implementation opportunities-Report from an ECVAM workshop. Mutat Res 2009;680(1–2):31–42.

[39] MacGregor JT, Tucker JD, Eastmond DA, Wyrobek AJ. Integration of cytogenetic assays with toxicology studies. Environ Mol Mutagen 1995;25(4):328–37.

[40] Hamada S, Sutou S, Morita T, Wakata A, Asanami S, Hosoya S, et al. Evaluation of the rodent micronucleus assay by a 28-day treatment protocol: summary of the 13th collaborative study by the collaborative study group for the micronucleus test (CSGMT)/Environmental Mutagen Society of Japan (JEMS)-Mammalian Mutagenicity Study Group (MMS). Environ Mol Mutagen 2001;37(2):93–110.

[41] Pascoe S, Gatehouse D. The use of a simple haematoxylin and eosin staining procedure to demonstrate micronuclei within rodent bone marrow. Mutat Res 1986;164(4):237–43.

[42] Romagna F, Staniforth CD. The automated bone marrow micronucleus test. Mutat Res 1989;213(1):91–104.

[43] Sun JT, Armstrong MJ, Galloway SM. Rapid method for improving slide quality in the bone marrow micronucleus assay; an adapted cellulose column procedure. Mutat Res 1999;439(1):121–6.

[44] Hayashi M, MacGregor JT, Gatehouse DG, Adler ID, Blakey DH, Dertinger SD, et al. In vivo rodent erythrocyte micronucleus assay. II. Some aspects of protocol design including repeated treatments, integration with toxicity testing, and automated scoring. Environ Mol Mutagen 2000;35(3):234–52.

[45] Dertinger SD, Bishop ME, McNamee JP, Hayashi M, Suzuki T, Asano N, et al. Flow cytometric analysis of micronuclei in peripheral blood reticulocytes: I. Intra- and inter-laboratory comparison with microscopic scoring. Toxicol Sci 2006;94(1):83–91.

[46] Davoli T, de Lange T. The causes and consequences of polyploidy in normal development and cancer. Annu Rev Cell Dev Biol 2011;27:585–610.

[47] Lucas JN, Tenjin T, Straume T, Pinkel D, Moore 2nd D, Litt M, et al. Rapid human chromosome aberration analysis using fluorescence in situ hybridization. Int J Radiat Biol 1989;56(1):35–44.

[48] Lucas JN, Awa A, Straume T, Poggensee M, Kodama Y, Nakano M, et al. Rapid translocation frequency analysis in humans decades after exposure to ionizing radiation. Int J Radiat Biol 1992;62(1):53–63.

[49] Tucker JD, Ramsey MJ, Lee DA, Minkler JL. Validation of chromosome painting as a biodosimeter in human peripheral lymphocytes following acute exposure to ionizing radiation in vitro. Int J Radiat Biol 1993;64(1):27–37.

[50] Matsuoka A, Tucker JD, Hayashi M, Yamazaki N, Sofuni T. Chromosome painting analysis of X-ray-induced aberrations in human lymphocytes in vitro. Mutagenesis 1994;9(2):151–5.

[51] Richold M. Practical application of new approaches in genetic toxicology. Toxicol Vitro 1990;4(4–5):644–5.

[52] Lovell DP. Editorial: systematic reviews and meta-analyses of gene association studies. Biomarkers 2016:1–5.

[53] Mitchell IG, Carlton JB, Gilbert PJ. The detection and importance of outliers in the in vivo micronucleus assay. Mutagenesis 1988;3(6):491–5.

[54] Hayes J, Doherty AT, Adkins DJ, Oldman K, O'Donovan MR. The rat bone marrow micronucleus test–study design and statistical power. Mutagenesis 2009;24(5):419–24.

[55] Kissling GE, Dertinger SD, Hayashi M, MacGregor JT. Sensitivity of the erythrocyte micronucleus assay: dependence on number of cells scored and inter-animal variability. Mutat Res 2007;634(1–2):235–40.

[56] Health Canada. Discussion document on assay sensitivity, sample size and statistical power for the erythrocytes micronucleus assay (TG474) produced for the OECD expert group meeting. 2012.

[57] Adler ID, Bootman J, Favor J, Hook G, Schriever-Schwemmer G, Welzl G, et al. Recommendations for statistical designs of in vivo mutagenicity tests with regard to subsequent statistical analysis. Mutat Res 1998;417(1):19–30.

[58] LeBaron MJ, Schisler MR, Torous DK, Dertinger SD, Gollapudi BB. Influence of counting methodology on erythrocyte ratios in the mouse micronucleus test. Environ Mol Mutagen 2013;54(3):222–8.

[59] Hayashi M, Dearfield K, Kasper P, Lovell D, Martus HJ, Thybaud V. Compilation and use of genetic toxicity historical control data. Mutat Res 2011;723(2):87–90.

[60] Organization for Economic Co-operation, Development (OECD). Guideline for testing chemicals 474: mammalian erythrocyte micronucleus test. 2014. http://www.oecd-ilibrary.org/environment/test-no-474-mammalian-erythrocyte-micronucleus-test_9789264224292-en.

[61] Organization for Economic Co-operation, Development (OECD). Guideline for testing chemicals 475: mammalian bone marrow chromosomal aberration test. 2014. http://www.oecd-ilibrary.org/environment/test-no-475-mammalian-bone-marrow-chromosome-aberration-test_9789264071308-en.

[62] Tweats DJ, Blakey D, Heflich RH, Jacobs A, Jacobsen SD, Morita T, et al. Report of the IWGT working group on strategies and interpretation of regulatory in vivo tests I. Increases in micronucleated bone marrow cells in rodents that do not indicate genotoxic hazards. Mutat Res 2007;627(1):78–91.

CHAPTER SIX

Mutagenicity and Genotoxicity Testing in Environmental Pollution Control

Amit Bafana, Kannan Krishnamurthi, Saravanadevi Sivanesan, Pravin K. Naoghare
CSIR–National Environmental Engineering Research Institute, Nagpur, India

1. INTRODUCTION

One of the greatest problems that the world is facing today is that of environmental pollution. Pollution may be defined as "the anthropogenic introduction of substances or energy into the environment, which may cause hazards to human health, living resources, ecological systems, or structures." This definition excludes the natural release of polluting materials to the environment, such as volcanic eruptions or geogenic release of metals to water bodies. This is because natural release of such materials to the environment is generally low. Besides, environmental processes can assimilate pollutants when the volume and rate of emission and toxicity of pollutants are low [1]. Evolution of human civilization, on other hand, has greatly increased the scale and range of pollutants not only on earth but also in near solar system. Many effects have become obvious now and many more will appear as the time and human spread increases. Several human activities, such as forest clearance, agriculture, construction, industrialization, energy generation, waste production and disposal, have increased the burden of pollutants. Ever increasing size of human population has made this burden disproportionate with the assimilation capacity of the nature. In other words, we have already crossed the carrying capacity of environment, and there is no immediate way that we can undo it. However, the focus needs to shift to the use of green and eco-friendly technologies to eventually achieve the balance with the environment [2].

A pollution event can be characterized by the source of pollutants (thermal power plant, industrial effluent, etc.), transport medium (air, water),

Mutagenicity: Assays and Applications
ISBN 978-0-12-809252-1
http://dx.doi.org/10.1016/B978-0-12-809252-1.00006-7

and target (ecosystems or individual organisms). The pollutants that can be dispersed to a larger area through air or water are expected to affect larger populations [3]. For example, several pesticides have been found in Antarctica, although they were never used there. It is proposed that these pesticides might have been transported by long-range transport through air from other continents. The severity of pollutants is determined by their properties such as toxicity, persistence, dispersion, chemical reactivity, bioaccumulation, and ease of control [4]. Based on these properties, regulatory agencies have set up the criteria for allowed limit of these pollutants in different media.

2. TYPES AND SOURCES OF POLLUTION

Pollution can be classified in several ways depending on the source (e.g., agricultural pollution, thermal power plant pollution, etc.), media affected (air pollution, water pollution, etc.), or nature of pollutant (radioactive pollution, heavy metal pollution, etc.). Broadly, it can be categorized in three main types: air, water, and soil pollutions [4]. Many other types such as noise, radiation, heat, and solid waste pollutions have also been described commonly in literature.

The environment is continuously being burdened with newer foreign organic chemicals (xenobiotics), such as polychlorinated biphenyls (PCBs), organochlorine pesticides, polycyclic aromatic hydrocarbons (PAHs), polychlorinated dibenzofurans, and dibenzo-p-dioxins [3]. Nanomaterials are relatively recent introduction, which are increasingly being used for commercial purposes such as fillers, catalysts, semiconductors, cosmetics, microelectronics, and drug carriers. The production, use, and disposal of nanomaterials will inevitably lead to their release into air, water, and soil [5]. Electronic waste (e-waste) includes disposed electronic products like computers, mobile phones, printers, television, and toys, which are made of sophisticated blends of plastics, metals, among other materials. E-waste is growing at about 4% per year and has become the fastest growing waste stream in the industrialized world. During recycling, incomplete combustion of e-waste results in the release of persistent organic pollutants such as polybrominated diphenyl ethers, dioxins/furans, PAHs, PCBs, and heavy metals/metalloid [6]. Radiation pollution includes atomic and nuclear radiation, which can be released accidentally into the environment and have catastrophic effect. The Fukushima Daiichi nuclear disaster (2011) and Chernobyl disaster (1986) are well-known examples of nuclear power plant accidents. Radioactive pollution is very dangerous as it is mutagenic and can persist in the environment for a very long time depending on the half-life of the material [7]. Use of X-rays is very common nowadays and repeated exposure can have mutagenic effects. Increasing chemical pollution has also increased exposure to harmful UV rays in recent past indirectly through damaging the ozone layer. The discussion on various types of pollution can be endless. Hence, this chapter will focus on the three main types of pollution, i.e., air, water, and soil pollutions.

2.1 Air Pollution

Air pollutants can be broadly categorized into gaseous pollutants and particulate matter (PM). Gaseous pollutants include SOx, NOx, CO, O_3, and volatile organic compounds (VOCs). PM consists of particles suspended in air, which vary in size and composition and are produced by a wide variety of natural and anthropogenic activities. Major PM sources are factories, power plants, motor vehicles, construction activity, and natural windblown dust. PM may contain PAHs, metals, and other organic components depending on the source. Based on the size, PM particles can be classified into PM_{10} (aerodynamic diameter smaller than $10\,\mu m$) and $PM_{2.5}$ (aerodynamic diameter smaller than $2.5\,\mu m$). Recently, other categories have also been defined such as ultrafine particles smaller than $0.1\,\mu m$ and fine particles smaller than $1\,\mu m$. Larger particles deposit mainly in the upper respiratory tract, while fine and ultrafine particles are able to reach lung alveoli. There is strong evidence that finer particles are more hazardous than larger ones. However, chemical composition of the particles also plays important role in determining its toxicity [8]. Photochemical reactions in the atmosphere can give rise to secondary pollutants. The oxidation of NOx and VOCs in the troposphere may lead to a variety of secondary pollutants (e.g., O_3, peroxy acetyl nitrate, secondary organic aerosols), many of which are potentially more harmful than their precursors [9].

2.2 Water Pollution

More than one-third of the Earth's accessible renewable freshwater is used for agricultural, industrial, and domestic purposes, and most of these activities lead to water contamination with numerous synthetic compounds. Widespread use of pesticides for agriculture has resulted in the presence of their residues in various environmental matrices. Pesticide contamination of surface waters has been well documented worldwide and constitutes a major issue. Pesticide residues reach the aquatic environment through direct runoff, leaching, careless disposal, equipment washing, etc. [10]. In an EU-wide reconnaissance of the occurrence of persistent organic pollutants in European river waters, only about 10% of the river water samples analyzed could be classified as "very clean" in terms of chemical pollution [11]. Heavy metal contamination of aquatic system has attracted the attention of several investigators both in the developed and developing countries of the world. Pollution of rivers is a very critical issue as they provide potable water to a large number of adjacent towns and villages. Rivers such as Ganga are polluted by a wide range of sources including domestic sewage; industrial wastewater; surface runoff containing fertilizers and pesticides, animal carcasses, and human corpses. Cleaning of rivers or other water bodies is a costly program. Ganga action plan is one such example undertaken by the Government of India [12]. In recent years, levels of contaminants in the marine environment have increased as a consequence of anthropogenic activities. It has now been fairly established that operational or accidental oil spills from tankers and other oil-related activities can lead to large-scale destruction of marine life [13]. Thus, pollution can lead to degradation of water and sediment quality and, consequently, a decrease in natural resources.

2.3 Soil Pollution

Ground sealing through expansion of residential and traffic areas has progressively led to reduction in the surface under vegetation and soils. This is coupled to contamination by wastes and atmospheric depositions, loss of organic matter, changes in soil reaction, structural degradation, and infection by pathogenic microorganisms. Heavy metals, radionuclides, chlororganic compounds, etc. are the most common contaminants, mainly arising from mining, industrial plants, thermal power stations, vehicular traffic, and road infrastructures [14]. Soil contamination has adversely affected the ecological functions of soils. Wastewater irrigation, solid waste disposal, sludge applications, vehicular exhaust, and industrial activities have led to soil contamination with heavy metals, and an increased metal uptake by food crops grown on such contaminated soils is often observed. Street dusts also often contain elevated concentrations of a range of toxic elements including heavy metals [15].

3. IMPACT OF ENVIRONMENTAL POLLUTION

Since the early 1960s, mankind has become aware of the potential long-term adverse effects of pollution in general and its potential risks. Environmental pollution has existed for centuries but only started to be significant following the industrial revolution in the 19th century. There is no doubt that excessive levels of pollution are causing a lot of damage to animals and plants. Pollution also disturbs our ecosystem and the balance in the environment. Pollution can exert negative effect on natural elements that are essential for life, such as water and air. The effects in living organisms may range from mild discomfort to serious diseases and physical deformities [16].

3.1 Effect of Pollution on Ecosystem

Some of the effects of air pollutants such as acid rain, global warming, and ozone layer depletion leading to increased UV penetration and health hazards are well recognized now [17]. Further, suspended particulate matter in the air affects vegetation and ecosystem both physically and chemically. Deposition of PM to vegetated surfaces may cause abrasion and radiative heating, and may reduce the photon flux reaching the photosynthetic tissues. The chemical pollutants of PM deposited in the rhizosphere may be taken up via roots. PM deposited to the soil can also influence nutrient cycling through its effects on the rhizosphere bacteria and fungi [18].

Many industrial and agricultural processes have contributed to the contamination of freshwater systems thereby causing adverse effects on aquatic biota. Emissions of NOx and SOx are playing an increasing role in the acidification of freshwater ecosystems. Release of nitrogen in aquatic ecosystem can enhance the development of primary producers, resulting in eutrophication. Eutrophication further leads to hypoxia and consequent mortality of invertebrates and fishes. The decline in dissolved oxygen concentrations can also promote formation of reduced compounds, such as hydrogen sulfide, resulting in toxic effects on aquatic animals. Additionally, the growth of toxic algae can significantly contribute to the death of aquatic animals [19].

Food chain contamination is one of the important pathways for the entry of toxic pollutants such as heavy metals into the human body [20]. Heavy metals cannot be destroyed through biological degradation and have the ability to accumulate in the environment, which makes these toxicants deleterious to the living organisms [21]. Simsek et al. reported the highest heavy metal content in the milk samples collected from industrial regions followed by traffic intensive regions and rural regions [22]. Among heavy metals, mercury represents a major hazard. Organomercury compounds, which can arise from microbial metabolism of inorganic mercury, are more toxic than the inorganic mercury. Mercury can affect not only lower and higher animals and plants but also disturb microorganisms leading to disruption of natural ecological cycles [23].

Antibiotics are widely used for preventing and treating animal and plant infections as well as for promoting growth in animal farming. Antibiotic release in environment leads to selection of resistant bacteria and perturbation of the environmental microbiota. Other pharmaceutical ingredients have also been shown to be ecotoxic. Estrogen analogs are well known for their endocrine disruptive and reproductive effects. Similarly, excessive use of diclofenac has been shown to be the cause of decline in vulture population [24]. Antibiotic resistance genes can also be considered as molecular pollutants, which can be released from clinical settings and disseminated among pristine ecosystems without any record of antibiotic contamination [25]. Pollution by antibiotic resistance genes can increase the chances of human pathogens for acquiring resistance [26]. There is an increasing amount of research on the toxicology of nanomaterials. Some researchers have shown the toxicity of nanoparticles, such as fullerene, carbon nanotubes, and metal oxides to human cells, bacteria, plants, and rodents. Reactive oxygen species generation and oxidative stress are proposed to be responsible for the toxicity of nanoparticles [27].

Construction activities have modified the environment in a significant way. Laying down of a large network of roads across the globe is associated with negative effects on both terrestrial and aquatic ecosystems. Apart from removal of trees, road construction kills sessile and slow-moving organisms and alters physical conditions beneath a road. Roads change the soil density, temperature, water content, light levels, surface waters, and sedimentation as well as add heavy metals, salts, and organic molecules to soil. Movement of vehicles on roads kills or injures animals by collisions, and also changes behavior and movement of animals [28].

3.2 Effect of Pollution on Humans

Pollutants may cause mortality and morbidity or may pose potential hazard to human health. Human health risk is determined from clinical, epidemiological, or animal studies. Sporadic air pollution events, like the London fog in 1952, and a number of epidemiological studies have reported that air pollutants contribute to increased mortality and hospital admissions. Air pollution–induced human health effects include nausea, difficulty in breathing, skin irritation, cardiovascular problems, birth defects, reduced activity of the immune system, cancer etc. [29]. The London smog of 1952 led to introduction of Clean Air Act in the United Kingdom in 1956. However, introduction of such regulations is being offset by increasing level and range of pollutants due to the shear increase in modernization. After World War II, increased exploitation of local brown coal (lignite) for electricity generation resulted in marked increase in atmospheric SOx pollution because of coal's high sulfur content in Czech Republic. Several epidemiological studies have found association between exposure to airborne PM and incidence of mortality, cardiovascular disease and lung cancer [30]. Children are particularly high-risk group for air pollution. Indeed, some studies suggest that early exposure during childhood can play an important role in the development of chronic diseases in adulthood [31].

Significant proportion of people in developing countries uses coal and biomass in simple stoves with very incomplete combustion. According to World Health Organization (WHO), indoor air pollution may be responsible for nearly 2 million deaths and about 4% of the global burden of disease [32]. Indoor air pollution has been shown to increase the risk of chronic obstructive pulmonary disease and acute respiratory infections. Evidence also exists for the association with low birth weight, increased infant mortality, pulmonary tuberculosis, cataract, and cancer [33].

Water being an essential molecule for life, water pollution leads to a variety of deadly health effects. MacDonald et al. reported drinking water pollution as the leading contributor to health-care facility visits in a United Arab Emirates–wide study [34]. Hendryx et al. proposed that permitted surface water chemical discharges in the United States were related to population mortality [35]. Heavy metal contamination of drinking water poses a serious problem. Once ingested, heavy metals can enter human bodies and combine with biomolecules such as proteins, resulting in interference with their normal functioning [23]. Minamata disease of Japan is a well-known example. It was found to be the result of methylmercury poisoning due to consumption of fish and shellfish contaminated by methylmercury discharged from a chemical plant [36]. Similarly, manganese exposure is associated with neurotoxicity and correlated with the development of neurological disorders such as Parkinson's disease. Tarale et al. highlighted oxidative stress and epigenetic deregulation as the basis of Mn toxicity [37].

Ingested nitrites and nitrates from polluted drinking water can induce methemoglobinemia in humans, by reducing the oxygen-carrying capacity of hemoglobin. They can also cause cancer of the digestive tract through formation of nitrosamines. There is some evidence that ingested nitrites and nitrates may present risk factors for teratogenicity, birth defects, bladder and ovarian cancers, diabetes, thyroid hypertrophy, and respiratory tract infections. Indirect health hazards can occur due to nitrate contamination of water bodies and subsequent algal growth and toxin production. Consumption of algal toxins can cause nausea, vomiting, diarrhea, gastroenteritis, muscular cramps, and several poisoning syndromes [19]. Groundwater and surface water are also contaminated by pesticide runoff or leaching and can cause variety of health effects. Fernández et al. reported leaching of pollutants from municipal landfill into groundwater [38]. They found that the leachate presented high concentration of organic matter, apart from metallic and nonmetallic ions, in spite of the presence of single liner in the landfill site. Fluoride is present at dangerous levels in groundwater in several areas. Continuous exposure to fluoride has been shown to induce chronic oxidative and inflammatory stress, resulting in impediment in osteoblast differentiation and bone development [39]. Daiwile et al. reported that fluoride exposure leads to epigenetic alteration and implicated the microRNAs miR-124 and miR-155 in development of fluorosis [40]. Dyeing and textile industries are significant contributors of mutagenic and toxic chemicals in the environment. A series of confirmatory investigations on toxicity of these compounds has led to ban on many synthetic dues, especially azo dyes in several countries [41]. Nitroaromatics presents another important class of pollutants widely used in synthesis of pesticides, drugs, dyes, and other industrial chemicals. They are highly toxic, teratogenic, and even carcinogenic. Hence, many nitroaromatics have been listed as priority pollutants by the US Environmental Protection Agency (US EPA) [42]. Contamination of water with pathogenic microorganisms such as bacteria, fungi, or protozoa can similarly lead to a variety of fatal and chronic diseases [43].

A wide variety of PAHs are found in the environment as a result of incomplete combustion of organic matter. Some PAHs are also used in the synthesis of pesticides and dyes. Many PAHs have toxic, mutagenic, and carcinogenic properties. They can have a detrimental effect on the flora and fauna, bioaccumulate in food chains, and cause serious health problems or genetic defects in humans [44]. Consequently, the US EPA has listed several PAHs as priority pollutants for remediation [45]. Perera et al. reviewed the effects of in utero exposure to common environmental contaminants, including PAH, PM, and environmental tobacco smoke [46]. They concluded that there is significant transplacental transfer of PAH and tobacco smoke constituents from mother to fetus. PAHs can form adducts with DNA and cause genetic damage to the fetus. Several pollutants, including PAH, are known to suppress the functioning of immune system through dysregulated cytokine and chemokine production, loss of migratory potential, or the inability to phagocytose pathogens by immune cells. This tips the balance between pathogen and host defense system to the benefit of the pathogen leading to increased incidence of infectious diseases [47].

Effects of soil pollution on human health have been demonstrated in several studies. Heavy metal pollution in soil can arise from mining, waste dumping, agrochemicals, vehicular emission, and road construction. Fly ash dust emission from thermal power plants in Odisha was found to lead to high pH and greater accumulation of heavy metals in surrounding soil and consequent decrease in the soil enzyme activities [48]. It presents high carcinogenic and noncarcinogenic risk to the public, especially to children and those living in the vicinity of polluted soil [49]. Li et al. reported that people living near industrial area were exposed to high levels of persistent toxic contaminants in soil and, consequently were at higher risk of breast cancer, stomach cancer, dermatitis, gastroenteritis, and pneumonia [50]. As children are often in contact with surface soil, Ottesen et al. studied soil contamination in day care centers and playgrounds in Norwegian cities [51]. They found high concentrations of Pb and PAHs in soil samples and recommended remediation in majority of day care centers. Pesticide application presents a global pollution threat. Effects of endosulfan application in Kasargod District of Kerala, India, in 1978 are still persistent after so many years. Gandhi et al. showed deregulation of several genes and proteins on endosulfan exposure and correlated them to the mechanism of neurotoxicity and hepatotoxicity of endosulfan [52,53]. About 1200 tonnes of ammunition were dropped on Iraq during the Gulf Wars of 1991 and 2003. Resulting contamination with depleted uranium has been correlated with increased incidence of various cancers and birth defects in Iraq [54].

3.3 Mutagenic Effects of Environmental Pollution

Many environmental pollutants have genotoxic effects and can cause mutations. Mutations are changes in the number or sequence of genetic material, generally DNA molecule (point mutation, chromosomal aberration, etc.). In contrast to general toxicity, genotoxicity can exert long-term effects, with serious health and environmental consequences. Mutations can affect, either somatic or germ cells. Germline mutations cause abnormal development of embryo, prenatal death, or genetically defective offspring. They may also influence the reproductive capacity or survival of species, which is of prime importance in population ecology. Somatic mutations, on the other hand, can lead to development of degenerative diseases including autoimmune defects and cancer [55].

Environmental mutagens such as PAH and heterocyclic amines are known to bind to nucleotides, resulting in the formation of DNA adducts. These DNA adducts lead to mutations during replication of DNA. Reactive oxygen species generated by pollutants can also damage DNA molecules. Several DNA adducts are used as marker for the exposure of humans to mutagens [56]. 8-OxodG (8-oxo-2'-deoxyguanosine) in lymphocyte DNA and markers of oxidative damage in plasma have been shown to be associated with $PM_{2.5}$ exposure [57]. Perera et al. found positive association between aromatic DNA adducts and somatic hypoxanthine–guanine phosphoribosyltransferase mutation frequency in the newborns [58]. Both human and animal studies provide strong evidence that air pollution, especially PM, is genotoxic to male germ cells. This can take the form of mutations, DNA strand breaks, and altered methylation pattern resulting in abnormal sperm morphology and reduced sperm performance in men [59]. Jurewicz et al. studied air pollution–induced aneuploidy in human sperm [60]. Positive associations were observed between exposure to $PM_{2.5}$ and disomy Y, sex chromosome disomy and disomy 21. Exposure to PM_{10} was associated with disomy 21.

Water and soil pollutions from industries have also been shown to produce pronounced genotoxicity. DNA damaging potential of pesticide-contaminated soil was demonstrated using fluorimetric analysis of DNA unwinding assay by Krishnamurthi et al. [61]. In another study, genotoxic potential of oily sludge collected from petroleum refineries was demonstrated. Exposure with sludge extracts caused DNA damage, chromosomal aberration, p53 protein induction, and apoptosis in CHO–K1 cells, and the genotoxicity was attributed to PAHs present in the sludge [62]. Ishikawa et al. reported higher micronuclei frequency in females residing in industrial area as compared to those residing in rural setting, and attributed this effect to exposure to industrial pollution [63]. In an interesting study, the genotoxicity removal potential of industrial wastewater plants of petrochemical, petroleum refinery, and steel industries was studied. It was concluded that the treatment plants of petrochemical and steel industries completely eliminated genotoxicity of the wastewater. However, the third plant in petroleum refinery could achieve a reduction in genotoxicity but significant genotoxic contaminations were still present [64]. This highlights the importance of genotoxicity assays in assessing the efficiency of wastewater treatment plants, especially those treating potentially mutagenic wastes.

Mutagenic effects of environmental pollution are also evident in other animals. Herring gulls (*Larus argentatus*) nesting near steel mills were shown to have higher germline mutation rates at minisatellite DNA loci than those at rural sites, and mutation frequency increased with colony proximity to integrated steel mills. It was proposed that inhalation of air pollutants emitted from steel mills, such as PAH, was responsible for mutation induction [55]. Eeva et al. studied nucleotide diversity in insectivorous passerines living near a smelter [65]. They concluded that passerines living near the smelter site possessed high concentrations of heavy metals, especially arsenic, coupled with significantly high nucleotide diversity, suggesting increased mutation rates in a polluted environment.

4. NEED OF MUTAGENICITY ASSAYS TO ASSESS ENVIRONMENTAL POLLUTION

Estimation of adverse effects of pollutants on ecosystem is an essential component of environmental risk assessment. The steps involved in risk assessment process include hazard identification, effect assessment, exposure assessment, and risk characterization [16]. In recent years, it has been recognized that risk assessment cannot be solely based on chemical analysis of environmental samples because this approach does not provide any indication of deleterious effects of contaminants on the biota. Therefore, the measurement of the biological effects of pollutants has major importance in the assessment of the quality of the environment. This has led to the development of several approaches to estimate the ecotoxicity of pollutants and polluted sites. Bioindicators are organisms that can be used for the identification and qualitative determination of pollutants, whereas biomonitors are organisms mainly used for the quantitative determination of contaminants [66]. Lichens, which are symbiotic associations between fungi and algae, are the most studied bioindicators of air quality in view of their sensitivity to various environmental factors [67]. Fish are largely being used for the assessment of the quality of aquatic environment and as such can serve as bioindicators of environmental pollution [21]. While it is easy to correlate the immediate adverse effects by acute exposure to environmental pollution, delayed effects are not easy to trace back. Because such long-term tests are difficult and expensive, alternative short-term measures are needed. Biomarkers of exposure have been proposed to bridge the gap between chemical analysis and late biological effects. This is done by correlating the biomarkers of exposure and effects in different organisms. In the European Union (EU) project BEEP (Biological Effects of Environmental Pollution on Marine Coastal Ecosystems), the biomarker data collected from flounder, eelpout, and blue mussel populations were developed into mathematical models. The resulting health status agreed well with the known contamination levels in the different study areas [68]. Biomarkers, such as metallothionein induction, acetylcholinesterase inhibition, cytochrome P450 induction, and lysosomal alterations, have been proposed to estimate exposure and effect of different contaminants such as metals and organic xenobiotics. Marine bivalve mollusks such as mussels have emerged as one of the candidate sentinel species for most of the biomarkers [69]. Recently the biomarker approach has been incorporated into several pollution monitoring programmes in Europe and the United States (e.g., the North Sea Task Force Monitoring Master Plan and the NOAA's National Status and Trends Program).

Among the several adverse health effects associated with exposure to pollutants, genetic damage has received a particular interest. Mutagenic pollution of the natural environment is undoubtedly a serious and general problem because such compounds are capable of inducing serious diseases, including cancer. As several of the chemical mutagens can elicit deleterious effects at extremely low concentrations, their detection in the environment is a challenge. Chemical methods are expensive and time consuming, and the sample needs to be concentrated to bring the contaminant within detection range of the method. Moreover, a single method cannot suffice, as there are hundreds of different mutagens with different physicochemical properties. Therefore, it has become desirable to perform a quick and preliminary biological mutagenicity assay with wide detection range for mutagenic agents. Subsequent chemical analysis can be performed to determine the specific type of mutagens, if necessary [70].

Mutagenic effects are relatively long-term results of exposure to pollutants. However, such long-term studies are costly and not always feasible. Mutagenicity in human beings can be established from epidemiologic data, linking the chemical exposure with heritable effects. However, it is difficult to obtain such data because of the low rate of mutation, human genetic variability, small numbers of offspring, and long generation times. In addition, conditions caused by autosomal recessive disorders or by polygenic traits may go unrecognized for many generations [71]. Therefore, experimental animals are widely used as surrogate model systems. However, animal studies are also constrained by ethical and cost considerations. Fortunately, there is a good correlation between the mutagenicity of chemicals in simple microbes such as yeast and bacteria, and mammals [70]. Differences do arise because the primary DNA damage is not genetically relevant until processed metabolically. Furthermore, the genetic change in higher organisms may not necessarily be tumorigenic. Also, there are numerous types of genetic alterations, such as point mutations, frame-shift mutations, structural aberrations to the chromosomes (e.g., deletions, duplications, insertions, inversions, and translocations), and changes in chromosome number (e.g., monosomy, trisomy, haploidy, polyploidy). Certain mutagens, such as alkylating agents, can directly induce DNA damage, while some mutagens can indirectly affect DNA (e.g., agents affecting DNA synthesis, DNA repair, DNA methylation). It necessitates the use of a battery of test organisms to cover the full spectrum of possible genetic effects [72]. Another problem in short-term mutagenicity testing arises because many mutagens are actually promutagens. In other words, they need to be activated or converted to mutagenic derivatives by various metabolic reactions. Therefore, the selection of mutagenicity assays must account for not only the entire range of genetic effects relevant to man, but also metabolic activation systems relevant to human exposure. At present, liver microsome preparations are widely used as activation agents. Hierarchical approach has been used to improve the accuracy of mutagenicity assessments. It involves genotoxicity testing in three phases. The initial phase is screening essentially involving quick microbial tests. The second phase is confirmatory involving in vivo and in vitro tests in higher eukaryotes, including mammals. The third phase is essential for evaluation of human risk and involves in vivo tests in mammals [73].

5. APPLICATION OF MUTAGENICITY ASSAYS TO CONTROL ENVIRONMENTAL POLLUTION

In general, genotoxic monitoring can be grouped into two major approaches. First approach involves in situ monitoring of genetic changes in organisms from polluted environment. This is similar to the bioindicator/ biomonitor approach used for environmental risk assessment as discussed in previous section. The second approach involves ex situ laboratory examination of the samples using validated mutagenicity assays and test organisms.

This approach is useful to generate databases of mutagenic chemicals, which are required for legislation purpose [73].

Biological mutagenicity assays are based on the change in specific phenotypes of tester organisms after contact with mutagens. Changes in phenotypes result from changes in genetic material, which is the target for mutagens. Many different mutagenicity tests employing genetically modified bacterial strains have been described. Much of the testing currently being done uses the Ames reverse mutation test. It uses *Salmonella typhimurium* strains that are unable to produce histidine. Reverse mutations that restore ability of these cells to produce and consequently grow in the absence of histidine are significantly increased in the presence of mutagenic agents [74]. Tryptophan auxotrophs of *Escherichia coli* are also used in Ames test. Appearance of neomycin resistance in *Vibrio harveyi* due to mutagenic exposure has also been described [75]. Other bioluminescence-based assays such as Vitotox (commercially available from Thermo Electron Corporation, United States) and Mutatox tests (commercially available from AZUR Environmental) are also available [70]. Apart from bacteria, eukaryotic microorganisms, higher plants, insects, in vitro mammalian cells, and in vivo animal models are also established for mutagenicity testing. The micronucleus test is widely used as a marker of early biological effects in eukaryotes due to its ability to detect both clastogens and aneuploidy-inducing chemicals [76]. A high frequency of micronuclei in peripheral blood lymphocytes has been found to predict cancer occurrence [31]. The primary DNA damage may also be studied using comet assay (single cell gel electrophoresis test) and chromosomal aberration assay. Comet assay is a very rapid method to detect single-stranded and double-stranded breaks, alkali labile sites, and incomplete repair sites [77].

Literature abounds with application of mutagenicity assays to assess environmental pollution. Bafana et al. studied changes in the mutagenicity during biodegradation of azo dyes, direct black 38 and Congo red, by using Ames test [78,79]. Both the dyes are well-known carcinogens and are banned in several countries. Results indicated that mutagenicity initially increased due to the formation of mutagenic intermediates, but then declined after further degradation of the mutagenic intermediates. Such studies emphasize the need of mutagenicity assays in environmental pollution control. Routine chemical analysis of wastewater may not be able to showcase the dynamic changes in toxicity/mutagenicity, which is ultimately the main parameter of concern. In another study, detoxification of azo dye during biodegradation was checked by a battery of genotoxicity assays such as comet assay, DNA ladder formation, and terminal deoxynucleotidyl transferase–mediated dUTP nick-end labeling assay [80]. Laboratory studies have demonstrated that murine ESTR (expanded simple tandem repeat) loci are susceptible to germline mutations induced by chemical or radioactive mutagens and therefore may be useful tools for environmental contamination studies. Rodent ESTR DNA consists of 4- to 6-bp repeat units in long-tandem arrays that are unstable in the germline and tend to mutate by insertion or deletion of a number of repeat units. Recently, transgenic animals have been developed in which a target gene for detecting mutations is integrated. These transgenic animals can detect both mutation frequency and mutation spectra [56]. One of the most extensively used test for the induction of heritable mutation is the mouse-specific locus test, which measures the induction of recessive mutations at seven loci concerned with coat color and ear morphology. Although tests for the detection of variation in the chromosome number are still at an early stage of development, systems exist in diverse organisms such as fungi, *Drosophila*, and mammalian cells (e.g., mouse X-chromosome loss assay).

Regulatory decisions on mutagens involve two components: risk assessment and risk management. Risk assessment estimates the potential adverse health consequences of exposure, whereas risk management enables the regulatory legislation to decide the need to control the exposure. As discussed in the above section, risk assessment is comprised of hazard identification, effect assessment, exposure assessment, and risk characterization. Hazard identification is the qualitative determination of inherent mutagenicity of a chemical, whereas effect assessment involves measurement of the extent of mutagenicity of the chemical. These components provide information on whether an agent is likely to be mutagenic and how much mutational damage is likely to be produced under particular exposure scenarios. The exposure assessment determines the frequency and reach of human exposure. In risk characterization, all the above outputs are combined to estimate the mutation risk. Such mutagen risk assessment studies have led to the development of several guidelines and legislations. Commission of the European Communities (CEC), Organization for Economic Co-operation and Development (OECD), European Environmental Mutagen Society (EEMS), American Environmental Mutagen Society (AEMS), and WHO have made guidelines for mutagenicity testing, which recommend the following tests: Ames test, *Bacillus subtilis* rec test, gene mutation test in *Saccharomyces cerevisiae*, sex-linked recessive lethal mutation test in *Drosophila melanogaster*, spot test in mice, in vivo and in vitro cytogenetic tests, micronucleus test, dominant lethal mutation test, heredity translocation test, unscheduled DNA synthesis, mitotic recombination test in *Saccharomyces cerevisiae*, and sister chromatid exchange [81]. US EPA has published the guidelines for mutagenicity risk assessment and guidelines for carcinogen risk assessment [71,82]. These guidelines can be consulted when conducting a risk assessment to ensure that mutagenic and/or carcinogenesis effects are considered together with other health effects in the overall characterization of risk. In 2012, EPA designated specialized industrial products as a distinct group of products that meet tailored criteria under the Safer Choice Program [83]. Safer Choice ensures that the allowed ingredients are the safest in their functional component class and no ingredient should be a listed carcinogen; mutagen; reproductive or developmental toxicant; or a persistent, bioaccumulative, and toxic chemical.

REFERENCES

[1] Hill MK. Understanding environmental pollution. 3rd ed. Cambridge: Cambridge University Press; 2010.
[2] Manahan SE. Fundamentals of environmental chemistry. 3rd ed. Florida: CRC Press; 2011.
[3] Alloway B, Ayres DC. Chemical principles of environmental pollution. 2nd ed. Florida: CRC Press; 1997.
[4] Shafi SM. Environmental pollution. New Delhi: Atlantic Publishers; 2005.
[5] Lutz C, Steevens JA. Nanomaterials: risks and benefits. New York: Springer; 2008.
[6] Wong MH, Wu SC, Deng WJ, Yu XZ, Luo Q, Leung AO, et al. Export of toxic chemicals – a review of the case of uncontrolled electronic-waste recycling. Environ Pollut 2007;149:131–40.
[7] Kritidis P, Florou H, Eleftheriadis K, Evangeliou N, Gini M, Sotiropoulou M, et al. Radioactive pollution in Athens, Greece due to the Fukushima nuclear accident. J Environ Radioact 2012;114:100–4.
[8] Kampa M, Castanas E. Human health effects of air pollution. Environ Pollut 2008;151:362–7.
[9] Jenkin ME, Clemitshaw KC. Ozone and other secondary photochemical pollutants: chemical processes governing their formation in the planetary boundary layer. Atmos Environ 2000;34:2499–527.
[10] Konstantinou IK, Hela DG, Albanis TA. The status of pesticide pollution in surface waters (rivers and lakes) of Greece. Part I. Review on occurrence and levels. Environ Pollut 2006;141:555–70.
[11] Loos R, Gawlik BM, Locoro G, Rimaviciute E, Contini S, Bidoglio G. EU-wide survey of polar organic persistent pollutants in European river waters. Environ Pollut 2009;157:561–8.
[12] Rai B. Pollution and conservation of Ganga river in modern India. Int J Sci Res Publ 2013;3:221–4.
[13] Ladwani KD, Ladwani KD, Ramteke DS. Assessment of poly aromatic hydrocarbon (PAH) dispersion in the near shore environment of Mumbai, India after a large scale oil spill. Bull Environ Contam Toxicol 2013;90:515–20.
[14] Imperato M, Adamo P, Naimo D, Arienzo M, Stanzione D, Violante P. Spatial distribution of heavy metals in urban soils of Naples city (Italy). Environ Pollut 2003;124:247–56.

[15] Banerjee AD. Heavy metal levels and solid phase speciation in street dusts of Delhi, India. Environ Pollut 2003;123:95–105.

[16] van der Oost R, Beyer J, Vermeulen NP. Fish bioaccumulation and biomarkers in environmental risk assessment: a review. Environ Toxicol Pharmacol 2003;13:57–149.

[17] Rosenzweig C, Karoly D, Vicarelli M, Neofotis P, Wu Q, Casassa G, et al. Attributing physical and biological impacts to anthropogenic climate change. Nature 2008;453:353–7.

[18] Grantz DA, Garner JHB, Johnson DW. Ecological effects of particulate matter. Environ Int 2003;29:213–39.

[19] Camargo JA, Alonso A. Ecological and toxicological effects of inorganic nitrogen pollution in aquatic ecosystems: a global assessment. Environ Int 2006;32:831–49.

[20] Khan S, Cao Q, Zheng YM, Huang YZ, Zhu YG. Health risks of heavy metals in contaminated soils and food crops irrigated with wastewater in Beijing, China. Environ Pollut 2008;152:686–92.

[21] Farombi EO, Adelowo OA, Ajimoko YR. Biomarkers of oxidative stress and heavy metal levels as indicators of environmental pollution in African cat fish (Clarias gariepinus) from Nigeria Ogun River. Int J Environ Res Public Health 2007;4:158–65.

[22] Simsek O, Gültekin R, Oksüz O, Kurultay S. The effect of environmental pollution on the heavy metal content of raw milk. Nahrung 2000;44:360–3.

[23] Bafana A, Chakrabarti T, Krishnamurthi K. Mercuric reductase activity of multiple heavy metal-resistant Lysinibacillus sphaericus G1. J Basic Microbiol 2015;55:285–92.

[24] Carlsson C, Johansson AK, Alvan G, Bergman K, Kühler T. Are pharmaceuticals potent environmental pollutants? Part I: environmental risk assessments of selected active pharmaceutical ingredients. Sci Total Environ 2006;364:67–87.

[25] Martinez JL. Environmental pollution by antibiotics and by antibiotic resistance determinants. Environ Pollut 2009;157:2893–902.

[26] Pauwels B, Verstraete W. The treatment of hospital wastewater: an appraisal. J Water Health 2006;2006(4):405–16.

[27] Soni D, Naoghare PK, Saravanadevi S, Pandey RA. Release, transport and toxicity of engineered nanoparticles. Rev Environ Contam Toxicol 2015;234:1–47.

[28] Trombulak SC, Frissell CA. Review of ecological effects of roads on terrestrial and aquatic communities. Conserv Biol 2000;14:18–30.

[29] Chuang KJ, Chan CC, Su TC, Lee CT, Tang CS. The effect of urban air pollution on inflammation, oxidative stress, coagulation, and autonomic dysfunction in young adults. Am J Respir Crit Care Med 2007;176:370–6.

[30] Anderson JO, Thundiyil JG, Stolbach A. Clearing the air: a review of the effects of particulate matter air pollution on human health. J Med Toxicol 2012;8:166–75.

[31] Wild CP, Kleinjans J. Children and increased susceptibility to environmental carcinogens: evidence or empathy? Cancer Epidemiol Biomarkers Prev 2003;2:1389–94.

[32] Ezzati M, Kammen DM. The health impacts of exposure to indoor air pollution from solid fuels in developing countries: knowledge, gaps, and data needs. Environ Health Perspect 2002;110:1057–68.

[33] Bruce N, Perez-Padilla R, Albalak R. Indoor air pollution in developing countries: a major environmental and public health challenge. Bull World Health Organ 2000;78:1078–92.

[34] MacDonald Gibson J, Thomsen J, Launay F, Harder E, DeFelice N. Deaths and medical visits attributable to environmental pollution in the United Arab Emirates. PLoS One 2013;8:e57536.

[35] Hendryx M, Conley J, Fedorko E, Luo J, Armistead M. Permitted water pollution discharges and population cancer and non-cancer mortality: toxicity weights and upstream discharge effects in US rural-urban areas. Int J Health Geogr 2012;2(11):9.

[36] Harada M. Minamata disease: methylmercury poisoning in Japan caused by environmental pollution. Crit Rev Toxicol 1995;25:1–24.

[37] Tarale P, Chakrabarti T, Sivanesan S, Naoghare P, Bafana A, Krishnamurthi K. Potential role of epigenetic mechanism in manganese induced neurotoxicity. Biomed Res Int 2016;2016:2548792.

[38] Fernández DS, Puchulu ME, Georgieff SM. Identification and assessment of water pollution as a consequence of a leachate plume migration from a municipal landfill site (Tucumán, Argentina). Environ Geochem Health 2014;36:489–503.

[39] Gandhi D, Naoghare PK, Bafana A, Kannan K, Sivanesan S. Fluoride-induced oxidative and inflammatory stress in osteosarcoma cells: does it affect bone development pathway?. Biol Trace Elem Res 2017;175:103–11.

[40] Daiwile AP, Sivanesan S, Izzotti A, Bafana A, Naoghare PK, Arrigo P, et al. Noncoding RNAs: possible players in the development of fluorosis. Biomed Res Int 2015:274852.

[41] Bafana A, Devi SS, Chakrabarti T. Azo dyes: past, present and the future. Environ Rev 2011;19:350–70.

[42] Tiwari J, Naoghare P, Sivanesan S, Bafana A. Biodegradation and detoxification of chloronitroaromatic pollutant by *Cupriavidus*. Bioresour Technol 2017;223:184–91.

[43] Wanjugi P, Sivaganesan M, Korajkic A, Kelty CA, McMinn B, Ulrich R, et al. Differential decomposition of bacterial and viral fecal indicators in common human pollution types. Water Res 2016;105:591–601.

[44] Samanta SK, Singh OV, Jain RK. Polycyclic aromatic hydrocarbons: environmental pollution and bioremediation. Trends Biotechnol 2002;20:243–8.

[45] Liu K, Han W, Pan W-P, Riley JT. Polycyclic aromatic hydrocarbon (PAH) emissions from a coal fired pilot FBC system. J Hazard Mater 2001;84:175–88.

[46] Perera FP, Jedrychowski W, Rauh V, Whyatt RM. Molecular epidemiologic research on the effects of environmental pollutants on the fetus. Environ Health Perspect 1999;107:451–60.

[47] Karavitis J, Kovacs EJ. Macrophage phagocytosis: effects of environmental pollutants, alcohol, cigarette smoke, and other external factors. J Leukoc Biol 2011;90:1065–78.

[48] Raja R, Nayak AK, Shukla AK, Rao KS, Gautam P, Lal B, et al. Impairment of soil health due to fly ash-fugitive dust deposition from coal-fired thermal power plants. Environ Monit Assess 2015;187:679.

[49] Singh S, Raju NJ, Nazneen S. Environmental risk of heavy metal pollution and contamination sources using multivariate analysis in the soils of Varanasi environs, India. Environ Monit Assess 2015;187:345.

[50] Li J, Lu Y, Shi Y, Wang T, Wang G, Luo W, et al. Environmental pollution by persistent toxic substances and health risk in an industrial area of China. J Environ Sci (China) 2011;23:1359–67.

[51] Ottesen RT, Alexander J, Langedal M, Haugland T, Høygaard E. Soil pollution in day-care centers and playgrounds in Norway: national action plan for mapping and remediation. Environ Geochem Health 2008;30:623–37.

[52] Gandhi D, Tarale P, Naoghare PK, Bafana A, Krishnamurthi K, Arrigo P, et al. An integrated genomic and proteomic approach to identify signatures of endosulfan exposure in hepatocellular carcinoma cells. Pestic Biochem Physiol 2015;125:8–16.

[53] Gandhi D, Tarale P, Naoghare PK, Bafana A, Kannan K, Sivanesan S. Integrative genomic and proteomic profiling of human neuroblastoma SH-SY5Y cells reveals signatures of endosulfan exposure. Environ Toxicol Pharmacol 2016;41:187–94.

[54] Fathi RA, Matti LY, Al-Salih HS, Godbold D. Environmental pollution by depleted uranium in Iraq with special reference to Mosul and possible effects on cancer and birth defect rates. Med Confl Surviv 2013;29:7–25.

[55] Somers CM, Yauk CL, White PA, Parfett CL, Quinn JS. Air pollution induces heritable DNA mutations. Proc Natl Acad Sci USA 2002;99:15904–7.

[56] Sato H, Aoki Y. Mutagenesis by environmental pollutants and bio-monitoring of environmental mutagens. Curr Drug Metab 2002;3:311–9.

[57] Sørensen M, Autrup H, Møller P, Hertel O, Jensen SS, Vinzents P, et al. Linking exposure to environmental pollutants with biological effects. Mutat Res 2003;544:255–71.

[58] Perera F, Hemminki K, Jedrychowski W, Whyatt R, Campbell U, Hsu Y, et al. In utero DNA damage from environmental pollution is associated with somatic gene mutation in newborns. Cancer Epidemiol Biomarkers Prev 2002;11:1134–7.

[59] Somers CM. Ambient air pollution exposure and damage to male gametes: human studies and in situ 'sentinel' animal experiments. Syst Biol Reprod Med 2011;57:63–71.

[60] Jurewicz J, Radwan M, Sobala W, Polańska K, Radwan P, Jakubowski L, et al. The relationship between exposure to air pollution and sperm disomy. Environ Mol Mutagen 2015;56:50–9.

[61] Krishnamurthi K, Saravana Devi S, Chakrabarti T. DNA damage caused by pesticide-contaminated soil. Biomed Environ Sci 2006;19:427–31.

[62] Krishnamurthi K, Devi SS, Chakrabarti T. The genotoxicity of priority polycyclic aromatic hydrocarbons (PAHs) containing sludge samples. Toxicol Mech Methods 2007;17:1–12.

[63] Ishikawa H, Tian Y, Piao F, Duan Z, Zhang Y, Ma M, et al. Genotoxic damage in female residents exposed to environmental air pollution in Shenyang city, China. Cancer Lett 2006;240:29–35.

[64] Krishnamurthi K, Saravana Devi S, Hengstler JG, Hermes M, Kumar K, Dutta D, et al. Genotoxicity of sludges, wastewater and effluents from three different industries. Arch Toxicol 2008;82:965–71.

[65] Eeva T, Belskii È, Kuranov B. Environmental pollution affects genetic diversity in wild bird populations. Mutat Res 2006;608:8–15.

[66] Chu EHY, Generoso WM. Mutation, cancer, and malformation. Environmental science research, vol. 31. New York: Springer; 2012.

[67] Conti ME, Cecchetti G. Biological monitoring: lichens as bioindicators of air pollution assessment — a review. Environ Pollut 2001;114:471–92.

[68] Broeg K, Lehtonen KK. Indices for the assessment of environmental pollution of the Baltic Sea coasts: integrated assessment of a multi-biomarker approach. Mar Pollut Bull 2006;53:508–22.

[69] Cajaraville MP, Bebianno MJ, Blasco J, Porte C, Sarasquete C, Viarengo A. The use of biomarkers to assess the impact of pollution in coastal environments of the Iberian Peninsula: a practical approach. Sci Total Environ 2000;247:295–311.

[70] Wegrzyn G, Czyz A. Detection of mutagenic pollution of natural environment using microbiological assays. J Appl Microbiol 2003;95:1175–81.

[71] US EPA (US Environmental Protection Agency). Guidelines for mutagenicity risk assessment. 1986. Available from: https://www.epa.gov/risk/guidelines-mutagenicity-risk-assessment.

[72] Valon CL. New developments in mutation research. New York: Nova Publishers; 2007.

[73] Waters MD. Short-term bioassays in the analysis of complex environmental mixtures III. Environmental science research, vol. 27. New York: Springer; 2012.

[74] Maron DM, Ames BN. Revised methods for the *Salmonella* mutagenicity test. Mutat Res 1983;113:173–215.

[75] Czyż A, Kowalska W, Węgrzyn G. *Vibrio harveyi* mutagenicity assay as a preliminary test for detection of mutagenic pollution of marine water. Bull Environ Contam Toxicol 2003;70:1065–70.

[76] Kirsch-Volders M, Plas G, Elhajouji A, Lukamowicz M, Gonzalez L, Vande Loock K, et al. The in vitro MN assay in 2011: origin and fate, biological significance, protocols, high throughput methodologies and toxicological relevance. Arch Toxicol 2011;85:873–99.

[77] Bajpayee M, Kumar A, Dhawan A. The comet assay: assessment of in vitro and in vivo DNA damage. Methods Mol Biol 2013;1044:325–45.

[78] Bafana A, Chakrabarti T, Devi SS. Azoreductase and dye detoxification activities of *Bacillus velezensis* strain AB. Appl Microbiol Biotechnol 2008;77:1139–44.
[79] Bafana A, Chakabarti T, Muthal P, Kanade G. Detoxification of benzidine-based azo dye by *E. gallinarum*: time-course study. Ecotoxicol Environ Saf 2009;72:960–4.
[80] Bafana A, Jain M, Agrawal G, Chakrabarti T. Bacterial reduction in genotoxicity of Direct Red 28 dye. Chemosphere 2009;74:1404–6.
[81] OECD. Introduction to the OECD guidelines on genetic toxicity and guidance on the selection and application of assays. Paris: Organisation for Economic Co-Operation and Development; 1986.
[82] US EPA (US Environmental Protection Agency). Guidelines for carcinogen risk assessment. 2005. Available from: https://www3.epa.gov/airtoxics/cancer_guidelines_final_3-25-05.pdf.
[83] US EPA (US Environmental Protection Agency). Safer choice criteria for specialized industrial products. 2012. Available from: https://www.epa.gov/saferchoice/safer-choice-criteria-specialized-industrial-products.

CHAPTER SEVEN

Mutagens in Food

Payal Mandal[a], Ankita Rai[a], Sakshi Mishra, Anurag Tripathi, Mukul Das

CSIR-Indian Institute of Toxicology Research, Lucknow, India

1. INTRODUCTION

Food is one of the most fundamental necessities of life required for nutritional fulfillment and proper growth. Whether obtained from plants or animal sources, food consists of essential biomolecules such as carbohydrates, fats, proteins, vitamins, and minerals [1]. With the development of modern agricultural tools and the changes in the life style, the need for food preservation and processing has been emphasized to ensure for adequate food supply during nonagricultural periods. Food processing involves several steps to enhance the aesthetic value, shelf life, elevate the flavors, and presenting the food product in the most lucrative form to the consumers. Most of the intentionally added additives are safe if consumed within the prescribed average daily intake (ADI). However, it becomes a matter of concern if these additives are used beyond the prescribed limits [2]. Consumption of contaminated food is at times lethal to humans and therefore contamination is of global concern [3]. Generally, most of the compounds in food sustain in low concentrations, but if any of them poses carcinogenic risk, then consumption of that food may be hazardous to human health [4]. Food preservatives such as sorbic acid and sodium benzoate, which are generally used to protect the food products from microbial deterioration, have been shown to be genotoxic at higher doses in root tip cells of *Allium cepa* [5] as well as human lymphocytes [6]. Because they can cause detrimental effect to human health, their use in food commodity is governed by regulatory bodies based on the intake pattern in a particular population. Food colors added to food for technological necessity is also one of the issues of food industry, as many of them are not recommended by regulatory standards [7,8]. Consumption

[a] Equal contribution

Mutagenicity: Assays and Applications
ISBN 978-0-12-809252-1
http://dx.doi.org/10.1016/B978-0-12-809252-1.00007-9

of sugar and sugar substitutes in higher quantity can cause dental cavities, obesity, allergic reactions, cardiovascular diseases, and can even turn out to be carcinogenic [9]. Adulteration of vegetable oils causes serious health problems, for example adulteration of mustard oil with highly toxic argemone oil leads to a disease called "epidemic dropsy" [10–12]. Mycotoxins are secondary metabolites produced by different species of fungus infested on several staple food commodities. Some of these compounds are highly mutagenic and carcinogenic in nature [13,14] to animals and humans. Chemicals such as pesticides are used on plant and animal farming to increase production and ample food supply by preventing the food from insects, rodents, weeds, and microorganisms [15,16]. Pesticides can be hazardous if not used appropriately, which can be the reason of severe health problems for the ecosystems and humans as well. During the preparation of foods, several compounds such as polycyclic aromatic hydrocarbons (PAHs)/heterocyclic amines (HCAs) are generated, which are well-known mutagens and carcinogens. Heavy metal contamination in food is a matter of global concern due to their toxic hazard for environment and human health [17].

Considering the toxicological studies of the compounds present in a variety of foodstuffs, this chapter mainly focuses on the genotoxic/mutagenic or carcinogenic substances in food. These compounds have been further classified into (1) additives, (2) contaminants, (3) adulterants, and (4) mutagens produced due to processing of food.

2. ADDITIVES

2.1 Preservatives

Preservatives are known as natural or artificial substances added to the food stuff, cosmetic, and pharmacy products to extend their shelf life and to prevent from the microbial growth, fermentation, and degradation [18]. Food preservation is one of the very old methods to increase the shelf life or freshness of food. Sugar, salts, vinegar, and spices are being used as natural preservatives from the ancient time. Chemical preservatives such as sulfur dioxides, nitrites, and benzoates like antimicrobial agents delay or diminish the microbial growth in food, while the antioxidants such as butylated hydroxyanisol (BHA), butylated hydroxytoluene (BHT), and propyl gallate are used to reduce or prevent the process of oxidation [19]. Antienzymatic substances, such as citric acid, are used to prevent the ripening of fruits [18]. However, many of these molecules pose genotoxic and mutagenic

Table 7.1 Genotoxic effects of some food preservatives

Food preservatives	Health risk	References
BHA, BHT	Carcinogenicity, hyperactivity	[20,143,144]
Nitrites, nitrates	Carcinogenicity, blue baby syndrome, chromosomal aberration	[109,145]
Sorbic acid	Chromosomal aberration, sister chromatid exchange	[146,147]
Propyl gallate	Cytotoxicity	[148]

BHA, butylated hydroxyanisol; *BHT*, butylated hydroxytoluene.

effects [20]. Excessive use of sugar and salt can also lead to different type of diseases such as diabetes, blood pressure, kidney failure, and heart stroke. Synthetic agents such as benzoates, nitrates, sulfites, BHA, BHT, and others can cause neurological diseases, allergic reactions, cancer, and even death in extreme cases. BHA has been declared as possible carcinogen to humans by IARC [21], whereas BHT has been found to be carcinogenic in mice [22]. BHA at a higher dose has been shown to cause gastrointestinal hemorrhage in Japanese house musk shrew, whereas at lower dose it causes lung hyperplasia [23].

To avoid the risk of these preservatives as indicated in Table 7.1, few suggestions can be recommended as nonessential preservatives should be labeled with appropriate warning about the mutagenic/carcinogenic risk and should not be added to the products consumed by children below 15 years.

2.2 Food Colors

Food colors are being used since 1500 BC in India, China, and Egypt. Colors used as food additives can be broadly divided as natural and synthetic colors. Synthetic colors are further subdivided into permitted and nonpermitted colors. In the current scenario, food colors are considered as an important aspect of food industries used to elevate the aesthetic quality and are rigorously used mainly in processed food items.

There are many natural pigments that are part of our daily food. These are β-carotene (orange/yellow), chlorophyll (green), saffron (orange), riboflavin (yellow), caramel (beige/dark brown), turmeric/curcumin (orange), and annatto (yellow/orange/red), which are prescribed in majority of the countries, including India [8]. However, European countries allow other natural colors such as beetroot (pink/blue/red), capsanthin (red/orange), carmine (red), lycopene (reddish orange), carminic acid (orange/red), lutein (yellow), anthocynin (red/ purple), and vegetable carbon (black) [24]. Although the natural colors are relatively safe compared with synthetic ones, higher amounts of natural or permitted synthetic colors may pose a genotoxic/mutagenic threat. A study conducted by Massachusetts Medical Society reported that β-carotene, when used by the subjects having smoking habits, showed increased incidence of cancer [25].

The global production of synthetic food colorants is about 80,00,000 tons per year. Synthetic colorants can be broadly classified into five major categories: the azo compounds (such as amaranth, sunset yellow (SY), ponceau 4R, carmoisine), the triarylmethane group, the indigo colorants, xanthenes (such as erythrosine), and the chinophthalon derivatives of quinoline yellow [26]. In India, eight of the synthetic colors are permitted to be added in food commodities, which are SY FCF, tartrazine, ponceau 4R, carmoisine, erythrosine, brilliant blue FCF, fast green FCF, and indigocarmine [8]. Sunset yellow (SY) (C.I. 15985) is a synthetic coloring dye belonging to a group of benzidine-based azo compounds with a revised acceptable daily intake (ADI) of 0–4 mg/kg body weight per day. It is formed by the reaction of arene diazonium ions and reactive aromatic hydrocarbons. Some of the azo dyes are known to be metabolized in the liver and intestine to produce free aromatic amines that are potentially mutagenic in nature [27]. Evidences suggest that SY inhibits the cell cycle in the root tips of *Brassica campestris* L. by decreasing the mitotic index [28] and a similar effect was reported in lymphocyte cell culture assay [29]. Study conducted on somatic and germ cells of male Swiss mice showed that oral dose of SY induced DNA fragmentation and both autosomal and X–Y univalent chromosomal aberrations in bone marrow as well as spermatocytes in a dose-dependent manner, indicating the genotoxic and mutagenic effects [27].

Tartrazine (C.I. 19140), a lemon yellow azo dye, is used individually and in combination with brilliant blue FCF for obtaining green shades. The binding studies of tartrazine by spectroscopic analysis showed distinct isosbestic point in the spectrum at 307 nm that gives a clear picture of its binding capability with DNA, which is of great clinical importance with regard to genotoxicity [26]. Table 7.2 enlists the permitted synthetic food colors and their mutagenic manifestations. Besides permitted synthetic food colors, many a times several nonpermitted synthetic colors viz rhodamine B, malachite green, metanil yellow, Orange II, etc., are used in foodstuffs, which are well-known mutagens and carcinogens [30–32].

Table 7.2 Mutagenic manifestations of permitted synthetic food colors

Color	Types of mutation caused	Health hazards
Erythrosine B	DNA damage, chromosomal aberrations, frame shift, mitotic recombination [149]	Thyroid cell follicular adenomas, follicular cystic hyperplasia, lymphocytic lymphoma
Allura red	DNA damage and base substitution [150,151]	Food intolerance and attention-deficit hypersensitivity disorder–like behavior in children
Tartrazine	DNA damage, chromosomal aberrations, frame shift [152]	Sensitivity, gastric carcinogen, immunosuppressive
Sunset yellow	Chromosomal aberrations, forward mutation [28]	Food intolerance and attention-deficit hypersensitivity disorder–like behavior in children
Brilliant blue FCF	Chromosomal aberrations [29]	Allergy
Fast green FCF	Base pair, chromosomal aberrations, malignant cell transformation [153,154]	Tumors, carcinogen

2.3 Sweeteners

Sweeteners are the intentional additives used in foods to keep the food calories low. Sugar is a natural sweetener obtained from sugarcane or sugar beet. Artificial sugar substitutes add similar sweetness as sugar to the food with very low calorie [33]. The most common artificial sweeteners are saccharine, aspartame, neotame, acesulfame K, stevia, and sucralose approved by FDA. These sweeteners are used in different processed foods such as bakery products, jam and jellies, canned food puddings, and soft drinks [34]. Artificial sweeteners are manyfold sweeter than natural sugar. These low–calorie sweeteners have known in the food sector for their proposed health benefits as reduced risk factors for obesity and management of diabetes. However, there are some reports concluding that high and continuous usage of these low-calorie sweeteners can induce various health problems as seizures, allergy, breathlessness, headaches, mood swings, and cancer [35]. The clastogenicity and genotoxicity of some of the sweeteners are listed in Table 7.3.

Steviol glycosides are biological sweetener extracted from the leaves of stevia plant. Recent studies on steviosides indicate that there is no potential threat of genotoxicity and allergenicity [36]. Therefore, these sweeteners should be used for the specific sector of population that should be labeled on the packaging.

Table 7.3 Potential toxic effects of artificial sweeteners

Sweeteners	ADI by FDA (mg/kg b.w./day)	Health risk	References
Saccharine	5	Bladder carcinoma, hepato-toxicity, low birth weight	[155,156]
Aspartame	40	Migraine, lymphoma, leukemia in rats, chromosomal aberration	[35,157,158]
Sucralose	5	Thymus shrinking, cell enlargement in rats, diarrhea, migraine	[159–161]
Acesulfame K	15	Clastogenicity, genotoxicity, thyroid tumors in rats	[162]
Neotame	2	Headache, low birth weight, hepatotoxicity	[163,164]

ADI, average daily intake.

3. CONTAMINANTS
3.1 Mycotoxins

Mycotoxins are secondary fungal metabolites of molds that cause adverse effects on animals, humans, and crops, resulting in illnesses and economic loss. From food safety perspective, global mycotoxin contamination of food and feed is of major concern. These secondary metabolites contaminate a variety of food crops and foodstuffs, including cereals, fruits, nuts, apple juice, spices, dried fruits, and coffee, often under warm and humid conditions. Many mycotoxins have been shown to be genotoxic that may cause cancer in animal species.

The International Agency for Research on Cancer (IARC) and World Health Organization (WHO) have suggested for a coordinated international response to the problem of mycotoxin contamination of food [13,37]. WHO report [38] urges that there is widespread mycotoxin contamination in developing countries. In sub-Saharan Africa, Southeast Asia, and parts of Latin America, the conditions are appropriate for the fungus to grow that produce the toxin. The problem is aggravated in those regions of the globe where the staple food is contaminated with a high level of mycotoxins.

Mycotoxins of concern are aflatoxins (B1, B2, G1, G2, and M1), ochratoxin A (OTA), patulin (PAT), Fusarium toxins, including fumonisins (B1, B2, and B3), trichothecenes including nivalenol (NIV), deoxynivalenol

(DON), T-2 and HT-2 toxins, and zearalenone. There are different regulatory limits prescribed for mycotoxin present in food. Aflatoxins in diet are considered an important risk factor as epidemiological studies from Philippines, China, Thailand, and many African countries suggest direct link of liver cancer incidence to aflatoxin consumption in the diet [39,40]. The major aflatoxins include B1, B2, G1, and G2, M1, and M2, where the last two are metabolized from AFB1 and secreted in the milk. Aflatoxins B2 and G2 are the dihydroxy derivatives of B1 and G1, respectively, whereas, aflatoxin M1 is 4-hydroxy aflatoxin B1 (AFB1) and aflatoxin M2 is 4-dihydroxy aflatoxin B2. Aflatoxin B1 causes chromosomal aberrations in animal and human cells after undergoing metabolic conversion to its *exo*-8,9-epoxide that induces damage to DNA [41]. Genotoxicity of AFB1 has been studied in prokaryotic and eukaryotic systems, including human cells, and in vivo in humans and in a variety of animal species [42]. It induces gene mutations and chromosomal alterations, including micronuclei, sister chromatid exchange, and mitotic recombination. Studies regarding carcinogenicity of aflatoxin have been conducted in at least 12 different animal species [43]. The carcinogenicity and mutagenicity of aflatoxins B1, G1, and M1 results by the formation of a reactive epoxide at the 8,9 position of the terminal furan ring and its subsequent covalent binding to nucleic acid and to albumin in the blood serum [4]. Although aflatoxins G1 and M2 have been tested less extensively, they appear to be toxicologically similar to aflatoxin B1 [44].

IARC [45] suggests that there is sufficient evidence for the carcinogenicity of aflatoxins B1, G1, and M1, limited evidence for the carcinogenicity of aflatoxin B2, and inadequate evidence for the carcinogenicity of aflatoxin G2. Furthermore, in 2002 IARC report [46] suggested that studies confirm that naturally occurring aflatoxins are carcinogenic to humans (Group 1), while aflatoxin M1 is possibly carcinogenic to humans (Group 2B).

Ochratoxins are produced by *Aspergillus* species of fungi such as *Aspergillus ochraceus* or *Aspergillus niger* as well as by some *Penicillium* species, especially *Penicillium verrucosum* [47]. Main three forms of ochratoxin include A, B, and C among which ochratoxin B (OTB) is a nonchlorinated form of OTA and ochratoxin C (OTC) is an ethyl ester of OTA. OTA is a renal toxin in various species and one of the potent carcinogens in rats [47]. OTA has been classified as a possible human carcinogen (group 2B) by IARC [37] and has been associated with nephropathies and urothelial tract tumors in the Balkans and in North African countries, as well as in skin tumor initiation and promotion [48–53]. OTA has also been suggested as a cause of testicular cancer in young men [54–56]. Synthetic ochratoxin hydroquinone, an OTA metabolite, which forms covalent DNA adducts, is also reported to be mutagenic [57].

Fumonisins are suspected human carcinogens produced by *Fusarium moniliforme* (*Fusarium verticillioides*) and *F. proliferatum*. Fumonisin B1 (FB1) is the most common mycotoxin of this group. Studies suggest that fumonisins are not genotoxic (WHO 2000), but are potent liver cancer promoters as they have the ability to induce glutamyl transpeptidase and glutathione-S-transferase positive foci in rat liver [58]. FB1 is the most commonly found in mycotoxin corn, corn-based foods, beer, rice, sorghum, triticale, cowpea seeds, beans, soybeans, and asparagus. FB1 is also nephrotoxic and embryotoxic in experimental animals. In humans, fumonisins are linked with esophageal cancer and neural tube defects [59,60]. IARC (1993) [13] has placed FB1 in Group 2B as "possibly carcinogenic to humans." Fumonisin B2 (FB2) and fumonisin B3 (FB3) are closely related metabolites of FB1, which may cooccur with FB1 in lower concentrations.

The tricothecenes are produced by several species of fungal molds viz *Fusarium, Myrotecium, Trichoderma, Cephalosporium, Verticimonosporium, and Stachybotrys* [61]. These are relatively stable under different environmental conditions, including typical cooking environments, irradiation, physiological saline, temperature, and pH [62–64]. Till date, 148 trichothecenes have been isolated, but only a few are found to contaminate food and feed. These include DON, also known as vomitoxin, NIV, diacetoxyscirpenol (DAS), and T-2 toxins. In vitro studies conducted on DON suggest that it has the potential to damage DNA. It induces chromosomal aberration in CHL, V79, and rat primary hepatocytes and acts as tumor initiator [65,66]. The incidence of esophageal cancer in China and Africa has been linked to DON contamination [67,68]. IARC (1993) has classified DON, NIV, and T-2 in group 3 "not classifiable as to its carcinogenicity to humans" due to lack of direct evidences [13]. T-2 toxin has been found to induce DNA damage in immune cells and human cervical cells [69,70]. DAS is a very toxic mycotoxin and is able to induce chromosomal abnormalities however lacks mutagenicity in *Salmonella typhimurium* [71,72].

PAT is a mycotoxin produced by *Penicillium* and *Aspergillus* species [73]. In a recent study DNA damage was observed in several organs of mice treated acutely with PAT [74]. PAT-induced chromosomal aberrations, DNA damage, and micronucleus formation have also been reported in different mammalian cells, which may lead to carcinogenicity [75–77]. Zearalenone produced by *Fusarium graminearium* causes DNA damage. However, the mutagenic properties of this toxin are controversial. A limited number of genotoxicity assays have been conducted with zearalenone. It was found to be negative in the *S. typhimurium* assay [78,79]. Citrinin, another mycotoxin produced by *Penicillium*, *Monascus*, and *Aspergillus*, is mostly found in wheat, maize, barley, oats, etc. Studies performed by Kumar et al. [80] suggest that citrinin induces DNA damage in mice skin and also has the potential for genotoxic risk and oxidative damage to mammalian cells. Although human data are lacking for several mycotoxins, many of these are of concern to public health.

3.2 Pesticides

Pesticides are chemicals used by humans to kill pests that damage crops. Pesticides are potentially toxic for other organisms, including humans, and should be used safely. Vegetables, fruits, grains, and other foods are the vulnerable commodities for pesticide residues. Residues of pesticides in food commodities may gradually decrease when the crops are harvested, washed, cleaned, transported, and cooked. The common pesticides that are found in food may have mutagenic/genotoxic properties including organophosphates such as parathion, methyl parathion, diazinon, malathion, dichlorvos, fenitrothion, chlorpyrifos, phosmet, tetrachlorvinphos, azamethiphos, and azinphos-methyl. Methyl parathion and triazophos have been found to have mutagenic properties in *Drosophila melanogaster*. Parathion (PT), methyl paraoxon (PO), and dimefox (DF) induce DNA damage and affect cell proliferation in human hepatoma HepG2 cells [81]. IARC has classified five organophosphate pesticides viz glyphosate, malathion, and diazinon, as probably carcinogens to humans (Group 2A), and tetrachlorvinphos and parathion, as possibly carcinogens to humans (Group 2B) [82].

Carbamates are derived from carbamic acid and used widely across the globe. The mode of action is similar to that of organophosphate insecticides, which primarily affects nerve impulse transmission. These include carbofuran, aldicarb, ethienocarb, carbaryl, oxamyl, and methomyl. Carbofuran exerts sister chromatid exchange and increases micronuclei frequencies along with DNA damage in mammalian cultured cells [83,84]. Mutagenicity and DNA damage in human lymphocyte have been observed for aldicarb and methomyl [85]. According to IARC, oxamyl is categorized in Group E "Evidence of Non-carcinogenicity for Humans" and nonmutagenic; however, few reports suggest that at higher doses it may produce genotoxicity [82].

Organochlorines contain minimum of one covalently bond chlorine atom. These include aldrin, chlordane, chlordecone, DDT, dieldrin, endosulfan, endrin, heptachlor, hexachlorobenzene (HCB), lindane (gamma-hexachlorocyclohexane; DCB), methoxychlor, and 1,1-dichloro-2,2-bis (p-bichlorophenyl) ethylene (DDE). Different organochlorine compounds have been associated with cancer in humans. DDT has been shown to be associated with the high occurrence of cancer among humans [85]. Genotoxic potential of DDE and HCB has been evaluated in vitro in human lymphocytes [86]. Another pesticide of this group, endosulfan and its metabolite induce DNA damage and mutation [87,88].

Direct studies on DNA-damaging effect of phenothrin are lacking in literature and need additional studies to reveal the genotoxic potential of the same. WHO (2004) reported that most genotoxicity tests, e.g., unscheduled DNA synthesis, reverse in vitro gene mutations, mutations in bacteria, cytogenetic tests in mammalian cells, and in vivo micronucleus (MN) tests in mice, had negative results [89]. However, later studies suggest that synthetic pyrethroid viz β-cyfluthrin, λ-cyhalothrin, and α-cypermethrin are genotoxic in fishes as well as in human blood lymphocyte [90–92]. These evidences also highlight the importance of additional studies regarding genotoxicity of phenothrin.

3.3 Metals

Food contains a wide range of metals such as Se, Mg, Cu, Fe, K, Na, Ca, Zn, I, and Bo, which are necessary in trace amounts. Deficiency of these metals cause low immunity (Zn) and lead to several other diseases like cancer (Fe). The essential metals necessary to support biological functions may exert toxic effects if present in excess amounts. Metal contamination in food can also occur by their presence in food, or during preparing equipment or kitchen preparation and storage. However, sometimes other heavy metals get mixed with the food that may exert toxic effects on health. Cadmium (Cd), mercury (Hg), arsenic (As), and lead (Pb) are heavy metals that exert toxic and carcinogenic effects if present in very low amounts [93,94]. These heavy metals can generate reactive oxygen species and nitrogen species and thereby causes oxidative damage DNA and lipid damage [95]. Thus heavy metals are well-known environmental health hazards due of their toxic potential [17]. Many organizations such as WHO, Centers for Disease Control and Prevention, Joint Food and Agricultural Organization (FAO)/WHO Expert Committee on Food Additives, and IARC are continuously updating the data on toxicity of these metals. The major heavy metals posing toxic manifestations are listed in Table 7.4.

Table 7.4 Toxic effects of heavy metals

Heavy metals	TDI by JECFA [165] µg/kg b.w./day	Health risk	References
Mercury	1.3	Neurotoxicity, kidney damage, cardiovascular toxicity, DNA damage	[166,167]
Lead	3.6	Neurological disorder, reproductive diseases, micronucleus formation	[168–171]
Arsenic	2.1	Lung and skin cancer, cardiovascular toxicity, micronucleus formation	[170–172]
Cadmium	1	Kidney failure, lung cancer, bone diseases, chromosomal aberrations, micronucleus formation	[173–177]

JECFA, Joint Food and Agricultural Organization/WHO Expert Committee on Food Additives; *TDI*, tolerable daily intake.

3.4 Intrinsic Dietary Components

Our diet harbors a battery of beneficial agents required to carry out various functions of daily life. However, there are some components that exist naturally or formed during the process of food and that have also been found to be harmful to health. For example, heterocyclic aromatic amines, which are produced during processed food, hydrazines in mushrooms, dioxins, and cycasin in cycad nuts are toxic in nature [4]. The toxicity of these molecules could be due to the parent form or the biotransformed product [96]. Many of these compounds possess mutagenic, genotoxic, and/or carcinogenic potential leading to disease conditions in consumers [4,96].

Hydrazines are mainly found in mushrooms particularly in commonly cultivated button mushroom (*Agaricus bisporus*), wild false morel (*Gyromitra esculent*), Japanese forest mushroom (*Cortinellus shiitake*). Agaritine (β-N-[γ-L(+)-glutamyl]-4-hydroxymethylphenylhydrazine) (found in *A. bisporus*, *C. shiitake*) and its breakdown products (N-acetyl derivative, tetrafluoroborate form, and 4-methylphenylhydrazine hydrochloride) covalently bind to DNA to initiate the process of mutagenicity that has been studied under in vitro and in vivo conditions [97]. *G. esculent* contains a compound called acetaldehyde formylmethylhydrazone that gets converted into methylhydrazine, in acidic environment of stomach, and N-methyl-N-formylhydrazine. These hydrazines have been shown to cause DNA damage via oxidative stress, which are responsible for neoplasms, angiosarcoma in blood vessels, adenocarcinoma of lungs, and hepatocarcinomas [4]. Besides the intrinsic components of food products posing mutagenicity on consumption, total calorie and sodium chloride intake are also important factors in aggravating the disease condition(s). There has been a close correlation between high sodium chloride intake and gastric cancers in Japanese population [98]. It has been observed that high ingestion of sodium chloride chronically depletes the mucin from the gastrointestinal tract and further destructs the epithelial cells because of enhanced osmotic pressure. This kind of damage exposes the epithelial stem cells for proliferation and the microenvironment becomes favorable for causing mutation in those proliferating stem cells [99]. Fig. 7.1 depicts the schematic representation of mutation and cancer caused by intrinsic factors.

High caloric intake/relative body weight increases the likelihood of cancer of the breast, colon, rectum, ovary, prostate, endometrium, kidney, gallbladder, cervix, and thyroid in the exposed subjects [100]. Evidences suggest that there has been increasing incidences of pancreatic duct

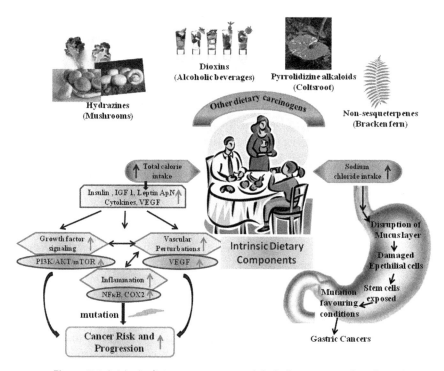

Figure 7.1 Intrinsic dietary mutagens and their downstream signaling.

adenocarcinomas that has been linked to obesity or increased calorie intake [101]. This phenomenon is characterized by increased levels of KRas (Kirsten rat sarcoma) (initial event), which is followed by mutations in the tumor suppressor genes initiating pancreatic cancer [100].

4. ADULTERANTS IN FOOD

4.1 Edible Oils

Edible oils, the source of essential fatty acids procured from plant, animal, or synthetic fat, is used in frying, baking, and other types of cooking/food preparations [102,103]. These are also used in various food preparations and flavorings, which do not involve any heat source, such as salad dressings and bread dips. From extraction, refinement and processing during various food preparations induce various alterations in the composition and physicochemical properties of the cooking medium. Therefore, choice of oil should be judiciously chosen for the purpose of fulfilling its specific property. Edible oils and fats is one commodity that is prone to various modes of adulterations [102]. The well-known oil adulterants are cheaper oils compromising the oil quality along with other nonedible oils or compounds such as argemone oil (AO), tricresyl phosphate (TCP), and butter yellow [8]. Studies suggest that consumption of mustard oil adulterated with AO leads to a clinical condition known as epidemic dropsy [102]. Sanguinarine, a benzophenanthridine alkaloid of AO, is an electrophilic molecule that intercalates in the GC-rich regions of DNA, thereby causing genotoxicity as evidenced in comet assay, chromosomal aberration, and micronuclei test [104]. AO has been shown to be carcinogenic in nature and it is feared that it is one of the etiological agent for gall bladder cancer (GBC) in the population of Indo-Gangetic Basin, where consumption of mustard oil is maximum [105,106]. Butter yellow, a fat-soluble azo nonpermitted dye, is often mixed with colorless cheap oil along with the mustard pungency factor, allyl isothiocyanate, to look and appear like mustard oil. Butter yellow interacts with macromolecules including DNA and proteins potentiating genotoxic and mutagenic responses [106,107]. It has been found as a contributory factor to cancer of respiratory tract, liver, and skin, due to which this color has been banned for food usage since 1950s [106]. Recently, this dye has also been shown to produce GBC in experimental animals [10]. TCP is an odorless, colorless, industrial, and organophosphorus chemical. It is used in lacquers and varnishes as a plasticizer and also in hydraulic fluid. TCP has been shown to cause paralysis of hands and feet in animals and humans [108]. However, its mode of action is not yet fully understood. It appears that ortho isomer of TCP is more toxic than its meta or para isomers. Outbreaks of TCP poisoning have occurred in past in the United States, Morocco, India, Durban, and Sri Lanka. In July 1988, an outbreak through the consumption of adulterated rapeseed oil took place at Behala area in Southwest outskirts of Kolkata, where about 600 victims of polyneuritis were reported to the hospital, of which 343 were admitted [108]. An 18-month follow-up study of these patients revealed that only 37.9% patients recovered from TCP-induced paralysis, whereas the majority (62.1%) still had neurological deficits. It was observed that alcoholics showed less response to recovery because of the formation of a metabolite saligenin cyclic-o-tolyl phosphate, which is five times more toxic than the parent compound [109]. Although saligenin cyclic-o-tolyl phosphate is an alkylating agent, no studies have been conducted on mutagenicity and genotoxicity of TCP. Another incident of mutagens in edible oil was recorded in the year 1981 in Spain, where consumption of contaminated colza oil led to the development of multisystem disorder, in which over 20,000 were affected and over 300 people lost their lives [10]. This was later called as toxic oil syndrome, in which prevalence of mutation was reported in the genes encoding for N-acetyl transferase 2 [110].

It has been observed that the generation of undesired process gives rise to contaminants possessing mutagenic and genotoxic potential such as trans-fatty acids, stigmasta-3,5-diene, and 3-monochloropropane-1,2-diol fatty acid esters. These are generated as a result of vegetable oils subjected to high temperature. These chemicals tend to generate oxidative stress in the system, thus forming DNA adducts, leading to chromosomal aberrations, and if these manifestations prevail for a longer period in the system, it may get converted into mutagenic outcome.

Cooking oil fumes (COF) has caused passive smoking in the nonsmoking individuals, increasing the risk of lung cancer [111]. Several epidemiological studies from China and Taiwan showed that subjects exposed to COF had major bulky DNA adducts (benzo[a]pyrene 7,8-diol 9,10-epoxide-N-2-deoxyguanosine (BPDE-N-2-dG) adduct), causing p53 hot spot mutations, leading to lung tumors. In another study, on exposure to COF, there was induction of 8-hydroxydeoxyguanosine in calf thymus as a result of oxidative DNA damage, which may also be linked to the etiology of lung cancer [111]. Fig. 7.2 shows the mutagens produced in oils and the respective effect on target organs.

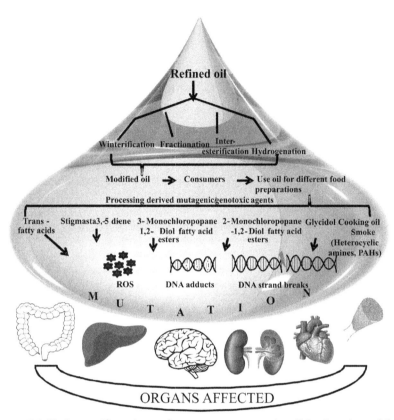

Figure 7.2 Toxic manifestations of mutagens present in oil in duration of food preparations.

4.2 Nonpermitted Food Colors

The requirement in addition of colors of food has been explained in Section 2.2. However, a number of colors/dyes that are banned for food usage because of their harmful effects still find their way in food. Reports prove that metanil yellow, Orange II, auramine, rhodamine B, blue VRS, malachite green, and Sudan III produce DNA damage and cause lesions in kidney, spleen, liver, and bone marrow [32,112]. Metanil yellow and its metabolites cause mutagenic effects in germ cells as well as disarrangement of mucosal folds in kidney and affect other organs such as spleen as well as reproductive system. Metanil yellow also induces cancers of lungs, intestine, ovary, liver, testis, and kidney [113].

5. MUTAGENS PRODUCED AS AN OUTCOME OF FOOD PROCESSING

Many undesired chemicals are formed during processing of foods as a result of reactions between the components of the food or any food additive that is added. This may happen mostly when the foods are heat processed, for example: baked, cooked on high temperatures, or deep-fried, etc. Some of these, involve components naturally occurring in the food, components involved in food packaging materials, food additives, or ingredients that are added intentionally. These reactions could generate potentially harmful compounds. Examples of chemicals produced during food processing are explained in the following sections.

5.1 Acrylamide

Many starchy food including potatoes contain reducing sugars and small amounts of amino acid, aspargine. When these vegetables are heated or fried at high temperatures, a well-known neurotoxin, acrylamide, is formed [114]. Frying, broiling, baking, or high-temperature cooking methods are also responsible to produce acrylamide formation. Potato chips and french fries have been reported to contain high levels of acrylamide compared with other foods [115]. IARC has classified acrylamide as a Group 2A carcinogen, i.e., "probable human carcinogen" based on studies conducted in laboratory animals [116]. The WHO/FAO suggests that the levels of acrylamide in foods pose a "major concern" and more research is needed globally to assess its dietary exposure in humans. Acrylamide has been shown to be metabolized by cytochrome P450 2E1 to glycidamide, which is responsible for in vivo genotoxicity and probable carcinogenicity [117]. However in vitro studies suggest that acrylamide is non-genotoxic [118].

5.2 Benzene

The presence of benzene has been detected in contaminated drinking water, packaging, storage conditions, degradation of food preservatives such as benzoates, cooking processes, and irradiation processes [119]. Benzene can also be introduced in foods by certain processes such as smoking, roasting, and ionizing radiation [120] and can also be formed when phenylalanine is broken down by ionizing radiation [118]. Lachenmeier et al. [121] had reported that β-carotene, terpenes, or phenylalanine during processing may decompose, leading to the formation of benzene. IARC has categorized benzene as a Group 1 carcinogen and its role as a leukemogen has been clearly established by a series of epidemiological studies [21]. After conducting genotoxic tests for benzene, it was found to be clastogenic and aneugenic and causes DNA strand breaks and lesions and chromosomal aberrations, producing micronuclei.

5.3 Heterocyclic Amines/Polycyclic Aromatic Hydrocarbons

HCAs and PAHs are formed when beef, pork, meat, fish, or poultry is cooked at higher temperature. Formation of HCAs occurs as a result of reaction between amino acids, sugars, and creatine at high temperatures, whereas PAHs are formed when meat, fat, and juices are grilled or roasted directly on fire [122]. Research conducted in laboratory experiments suggest that HCAs and PAHs are mutagenic because they cause damage to DNA that may increase the risk of cancer [123]. Studies have shown that rodents fed with HCAs developed tumors of breast, colon, liver, skin, lung, and prostate [124]. The mechanism of mutagenicity and carcinogenicity by HAA is shown in Fig. 7.3. Committee on Diet, Nutrition, and Cancer suggested that PAHs are able to develop tumors of gastrointestinal tract, urinary

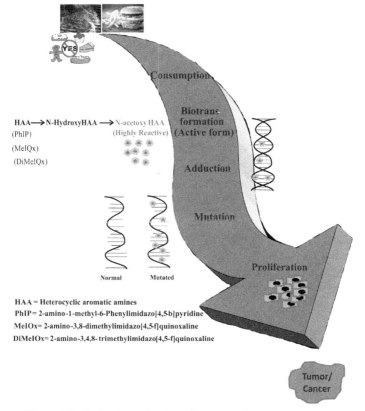

Figure 7.3 Mechanism of action of heterocyclic aromatic amines.

bladder, and lungs [4]. Repeated frying of foodstuffs, including fish in oil, generates considerable amount of PAHs that has been shown to have genotoxic and carcinogenic potential [125–127]. Thus, consumption of well-fried or barbecued meat is associated with increased risks of pancreatic, colorectal, lungs, and prostate cancer [122,128].

5.4 Chloropropanols

Chloropropanols are well-known contaminants of acid-hydrolyzed vegetable protein (acid-HVP), a frequently used ingredient of savory foods such as soups, prepared meals, gravy mixes, savory, and snacks. The major chloropropanols that are found in foods include 3-chloro-1,2-propanediol (3-MCPD) and 1,3-dichloro-2-propanol (1,3-DCP). Studies have shown that 3-MCPD may also be produced in thermally processed foods such as bread, biscuits, malt-derived products, coffee, cheese, smoked food, meat, and salted fish [129,130]. In vitro studies have shown that 3-MCPD induces reverse mutations [131] and DNA strand breaks [132]; however, no genotoxic effects of 3-MCPD have been observed under in vivo conditions [132]· It is interesting to note that 3-MCPD exerts its carcinogenicity without mutagenic pathway, with 1,3-DCP has been shown to exert genotoxicity in S. typhimurium and various mammalian cell lines [133].

5.5 Nitrosamines

Nitrosamines are produced from nitrites and secondary amines of food under certain conditions, which include strong acidic conditions, such as that of the stomach and high temperatures used for frying. Studies report that dimethylnitrosamine causes liver tumors in rats and approximately 90% of nitrosamine compounds were found to be carcinogenic [134]. Study conducted on 61,433 Sweden women suggested that high consumption of processed meat might be associated with increased risk of stomach cancer [135]. In vivo and in vitro genotoxicity has been reported for several N-nitrosamines in extra hepatic tissues of the rat, which may be responsible for cancer [136].

5.6 Furan

Furan has been identified in a number of food products [137], especially in jarred and canned food items that undergo heat treatment within a sealed container. Furthermore, furan has been detected in food items such as meats, beer, and nuts [138]. It is classified as "possible human carcinogen (Group 2B)" by the IARC on Cancer [139]. Furan is a potent hepatotoxicant, which is responsible for producing liver cancer in rodents [133]. However, short-term tests for genotoxicity of furan are inconclusive [140]. Various pathways have been reported for the formation of furan, that is, thermal degradation and/or thermal rearrangement of carbohydrates in the presence of amino acids. In particular, the role of genotoxicity in inducing furan carcinogenicity is still not clear. Studies indicate that the in vivo exposure to furan gives rise to premutagenic DNA damage in resting splenocytes [141]. Studies on male Balb/C mice suggested that neither the in vivo nor the in vitro studies had any significant increase in the micronucleus frequency after treatment with furan [142]. Furan did not induce unscheduled DNA synthesis in hepatocytes of F344 mice [143]. Thus, it has been postulated that carcinogenicity of furan may be caused by a nongenotoxic mechanism.

6. SUMMARY

A relationship between diet and cancer is gaining considerable importance in the last couple of decades. Attempts have been made to detect and determine these carcinogens/mutagens in food items. The food that animals or humans consume may contain contaminants, intentionally or unintentionally added additives, and adulterants, which may have genotoxic and mutagenic potential and thus may pose the hazard to the consumers. This chapter has mainly focused on the widely found mutagens present in food. Many new sensitive tools are being developed by researchers to study and identify the mutagens in raw and processed food. To protect consumer safety, strict legislative rules should be implemented for foodstuffs. Further research needs to be carried out for the analysis and risk assessment of mutagenic, genotoxic, and carcinogenic compounds present in food.

ACKNOWLEDGMENTS

We thank the Director, CSIR-IITR, Lucknow, for his support and encouragement. P.M. and A.R. are thankful to University Grant Commission for financial assistance, while S.M. acknowledges DST/WOS (A) fellowship award. This manuscript is CSIR-IITR communication #3447.

REFERENCES

[1] Francis F. Pioneer in food sciences and quality. Century Food Sci Inst Food Technol 2000.
[2] Goldman R, Shields PG. Food mutagens. J of Nutr 2003;133:965S–73S.

[3] Fukuda K. Food safety in a globalized world. Bull World Health Organ 2015;93:212.
[4] Peter FM. Diet, nutrition, and cancer. In: Committee on diet, nutrition, and cancer. National Research Council; 1982. ISBN: 0-309-56792-0. 496 pages.
[5] Türkoğlu Ş. Genotoxicity of five food preservatives tested on root tips of *Allium cepa* L. Mutat Res 2007;626:4–14.
[6] Mamur S, Yüzbaşıoğlu D, Ünal F, Aksoy H. Genotoxicity of food preservative sodium sorbate in human lymphocytes in vitro. Cytotechnology 2012;64:553–62.
[7] Silva NO, Reed E. Estudo sobre corantes artificiais em alimentos: quais os riscos mais comuns pelo consumo excessivo. Cadernos de Educação Tecnologia e Sociedade 2011;2:83–6.
[8] Das M. Food contamination and adulteration. Environmental biochemistry. New Delhi: E-book, National Science Digital Library, National Institute of Science Communication & Information Resources; 2008. p. 1–32.
[9] Zygler A, Wasik A, Namieśnik J. Analytical methodologies for determination of artificial sweeteners in foodstuffs. TrAC 2009;28:1082–102.
[10] de la Paz MP, Philen R, Borda IA, Socias JS, de la Cámara AG, Kilbourne E. Toxic oil syndrome: traceback of the toxic, oil and evidence for a point source epidemic. Food Chem Toxicol 1996;34:251–7.
[11] Babu CK, Khanna SK, Das M. Adulteration of mustard cooking oil with argemone oil: do Indian food regulatory policies and antioxidant therapy both need revisitation? Antioxid Redox Signal 2007;9:515–25.
[12] Clemente TE, Cahoon EB. Soybean oil: genetic approaches for modification of functionality and total content. Plant Physiol 2009;151:1030–40.
[13] International Agency for Research on Cancer (IARC). Toxins derived from *Fusarium moniliforme*: fumonisins B1 and B2 and fusarin C. IARC Monogr Eval Carcinog Risks Hum 1993;56:445–62.
[14] World Health Organization. Safety evaluation of certain mycotoxins in food. In: Prepared by 49th meeting of JECFA, food additives series, WHO, Geneva. 1998. p. 47.
[15] Donaldson D, Kiely T, Grube A. Pesticide's industry sales and usage 1998-1999 market estimates. Washington (DC): US Environmental Protection Agency; 1997. Report No. EPA-733-R-02-OOI. Available from: http://www.epa.gov/oppbead/pesticides/99pestsales/market-estimates.
[16] Alavanja MC. Introduction: pesticides use and exposure, extensive worldwide. Rev Environ Health 2009;24:303–10.
[17] Chandorkar S, Deota P. Heavy metal content of foods and health risk assessment in the study population of vadodara. Curr World Environ 2013;8:291–7.
[18] Anand S, Sati N. Artificial preservatives and their harmful effects: looking toward nature for safer alternatives. Inter J Pharma Sci Res 2013;4:2496.
[19] Elgazar AF. Effects of butylated hydroxytoluene and butylated hydroxyanisole against hepatotoxicity induced by carbon tetrachloride in rats. World Appl Sci J 2013;22:63–9.
[20] Silva MM, Lidon FC. Food preservatives-An overview on applications and side effects. Emir J Food Agric 2016;28:366.
[21] International Agency for Research on Cancer. Overall evaluations of carcinogenicity. An updating of IARC monographs volumes 1 to 42. Lyon: IARC; 1987.
[22] Clapp N, Tyndall R, Cumming R, Otten J. Effects of butylated hydroxytoluene alone or with diethylnitrosamine in mice. Food Cosmet Toxicol 1974;12:367–71.
[23] Amo H, Kubota H, Lu J, Matsuyama M. Adenomatous hyperplasia and adenomas in the lung induced by chronic feeding of butylated hydroxyanisole of Japanese house musk shrew (*Suncus murinus*). Carcinogenesis 1990;11:151–4.
[24] Kapoor V. Food colours: concern regarding their safety and toxicity. Inter Soc Environ Botanist 2006;12:7–8.

[25] Vrolijk MF, Opperhuizen A, Jansen EH, Godschalk RW, Van Schooten FJ, Bast A, et al. The shifting perception on antioxidants: the case of vitamin E and β-carotene. Redox Biol 2015;4:272–8.

[26] Mpountoukas P, Pantazaki A, Kostareli E, Christodoulou P, Kareli D, Poliliou S, et al. Cytogenetic evaluation and DNA interaction studies of the food colorants amaranth, erythrosine and tartrazine. Food Chem Toxicol 2010;48:2934–44.

[27] Sayed HM, Fouad D, Ataya FS, Hassan NH, Fahmy MA. The modifying effect of selenium and vitamins A, C, and E on the genotoxicity induced by sunset yellow in male mice. Mutat Res 2012;744:145–53.

[28] Dwivedi K, Kumar G. Genetic damage induced by a food coloring dye (sunset yellow) on meristematic cells of Brassica campestris L. J Environ Public Health 2015:2015.

[29] Kus E, Eroglu HE. Genotoxic and cytotoxic effects of sunset yellow and brilliant blue, colorant food additives, on human blood lymphocytes. Pak J Pharm Sci 2015;28:227–30.

[30] Tripathi M, Khanna SK, Das M. Surveillance on use of synthetic colours in eatables vis a vis prevention of food adulteration act of India. Food Control 2007;18:211–9.

[31] Dixit S, Purshottam S, Khanna SK, Das M. Surveillance on quality of turmeric powders vis a vis curcumin content and presence of extraneous color s from city markets of India. Food Addit Contam 2009;26:1227–31.

[32] Khanna S, Das M. Toxicity, carcinogenic potential and clinico-epidemiological studies on dyes and dye intermediates. J Sci Ind Res 1991;50:965–74.

[33] Wiebe N, Padwal R, Field C, Marks S, Jacobs R, Tonelli M. A systematic review on the effect of sweeteners on glycemic response and clinically relevant outcomes. BMC Med 2011;9:1.

[34] Tripathi M, Khanna SK, Das M. Usage of saccharin in food products and its intake by the population of Lucknow, India. Food Addit Contam 2006;23:1265–75.

[35] Soffritti M, Belpoggi F, Tibaldi E, Esposti DD, Lauriola M. Life-span exposure to low doses of aspartame beginning during prenatal life increases cancer effects in rats. Environ Health Perspect 2007:1293–7.

[36] Yang Q. Gain weight by "going diet?" Artificial sweeteners and the neurobiology of sugar cravings. Yale Biol Med 2010;83:101–8.

[37] International Agency for Research on Cancer (IARC), Ochratoxin A. IARC Monogr Eval Carcinog Risks Hum 1993;56:489–521.

[38] Wild C, Miller JD, Groopman JD. Mycotoxin control in low-and Middle-income countries. World Health Organization; 2016.

[39] Peers F, Linsell C. Dietary aflatoxins and liver cancer–a population based study in Kenya. Br J Cancer 1973;27:473.

[40] Van Rensburg S, Cook-Mozaffari P, Van Schalkwyk D, Van der Watt J, Vincent T, Purchase I. Hepatocellular carcinoma and dietary aflatoxin in Mozambique and Transkei. Br J Cancer 1985;51:713.

[41] Some current perspectives on chemical carcinogenesis in humans and experimental animals: presidential address. In: Miller EC, editor. Miller EC, editor. Cancer research, vol. 38. 1978. p. 1479–96.

[42] Rastogi S, Dogra R, Khanna SK, Das M. Skin tumorigenic potential of aflatoxin B1 in mice. Food Chem Toxicol 2006;44:670–7.

[43] Smela ME, Currier SS, Bailey EA, Essigmann JM. The chemistry and biology of aflatoxin B1: from mutational spectrometry to carcinogenesis. Carcinogenesis 2001;22:535–45.

[44] National Toxicology Program. Report on carcinogens. Washington (DC): US Department of Health and Human Services. Public Health Service; 1980.

[45] International Agency for Research on Cancer. Aflatoxins: natural occurring aflatoxinas (Group 1), aflatoxin M1 (Group 2B). IARC Scientific Publications; 1993.

[46] International Agency for Research on Cancer. Aflatoxins. IARC Monogr Eval Carcinog Risks Hum 2002;82:171–300.

[47] National Toxicology Program. Toxicology and carcinogenesis studies of ochratoxin a (CAS No. 303-47-9) in F344/N rats (Gavage studies). Natl Toxicol Program Tech Rep Ser 1989;358:51.

[48] Pfohl-Leszkowicz A, Manderville RA. Ochratoxin A: an overview on toxicity and carcinogenicity in animals and humans. Mol Nutr Food Res 2007;51:61–99.

[49] Pfohl-Leszkowicz A, Petkova-Bocharova T, Chernozemsky I, Castegnaro M. Balkan endemic nephropathy and associated urinary tract tumours: a review on aetiological causes and the potential role of mycotoxins. Food Addit Contam 2002;19:282–302.

[50] Stoev S. The role of ochratoxin A as a possible cause of Balkan endemic nephropathy and its risk evaluation. Vet Hum Toxicol 1998;40:352–60.

[51] Stoev SD. Studies on carcinogenic and toxic effects of ochratoxin A in chicks. Toxins 2010;2:649–64.

[52] Grosso F, Saïd S, Mabrouk I, Fremy J, Castegnaro M, Jemmali M, et al. New data on the occurrence of ochratoxin A in human sera from patients affected or not by renal diseases in Tunisia. Food Chem Toxicol 2003;41:1133–40.

[53] Kumar R, Ansari KM, Chaudhari BP, Dhawan A, Dwivedi PD, Jain SK, Das M. Topical application of ochratoxin A causes DNA damage and tumor initiation in mouse skin. PLoS One 2012;7:e47280.

[54] Kumar R, Alam S, Chaudhari BP, Dwivedi PD, Jain SK, Ansari KM, Das M. Ochratoxin A-induced cell proliferation and tumor promotion in mouse skin by activating the expression of cyclin D1 and cyclooxygenase-2 through nuclear factor-kappa B and activator protein-1. Carcinogenesis 2013;34:647–57.

[55] Schwartz GG. Hypothesis: does ochratoxin A cause testicular cancer? Cancer Causes Control 2002;13:91–100.

[56] Jennings-Gee JE, Tozlovanu M, Manderville R, Miller MS, Pfohl-Leszkowicz A, Schwartz GG, Ochratoxin A. In utero exposure in mice induces adducts in testicular DNA. Toxins 2010;2:1428–44.

[57] Tozlovanu M, Faucet-Marquis V, Pfohl-Leszkowicz A, Manderville RA. Genotoxicity of the hydroquinone metabolite of ochratoxin A: structure-activity relationships for covalent DNA adduction. Chem Res Toxicol 2006;19:1241–7.

[58] Gelderblom W, Kriek N, Marasas W, Thiel P. Toxicity and carcinogenicity of the Fusanum monilzforine metabolite, fumonisin B1, in rats. Carcinogenesis 1991;12:1247–51.

[59] Marasas W. Discovery and occurrence of the fumonisins: a historical perspective. Environ Health Perspect 2001;109:239.

[60] Missmer SA, Suarez L, Felkner M, Wang E, Merrill Jr AH, Rothman KJ, et al. Exposure to fumonisins and the occurrence of neural tube defects along the Texas-Mexico border. Environ Health Perspect 2006:237–41.

[61] Ueno Y. Trichothecene mycotoxins mycology, chemistry, and toxicology. In: Draper HH, editor. Advances in nutritional research. Boston (MA): Springer US; 1980. p. 301–53.

[62] Samar M, Neira M, Resnik S, Pacin A. Effect of fermentation on naturally occurring deoxynivalenol (DON) in Argentinean bread processing technology. Food Addit Contam 2001;18:1004–10.

[63] Mishra S, Dixit S, Dwivedi PD, Pandey HP, Das M. Influence of temperature and pH on the degradation of deoxynivalenol (DON) in aqueous medium: comparative cytotoxicity of DON and degraded product. Food Addit Contam Part a 2014;31:121–31.

[64] Lauren D, Smith W. Stability of the Fusarium mycotoxins nivalenol, deoxynivalenol and zearalenone in ground maize under typical cooking environments. Food Addit Contam 2001;18:1011–6.

[65] Ma YY, Guo HW. Mini-review of studies on the carcinogenicity of deoxynivalenol. Environ Toxicol Pharmacol 2008;25:1–9.

[66] Mishra S, Tewari P, Chaudhari BP, Dwivedi PD, Pandey HP, Das M. Deoxynivalenol induced mouse skin tumor initiation: elucidation of molecular mechanisms in human HaCaT keratinocytes. Inter J Cancer 2016;139:2033–46.

[67] Hsia C, Wu J, Lu XQ, Li Y. Natural occurrence and clastogenic effects of nivalenol, deoxynivalenol, 3-acetyl-deoxynivalenol, 15-acetyl-deoxynivalenol, and zearalenone in corn from a high-risk area of esophageal cancer. Cancer Detect Prev 1987;13:79–86.

[68] Luo Y, Yoshizawa T, Katayama T. Comparative study on the natural occurrence of Fusarium mycotoxins (trichothecenes and zearalenone) in corn and wheat from high- and low-risk areas for human esophageal cancer in China. Appl Environ Microbiol 1990;56:3723–6.

[69] Frankič T, Pajk T, Rezar V, Levart A, Salobir J. The role of dietary nucleotides in reduction of DNA damage induced by T-2 toxin and deoxynivalenol in chicken leukocytes. Food Chem Toxicol 2006;44:1838–44.

[70] Chaudhari M, Jayaraj R, Bhaskar A, Rao PL. Oxidative stress induction by T-2 toxin causes DNA damage and triggers apoptosis via caspase pathway in human cervical cancer cells. Toxicology 2009;262:153–61.

[71] Hassanane M, Abdalla E, El-Fiky S, Amer M, Hamdy A. Mutagenicity of the mycotoxin diacetoxyscirpenol on somatic and germ cells of mice. Mycotoxin Res 2000;16:53–64.

[72] Ueno Y. Mode of action of trichothecenes. Pure Appl Chem 1977;49:1737–45.

[73] Zhou S-m, Jiang L-p, Geng C-y, Cao J, Zhong L-f. Patulin-induced oxidative DNA damage and p53 modulation in HepG2 cells. Toxicon 2010;55:390–5.

[74] de Melo FT, de Oliveira IM, Greggio S, Dacosta JC, Guecheva TN, Saffi J, et al. DNA damage in organs of mice treated acutely with patulin, a known mycotoxin. Food Chem Toxicol 2012;50:3548–55.

[75] Alves I, Oliveira N, Laires A, Rodrigues A, Rueff J. Induction of micronuclei and chromosomal aberrations by the mycotoxin patulin in mammalian cells: role of ascorbic acid as a modulator of patulin clastogenicity. Mutagenesis 2000;15:229–34.

[76] Saxena N, Ansari KM, Kumar R, Chaudhari BP, Dwivedi PD, Das M. Role of mitogen activated protein kinases in skin tumorigenicity of patulin. Toxicol Appl Pharmacol 2011;257:264–71.

[77] Saxena N, Ansari KM, Kumar R, Dhawan A, Dwivedi PD, Das M. Patulin causes DNA damage leading to cell cycle arrest and apoptosis through modulation of Bax, p 53 and p 21/WAF1 proteins in skin of mice. Toxicol Appl Pharmacol 2009;234:192–201.

[78] Pfohl-Leszkowicz A, Chekir-Ghedira L, Bacha H. Genotoxicity of zearalenone, an estrogenic mycotoxin: DNA adduct formation in female mouse tissues. Carcinogenesis 1995;16:2315–20.

[79] Ouanes-Ben Othmen Z, Essefi S, Bacha H. Mutagenic and epigenetic mechanisms of zearalenone: prevention by Vitamin E. World Mycotoxin J 2008;1:369–74.

[80] Kumar R, Dwivedi PD, Dhawan A, Das M, Ansari KM. Citrinin-generated reactive oxygen species cause cell cycle arrest leading to apoptosis via the intrinsic mitochondrial pathway in mouse skin. Toxicol Sci 2011;122:557–66.

[81] Hreljac I, Zajc I, Lah T, Filipič M. Effects of model organophosphorous pesticides on DNA damage and proliferation of HepG2 cells. Environ Mol Mutagen 2008;49:360–7.

[82] Anttila A, Bhat RV, Bond JA, Borghoff SJ, Bosch FX, Carlson GP, Castegnaro M, Cruzan G, Gelderblom WC, Hass U, Henry SH. Some traditional herbal medicines, some mycotoxins, Naphthalene and Styrene. IARC Monogr Eval Carcinog Risks Hum 2002;82.

[83] Saxena S, Ashok B, Musarrat J. Mutagenic and genotoxic activities of four pesticides: captan, foltaf, phosphamidon and furadan. Biochem Mol Biol Inter 1997;41:1125–36.

[84] Sharma RK, Sharma B. In-vitro carbofuran induced genotoxicity in human lymphocytes and its mitigation by vitamins C and E. Dis Markers 2012;32:153–63.

[85] Sun X, Jin Y, Wu B, Wang W, Pang X, Wang J. Study on genotoxicity of aldicarb and methomyl. Huan Jing Ke Xue 2010;31:2973–80.

[86] Ennaceur S, Ridha D, Marcos R. Genotoxicity of the organochlorine pesticides 1, 1-dichloro-2, 2-bis (p-chlorophenyl) ethylene (DDE) and hexachlorobenzene (HCB) in cultured human lymphocytes. Chemosphere 2008;71:1335–9.

[87] Pandey N, Gundevia F, Prem A, Ray P. Studies on the genotoxicity of endosulfan, an organochlorine insecticide, in mammalian germ cells. Mutat Res 1990;242:1–7.

[88] Bajpayee M, Pandey AK, Zaidi S, Musarrat J, Parmar D, Mathur N, et al. DNA damage and mutagenicity induced by endosulfan and its metabolites. Environ Mol Mutagen 2006;47:682–92.

[89] Joint FAO/WHO Expert Committee on food additives. Evaluation of certain veterinary drug residues in food. WHO Tech Rep Ser 2004;9:9.

[90] Çelik A, Mazmanci B, Çamlica Y, Aşkin A, Çömelekoğlu Ü. Cytogenetic effects of lambda-cyhalothrin on Wistar rat bone marrow. Mutat Res 2003;539:91–7.

[91] Çelik A, Mazmanci B, Çamlica Y, Aşkin A, Çömelekoğlu Ü. Induction of micronuclei by lambda-cyhalothrin in Wistar rat bone marrow and gut epithelial cells. Mutagenesis 2005;20:125–9.

[92] Marinowic DR, Mergener M, Pollo TA, Maluf SW, da Silva LB. In vivo genotoxicity of the pyrethroid pesticide β-cyfluthrin using the comet assay in the fish Bryconamericus iheringii. Z Naturforschung C 2012;67:308–11.

[93] Zuo P, Yin B-C, Ye B-C. DNAzyme-based microarray for highly sensitive determination of metal ions. Biosens Bioelectron 2009;25:935–9.

[94] Aragay G, Pons J, Merkoçi A. Recent trends in macro-, micro-, and nanomaterial-based tools and strategies for heavy-metal detection. Chem Rev 2011;111:3433–58.

[95] Flora S, Mittal M, Mehta A. Heavy metal induced oxidative stress & its possible reversal by chelation therapy. Indian J Med Res 2008;128:501.

[96] Ho V, Peacock S, Massey TE, Ashbury JE, Vanner SJ, King WD. Meat-derived carcinogens, genetic susceptibility and colorectal adenoma risk. Genes Nutr 2014;9:1–12.

[97] Kondo K, Watanabe A, Akiyama H, Maitani T. The metabolisms of agaritine, a mushroom hydrazine in mice. Food Chem Toxicol 2008;46:854–62.

[98] Inoue M, Tsugane S. Epidemiology of gastric cancer in Japan. Postgrad Med J 2005;81:419–24.

[99] Sugimura T. Nutrition and dietary carcinogens. Carcinogenesis 2000;21:387–95.

[100] Dawson DW, Hertzer K, Moro A, Donald G, Chang H-H, Go VL, et al. High-fat, high-calorie diet promotes early pancreatic neoplasia in the conditional KrasG12D mouse model. Cancer Prev Res 2013;6:1064–73.

[101] Albanes D. Caloric intake, body weight, and cancer: a review. Nutr Cancer 1987;9:199–217.

[102] Das M, Khanna SK. Clinicoepidemiological, toxicological, and safety evaluation studies on argemone oil. Crit Rev Toxicol 1997;27:273–97.

[103] Evangelista CMW, Antunes LMG, Bianchi MdLP. In vivo cytogenetic effects of multiple doses of dietary vegetable oils. Genet Mol Biol 2006;29:730–4.

[104] Ansari KM, Chauhan LK, Dhawan A, Khanna SK, Das M. Unequivocal evidence of genotoxic potential of argemone oil in mice. Inter J Cancer 2004;112:890–5.

[105] Das M, Ansari KM, Dhawan A, Shukla Y, Khanna SK. Correlation of DNA damage in epidemic dropsy patients to carcinogenic potential of argemone oil and isolated sanguinarine alkaloid in mice. Inter J Cancer 2005;117:709–17.

[106] Mishra V, Mishra M, Ansari KM, Chaudhari BP, Khanna R, Das M. Edible oil adulterants, argemone oil and butter yellow, as aetiological factors for gall bladder cancer. Eur J Cancer 2012;48:2075–85.

[107] Arcos JC, Argus MF, Wolf G. Chemical induction of cancer: structural bases and biological mechanisms. Elsevier; 2013.

[108] Srivastava A, Das M, Khanna S. An outbreak of tricresyl phosphate poisoning in Calcutta, India. Food Chem Toxicol 1990;28:303–4.

[109] Srivastava AK, Das M, Khanna SK. An investigation of factors affecting recovery in victims of tricresyl phosphate induced polyneuorpathy: an 18 months followup study. Inter J Toxicol Occupat Environ Health 1992;1:41–5.

[110] de la Paz MP, Philen RM, Borda IA. Toxic oil syndrome: the perspective after 20 years. Epidemiol Rev 2001;23:231–47.

[111] Lin S-Y, Tsai S-J, Wang L-H, Wu M-F, Lee H. Protection by quercetin against cooking oil fumes-induced DNA damage in human lung adenocarcinoma CL-3 cells: role of COX-2. Nutr Cancer 2002;44:95–101.

[112] Zanoni TB, Lizier TM, das Dores Assis M, Zanoni MVB, De Oliveira DP. CYP-450 isoenzymes catalyze the generation of hazardous aromatic amines after reaction with the azo dye Sudan III. Food Chem Toxicol 2013;57:217–26.

[113] Sarkar R, Ghosh AR. Metanil yellow-an azo dye induced histopathological and ultrastructural changes in albino rat (Rattus norvegicus). Bioscan 2012;7:424–32.

[114] Wikkelsö C, Ekberg K, Lillienberg L, Wetterholm B, Karlsson B, Blomstrand C, Johansson BB. Cerebrospinal fluid proteins and cells in men subjected to long-term exposure to organic solvents. Acta Neurol ScSuppl 1983;100:113–9.

[115] Organization WH. Summary report of the sixty-fourth meeting of the Joint FAO/WHO Expert Committee on food additive (JECFA). Rome, Italy. Washington (DC): The ILSI Press International Life Sciences Institute; 2005.

[116] International Agency for Research on Cancer (WHO-IARC). Re-evaluation of some organic chemicals, hydrazine and hydrogen peroxide. IARC Monogr Eval Carcinog Risks Hum 1998:17–24.

[117] Carere A. Genotoxicity and carcinogenicity of acrylamide: a critical review. Ann Ist Super Sanita 2006;42:144.

[118] Sommers CH, Delincée H, Smith JS, Marchioni E. Toxicological safety of irradiated foods. Food Irradiat Res Technol 2012;2:53–74.

[119] Salviano dos Santos VP, Medeiros Salgado A, Guedes Torres A, Signori Pereira K. Benzene as a chemical hazard in processed foods. Inter J Food Sci 2015;2015.

[120] Stadler RH, Lineback DR. Process-induced food toxicants: occurrence, formation, mitigation, and health risks. John Wiley & Sons; 2008.

[121] Lachenmeier DW, Reusch H, Sproll C, Schoeberl K, Kuballa T. Occurrence of benzene as a heat-induced contaminant of carrot juice for babies in a general survey of beverages. Food Addit Contam 2008;25:1216–24.

[122] Cross AJ, Sinha R. Meat-related mutagens/carcinogens in the etiology of colorectal cancer. Environ Mol Mutagen 2004;44:44–55.

[123] Sugimura T, Wakabayashi K, Nakagama H, Nagao M. Heterocyclic amines: mutagens/carcinogens produced during cooking of meat and fish. Cancer Sci 2004;95:290–9.

[124] Ohgaki H, Kusama K, Matsukura N, Morino K, Hasegawa H, Sato S, et al. Carcinogenicity in mice of a mutagenic compound, 2-amino-3-methylimidazo [4,5-f] quinoline, from broiled sardine, cooked beef and beef extract. Carcinogenesis 1984;5:921–4.

[125] Pandey MK, Yadav S, Parmar D, Das M. Induction of hepatic cytochrome P450 isozymes, benzo (a) pyrene metabolism and DNA binding following exposure to polycyclic aromatic hydrocarbon residues generated during repeated fish fried oil in rats. Toxicol Appl Pharmacol 2006;213:126–34.

[126] Pandey MK, Dhawan A, Das M. Induction of P53, P21Waf1, orinithine decorboxylase activity, and DNA damage leading to cell-cycle arrest and apoptosis following topical application of repeated fish fried oil extract to mice. Mol Carcinogen 2006;45:805–13.

[127] Pandey MK, Das M. Assessment of carcinogenic potential of repeated fish fried oil in mice. Mol Carcinogen 2006;45:741–51.

[128] Sinha R, Park Y, Graubard BI, Leitzmann MF, Hollenbeck A, Schatzkin A, et al. Meat and meat-related compounds and risk of prostate cancer in a large prospective cohort study in the United States. Am J Epidemiol 2009;280.

[129] Baer I, de la Calle B, Taylor P. 3-MCPD in food other than soy sauce or hydrolysed vegetable protein (HVP). Anal Bioanal Chem 2010;396:443–56.

[130] Crews C, Brereton P, Davies A. The effects of domestic cooking on the levels of 3-monochloropropanediol in foods. Food Addit Contam 2001;18:271–80.

[131] Šilhánková L, Šmíd F, Černá M, Davídek J, Velíšek J. Mutagenicity of glycerol chlorohydrines and of their esters with higher fatty acids present in protein hydrolysates. Mutat Res Lett 1982;103:77–81.

[132] El Ramy R, Elhkim MO, Lezmi S, Poul J. Evaluation of the genotoxic potential of 3-monochloropropane-1, 2-diol (3-MCPD) and its metabolites, glycidol and β-chlorolactic acid, using the single cell gel/comet assay. Food Chem Toxicol 2007;45:41–8.

[133] National Toxicology Program. Toxicology and carcinogenesis studies of furan in F344/N rats and B6C3F1 mice. Research Triangle Park (NC): US Department of Health and Human Services. Public Health Service, National Institutes of Health; 1993. [NTP Technical Report No 402].

[134] Jakszyn P, González CA. Nitrosamine and related food intake and gastric and oesophageal cancer risk: a systematic review of the epidemiological evidence. World J Gastroenterol 2006;12:4296–303.

[135] Larsson SC, Bergkvist L, Wolk A. Processed meat consumption, dietary nitrosamines and stomach cancer risk in a cohort of Swedish women. Inter J Cancer 2006;119:915–9.

[136] Brendler S, Tompa A, Hutter K, Preussmann R, Pool-Zobel B. In vivo and in vitro genotoxicity of several N-nitrosamines in extrahepatic tissues of the rat. Carcinogenesis 1992;13:2435–41.

[137] Maga JA, Katz I. Furans in foods. Crit Rev Food Sci Nutr 1979;11:355–400.

[138] US Food and Drug Administration. Exploratory data on furan in food. Washington (DC): FDA; 2004.

[139] International Agency for research on Cancer. IARC monographs on the evaluation of carcinogenic risks to humans. Lyon, France: IARC; 2014.

[140] McDaniel LP, Ding W, Dobrovolsky VN, Shaddock JG, Mittelstaedt RA, Doerge DR, et al. Genotoxicity of furan in big blue rats. Mutat Res 2012;742:72–8.

[141] Leopardi P, Cordelli E, Villani P, Cremona TP, Conti L, De Luca G, et al. Assessment of in vivo genotoxicity of the rodent carcinogen furan: evaluation of DNA damage and induction of micronuclei in mouse splenocytes. Mutagenesis 2010;25:57–62.

[142] Wilson DM, Goldsworthy TL, Popp JA, Butterworth BE. Evaluation of genotoxicity, pathological lesions, and cell proliferation in livers of rats and mice treated with furan. Environ Mol Mutagen 1992;19:209–22.

[143] Tran AV. Do BHA and BHT induce morphological changes and DNA double-strand breaks in Schizosaccharomyces pombe? Scripps Senior Theses 2013;152.

[144] Williams G, Iatropoulos M, Whysner J. Safety assessment of butylated hydroxyanisole and butylated hydroxytoluene as antioxidant food additives. Food Chem Toxicol 1999;37:1027–38.

[145] Sanchez-Echaniz J, Benito-Fernández J, Mintegui-Raso S. Methemoglobinemia and consumption of vegetables in infants. Pediatrics 2001;107:1024–8.

[146] Hasegawa M, Nishi Y, Ohkawa Y, Inui N. Effects of sorbic acid and its salts on chromosome aberrations, sister chromatid exchanges and gene mutations in cultured Chinese hamster cells. Food Chem Toxicol 1984;22:501–7.

[147] Mukherjee A, Giri AK, Talukder G, Sharma A. Sister chromatid exchanges and micronuclei formations induced by sorbic acid and sorbic acid-nitrite in vivo in mice. Toxicol Lett 1988;42:47–53.

[148] Van der Heijden C, Janssen P, Strik J. Toxicology of gallates: a review and evaluation. Food Chem Toxicol 1986;24:1067–70.

[149] Hamdy A, Mekkawy A, Massoud A, El-zawahry M. Mutagenic effects of the food color erythrosine in rats. Probl Forensic Sci 2000;43:184–91.

[150] Tsuda S, Murakami M, Matsusaka N, Kano K, Taniguchi K, Sasaki YF. DNA damage induced by red food dyes orally administered to pregnant and male mice. Toxicol Sci 2001;61:92–9.

[151] Shimada C, Kano K, Sasaki YF, Sato I, Tsuda S. Differential colon DNA damage induced by azo food additives between rats and mice. J Toxicol Sci 2010;35:547–54.

[152] Hassan G. Effects of some synthetic coloring additives on DNA damage and chromosomal aberrations of rats. Arab J Biotechnol 2010;13:13–24.

[153] Roychoudhury A, Giri AK. Effects of certain food dyes on chromosomes of *Allium cepa*. Mutat Res 1989;223:313–9.

[154] Swaroop VR, Roy DD, Vijayakumar T. Genotoxicity of synthetic food colorants. J Food Sci Eng 2011;1:128.

[155] Negro F, Mondardini A, Palmas F. Hepatotoxicity of saccharin. N Engl J Med 1994;331:134–5.

[156] Weihrauch M, Diehl V. Artificial sweeteners—do they bear a carcinogenic risk? Ann Oncol 2004;15:1460–5.

[157] Mukhopadhyay M, Mukherjee A, Chakrabarti J. In vivo cytogenetic studies on blends of aspartame and acesulfame-K. Food Chem Toxicol 2000;38:75–7.

[158] Roberts H. Aspartame-induced thrombocytopenia. South Med J 2007;100:543–4.

[159] Sasaki YF, Kawaguchi S, Kamaya A, Ohshita M, Kabasawa K, Iwama K, et al. The comet assay with 8 mouse organs: results with 39 currently used food additives. Mutat Res 2002;519:103–19.

[160] Bigal ME, Krymchantowski AV. Migraine triggered by sucralose—a case report. Headache 2006;46:515–7.

[161] Bowen J. Splenda is not splendid. World National Health Organization; 2003. Available online at: http://www.wnho.net/splenda.htm.

[162] Mukherjee A, Chakrabarti J. In vivo cytogenetic studies on mice exposed to acesulfame-K—a non-nutritive sweetener. Food Chem Toxicol 1997;35:1177–9.

[163] Flamm WG, Blackburn GL, Comer CP, Mayhew DA, Stargel WW. Long-term food consumption and body weight changes in neotame safety studies are consistent with the allometric relationship observed for other sweeteners and during dietary restrictions. Regul Toxicol Pharmacol 2003;38:144–56.

[164] Mayhew DA, Comer CP, Stargel WW. Food consumption and body weight changes with neotame, a new sweetener with intense taste: differentiating effects of palatability from toxicity in dietary safety studies. Regul Toxicol Pharmacol 2003;38:124–43.

[165] Joint FAO/WHO Expert Committee on Food Additives. Evaluation of certain food additives and contaminants: sixty-first report of the Joint FAO/WHO Expert Committee on Food Additives. World Health Organization; 2004.

[166] Guallar E, Sanz-Gallardo MI, Veer P, Bode P, Aro A, Gómez-Aracena J, et al. Mercury, fish oils, and the risk of myocardial infarction. N Engl J Med 2002;347:1747–54.

[167] Yoshizawa K, Rimm EB, Morris JS, Spate VL, Hsieh C-c, Spiegelman D, et al. Mercury and the risk of coronary heart disease in men. N Engl J Med 2002;347:1755–60.

[168] Steenland K, Boffetta P. Lead and cancer in humans: where are we now? Am J Indus Med 2000;38:295–9.

[169] Patrick L. Lead toxicity part II: the role of free radical damage and the use of antioxidants in the pathology and treatment of lead toxicity. Altern Med Rev 2006;11:114.

[170] World Health Organization. Lead. Environmental health criteria. Geneva: WHO; 1995. 165.

[171] Steinkellner H, Mun-Sik K, Helma C, Ecker S, Ma TH, Horak O, et al. Genotoxic effects of heavy metals: comparative investigation with plant bioassays. Environ Mol Mutagen 1998;31:183–91.

[172] Chen Y, Wu F, Liu M, Parvez F, Slavkovich V, Eunus M, et al. A prospective study of arsenic exposure, arsenic methylation capacity, and risk of cardiovascular disease in Bangladesh. Environ Health Perspect 2013;121:832.

[173] Järup L, Hellström L, Alfvén T, Carlsson MD, Grubb A, Persson B, et al. Low level exposure to cadmium and early kidney damage: the OSCAR study. Occupat Environ Med 2000;57:668–72.

[174] Järup L, Berglund M, Elinder CG, Nordberg G, Vanter M. Health effects of cadmium exposure–a review of the literature and a risk estimate. ScJ Work Environ Health 1998:1–51.

[175] Ikeda M, Ezaki T, Tsukahara T, Moriguchi J. Dietary cadmium intake in polluted and non-polluted areas in Japan in the past and in the present. Inter Arch Occup Environ Health 2004;77:227–34.

[176] World Health Organization. Cadmium. Environmental health criteria. Geneva: WHO; 1992. 134.

[177] Zhang Y, Xiao H. Antagonistic effect of calcium, zinc and selenium against cadmium induced chromosomal aberrations and micronuclei in root cells of *Hordeum vulgare*. Mutat Res 1998;420:1–6.

FURTHER READING

[1] Cantoni O, Christie N, Swann A, Drath D, Costa M. Mechanism of HgCl2 cytotoxicity in cultured mammalian cells. Mol Pharmacol 1984;26:360–8.

[2] Aitio A, Becking G. Arsenic and arsenic compounds. Environmental health criteria. WHO; 2001. 224.

CHAPTER EIGHT

Emerging Computational Methods for Predicting Chemically Induced Mutagenicity

Shraddha Pandit, Alok Dhawan, Ramakrishnan Parthasarathi
CSIR-Indian Institute of Toxicology Research, Lucknow, India

1. INTRODUCTION

The rapid transition of chemical industry, medical and pharmaceutical sectors makes use of in vitro toxicity screening, in vivo toxicological experiments, and computational predictive methods to lower the failure rate in the product development (chemicals, drugs, etc.), decrease the cost, time, and human safety concerns [1–3]. Testing toxicity is about the determination of adverse outcome, which a chemical is likely to induce by its mode of action. Toxicity can be defined as the series of events initiated by exposure of a chemical, its progression through distribution and metabolism, and, ultimately, its interactions with different macromolecules of a cell, i.e., DNA or protein resulting in expressing its outcome in the form of various toxicity endpoints [4]. Mutagenicity represents a key toxicity endpoint for the safety evaluation of chemicals. The ability of a chemical to cause a change in DNA or RNA sequence is termed as mutagenicity and such mutations can be fixed or transmissible in nature [5]. DNA interacts with various environmental agents such as industrial chemicals, ultraviolet radiation, pesticides, and by-products from the combustion of fossil fuels [6,7]. The outcomes of mutagenic effects are huge such as (1) the genetic alteration can be heritable; (2) mutation in somatic cells can lead toward developing cancer; and (3) mutation in germ cells may cause infertility or abnormal kids. Mutation caused by a chemical depends on various parameters such as how often, how long, how much a cell or body is exposed to those chemicals, which are often found in paints, cosmetics, insecticides, pesticides, cleaning products, drugs, etc. All such commonly used products

Mutagenicity: Assays and Applications
ISBN 978-0-12-809252-1
http://dx.doi.org/10.1016/B978-0-12-809252-1.00008-0

161

contain many constituents that do not have comprehensive safety data about the impact on human health and the environment.

2. RELATIONSHIP BETWEEN MUTAGENS AND THEIR ABILITY TO CAUSE CANCER

Accurate identification of potentially toxic chemicals denotes a major problem. A chemical is mainly proved to be a mutagen through detailed battery of tests, which can be carried out through in vitro, in vivo, and in silico methods as shown in Fig. 8.1. The proper study of toxicological endpoints becomes necessary as it has highest concern on human health and the environmental toxicology. Mutagenicity and carcinogenicity are among the crucial endpoints of toxicology, which is known to have a strong interrelation. Although different in exact meaning, the ability of mutagens to cause cancer is evident because most mutations result in the abnormal cellular growth [8].

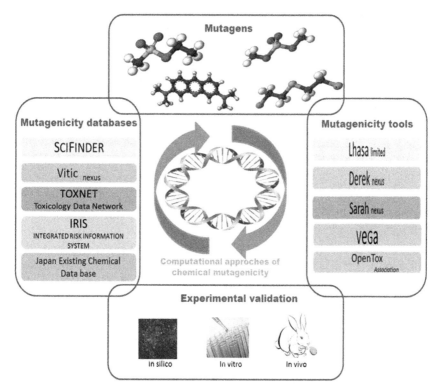

Figure 8.1 A compilation of in silico, in vitro, and in vivo methodology for mutagenicity prediction.

Miller and Miller's electrophilic theory of chemical carcinogen [9,10] states that most of the chemical carcinogens are either electrophilic or are converted into one where they further react with nucleophilic groups such as nucleic acids and proteins. This theory leads to the fact that electrophilic nature is a common point between mutagens and carcinogens [7]. Experimental evidence was provided by Heinrich Malling which showed that S 30 liver extract can cause metabolic transformation of a nonmutagenic chemical into mutagen [5]. Bruce Ames used liver extract in a bacterial assay to identify potential mutagens that may potentially induce cancer [11].

3. CRUCIAL ASPECTS OF COMPUTATIONAL PREDICTIVE MODELING

A comprehensive understanding of toxicity properties relevant to a chemical compound is crucial for its safe applicability and development of different products. Various in vitro/in vivo assays have been developed to evaluate mutagenicity assays, such as the Ames test for bacterial mutagenicity, chromosome aberration assays using human lymphocytes or other mammalian cells in culture and in vivo cytogenetics studies [1,2,12]. Some in vitro assays such as Ames test (a reverse mutation assay for bacterial strains), HGPRT gene mutation assay or TK gene mutation assays (mammalian assays) are routinely used for chemical toxicity due to their reproducibility, cost-effectiveness, and structure–activity relationships (SAR) [13]. In vivo assays on transgenic animal models also help in identification of gene mutation. Although such assays are carried out within a short span of time, their results do not evidently show applicability to evaluate large number of existing chemicals. International Conference on Harmonisation (ICH) M7 guideline has provided a framework that is applicable to the identification of mutagenic impurities, their categorization, qualification, and control to limit the risk of potential carcinogens [14]. One of the M7 guideline focuses on identification of mutagens using in silico predictions instead of in vitro studies [15]. So to assess large number of chemicals and also to reduce the experimental dependence and time, in silico approach is decisive as quantitative structure–activity relationship (QSAR) models have long been used to predict mutagenicity of chemical compounds [16]. Genotoxicity of chemical is detected earlier supported by SAR can generally guide subsequent chemical utilization prior to synthesis to avoid genotoxicity. It is also necessary to characterize the genotoxic potential of metabolites, degradants, impurities, and process intermediates involved in the synthetic reactions pathway of chemical synthesis [17]. Leading pharma industries widely use computational program to aid prediction of genotoxicity utilizing a combination of in vitro screening and in silico predictions [18]. The correct prediction depends on two main factors, i.e., sensitivity and specificity. Sensitivity is the ability of a program to correctly identify true positive mutagens, i.e., the chemicals that have the potential to cause mutations, and specificity is the ability to identify true negative mutagens, i.e., safe chemicals that do not have any mutagenic potential [19]. The available in silico tools for mutagenicity prediction add value to the process of identification of mutagens.

4. IN SILICO TOOLS AND TECHNIQUES FOR CHEMICAL MUTAGENIC PREDICTION

This section presents a number of computational approaches to understand and predict chemically induced mutagenicity and toxicology.

4.1 Virtual Screening

In the process of drug discovery and development, large combinatorial libraries of chemical compounds are searched to identify/analyze suitable molecule for a potential target [20]. Similarly, in silico screening technique can be applied for the evaluation of toxicity of a chemical, and number of chemicals can be filtered by evaluating large libraries of compounds. Virtual screening is a cheminformatics technique applied using two categories, i.e., ligand-based and structural-based screening. The approaches of virtual screening also utilize target and ligand information to predict ADMET (absorption, distribution, metabolism, excretion, and toxicity) properties, drug-likeness estimates, and other molecular property data [21]. This technique is used to screen chemicals for undesirable properties including mutagenic ability.

4.2 Quantitative Structure–Activity Relationship

The QSAR is an important tool of bio/chemoinformatics, which can be constructed primarily based on the data generated from the molecular modeling and computational chemistry to predict genotoxicity and mutagenicity of chemicals [22]. With the help of QSAR a correlation is set between the biological activity of a molecule, which is obtained by experimentation and its various molecular properties [23–25]. In mathematical form, QSAR can be expressed as activity $= f\,(physiochemical\ and/or\ structural\ properties)$.

In general, QSAR framework attempts to find a mathematical relationship between chemical structure and biological activity or chemical property including toxicity for a series of compounds. These series of compounds are called the training set. The generated mathematical equation can be used to predict the activity or property of any new compound, which has been built from the chosen training set. Here, the emphasis is on the current state-of-the-art QSAR methodology and its potential utility in chemical mutagenicity prediction. The fundamental approach for developing, validating, and using QSAR models for toxicity prediction are moderately consistent. QSAR is a technique, which tries to predict the activity and properties of an unknown set of molecules based on analysis of an equation connecting the structures of molecules to their respective measured activity and property. Generally, at first, the variation in property and/or toxicity of a known set of structurally analogue samples is studied with the changes in their molecular frameworks. The trends obtained from the study are then transformed into the form of model equations, and, further they are applied to determine the property and/or toxicity of chemical compounds that are not tested experimentally or new structurally similar set of systems. Because mathematical models are utilized to develop equations, the quantification of different properties of a set of molecules is also readily carried out. Thus, the QSAR-based models become useful toward predicting toxicity of chemicals. They act as a novel technique to build models correlating structure, activity, and toxicity. QSAR-based studies have shown their applications in many fields such as ecotoxicology, drug discovery, antitumor, biotoxicity, chemical–biological interactions, drug metabolism, anticancer drugs, pesticide toxicity, and in many other fields [26–29].

4.3 Molecular Docking

Molecular docking is a high-throughput tool that facilitate to model ligand binding sites and conformations within a target compound of interest. Docking studies help in identifying the orientation, mode of interaction as well as the conformations of a ligand in complex. Docking methods are applied for determining [1] the orientation of a ligand binding and [2] the affinity of the particle interaction mode [30]. This is considered as a scoring function, which comprises mathematical models that approximate the noncovalent binding energy of a ligand pose within the binding cavity and sampling of multiple ligand binding orientation (various conformations to obtain the most favorable mode and site of interactions). Sampling techniques such as genetic algorithms, Monte Carlo techniques, and matching algorithms are most frequently used in docking. Molecular docking offers a fast, computationally efficient method to screen receptor–ligand interactions. Mostly the docking studies report protein–ligand interaction. DNA is an important target for large number of chemicals and drugs. Holt et al. [31] reported that Surflex and AutoDock can precisely duplicate the crystal structure of bound ligand to DNA. Interaction of DNA with various chemicals and its stability can be predicted and can be applied for studying mutagenicity. Mostly, docking is used at the initial stages followed by detailed molecular dynamics (MD) simulation.

4.4 Molecular Dynamics Simulation

Molecular interaction is the basis of biological function. Considering the dynamics of biological molecules, it is important to investigate the motions of biomolecule (host) and biomolecule–ligand (host-guest) complexes in addition to their static structures. Elucidations of the atomic and molecular motions of biomolecules are investigated using computational methods via MD simulations. The foundation of MD simulations was initiated in the mid-1970s [32], and currently the field has advanced significantly due to sophisticated algorithm development, advancement in computational resource architecture and power. Atomistic MD simulations are based on the principles of Newtonian mechanics to simulate the motions of individual atoms over a simulated time period (trajectory). MD simulations provide a mechanism for chemical induced motion and conformational dynamics of macromolecular simulated system over time. Simulations are carried out to understand the structures and microscopic interaction between molecules in the complexes. A system can be represented at various levels including solvent effects and mimicking the environment in which the interaction/ complex formation is taking place. Once the system is built, with the help of derived equation, forces acting on each and every molecule are obtained to probe the host–guest (of various chemicals) interaction phenomena. Commonly used and most popular MD simulation codes are AMBER [33], CHARMM [34], GROMACS [35], NAMD [36], etc.

5. FRAMEWORK OF COMPUTATIONAL APPROACH FOR UNDERSTANDING MUTAGENICITY

The accurate prediction of mutagenicity is crucial because mutation generally occurs from direct covalent–noncovalent interaction of

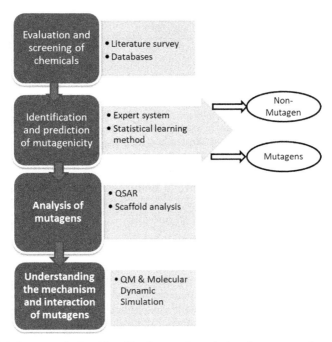

Figure 8.2 Framework for identification and analysis of mutagens/mutagenicity prediction.

chemicals with DNA. A general framework is highlighted in this section to identify and predict chemical induced mutagenicity using computational approaches. Fig. 8.2 is showing the workflow of the steps involved in probing mutagenicity.

5.1 Evaluation and Screening of Chemicals

The initial step in the identification of a mutagen is to analyze the structure and chemical composition. In case of a drug/active pharmaceutical ingredient (API), all the compounds involved in the synthesis of drugs are assessed based on the structure using high-throughput manner utilizing various cheminformatics approaches. When working with the in silico prediction of mutagenicity, it is always significant to check the availability of experimental mutagenicity data for the compound of interest. Scopus is one example of a commercial literature search engine that searches a molecule based on the CAS number or the name of the compound. The list of available databases is given in Table 8.1.

Table 8.1 Selected list of common mutagenicity databases[a]

Resources	Accessibility
BG-Chemie	http://www.bgchemie.de
MutAIT	http://www.mutait.org
1. TransgenicDB	
2. MGMD [8]	
EPA IRIS database	http://www.epa.gov/iris
INCHEM	http://www.inchem.org
Japan Existing Chemical Database	http://dra4.nihs.go.jp/mhlw_data/jsp/
Leadscope Model Applier	http://www.leadscope.com/
Scifinder	http://scifinder.cas.org
SciQSAR	http://www.scimatics.com/jsp/qsar/ key_features.jsp
Scopus	http://www.scopus.com
Commercially available databases	
TOXNET	http://toxnet.nlm.nih.gov
VITIC	https://viticnexus.lhasalimited.org/Vitic1-0/
Leadscope Toxicity Databases	http://www.leadscope.com/toxicity_databases/
BG-Chemie	http://www.bgchemie.de

[a]As on July 2017.

5.2 Identification and Prediction of Mutagenicity

There is a wide variety of in silico approaches to identify and predict chemically induced mutagenicity. For example, the following two in silico methods can be performed in coordination.

5.2.1 Expert System

A computational system which can mimic the ability of decision making of a human expert is known as Expert system [37]. Such systems solve complex problems through "if–then rules" rather than by conventional procedures. It contains all the approaches to supplement the SAR analysis performed by in silico tools. The expert system reviews already existing information such as mechanistic approach, physiochemical properties, reliability of specific structural alerts (SA), analysis of the applicability domain of a particular SA, evaluation of published (Ames) test data (e.g., lack of reproducibility, dose check, etc.), biophores/QSAR alerts designed to stimulate sensitivity, etc. The outcome of expert rule–based system depends on human written rules and other concerns such as data are analyzed manually and prediction is qualitative. Derek nexus is one such software that generates prediction of mutagenicity by applying expert system. It provides an inexpensive and quick way to identify potential mutagens [38].

5.2.2 Statistical Learning Methods

When prediction is done using statistical data and field analysis, such methods are statistical learning methods. Statistical learning mainly depends on two factors: prediction and inference (response). It is a basic framework for machine learning. Model developed using these methods relates the response to the predictors. Machine learning methods are categorized as supervised or unsupervised learning. For statistical learning method, the best method to be considered is supervised learning, which involves training data set–based learning approach. This method is based on structural and physiochemical properties of a molecule that is classified into two categories: (1) mutagens and (2) nonmutagens (Fig. 8.2). This approach for predicting mutagenicity has no restrictions on the type of the molecule or the features of its structure. Various types of algorithms (artificial neural network (ANN), k-nearest neighbor (kNN), support vector machine (SVM), naiv̈e Bayes (NB), etc.) are used for the prediction of mutagenicity of a compound to obtain reliable results. To recognize a mutagenic compound, the model is trained using specific characteristics such as electrophilicity of a compound. Once the model is curated, test set can be used to obtain mutagenic predictions. Main features of statistical learning methods are construction of fully automated hierarchical model and quantitative prediction. Sarah nexus is a unique machine learning tool, which uses this statistical based methodology for mutagenicity prediction [39]. Application of this tool to build Ames mutagenicity is demonstrated [40]. Sarah nexus is utilized to assess mutagenic impurities in regulation with ICH M7 guideline [15].

5.3 Analysis of Mutagens

Analysis of the chemical structure is a crucial step to identify that specific part of compound, which is potentially responsible for causing mutagenicity. QSAR is one such technique, which can be applied on various levels of predicting mutagenicity. Even in the analysis of a chemical, it plays a major role using various descriptors. Various softwares are currently available for mutagenicity prediction and analysis of mutagens. A list of available resources is given in Table 8.2. Leadscope Model Applier and MultiCASE (MC4PC) use QSAR-based approach for analyzing mutagenicity of a chemical [41]. Some of the descriptors used for the analysis of mutagens are discussed in this section.

Table 8.2 List of available software resources[a] used for predicting mutagenicity

Available resources	Description	Accessibility
Derek Nexus v 2.0	Expert rule–based and statistical based predictions for mutagenicity based on in vitro endpoints	www.lhasalimited.org
TEST v 4.0.1/Ames test	Toxicity Estimation Software Tool (TEST) evaluates the toxicity including mutagenicity using QSAR methods	www.epa.gov/chemical-research/toxicity-estimation-software-tool-test
TOPKAT (Accelrys discovery studio v 3.1)	TOPKAT predicts various ranges of toxicological endpoints such as mutagenicity, developmental toxicity, rodent carcinogenicity, rat chronic lowest observed adverse effect level (LOAEL), etc.	www.omictools.com/toxicity-prediction-by-komputer-assisted-technology-tool
Toxtree v 2.5.0	Toxtree estimates various toxic hazards by the structural rules	www.toxtree.sourceforge.net/index.html
VEGA Caesar v 2.1.10	Statistically based models, i.e., CEASAR which predicts five endpoints: developmental toxicity, skin sensitization, mutagenicity (Ames), carcinogenicity, and the bioconcentration factor	www.caesar-project.eu/
VEGA SARpy v 1.0.5–Beta	VEGA SARpy uses rule–based approach to predict both mutagenicity and nonmutagenicity	www.vegahub.eu
Mutagenicity (Ames test) model (ISS)—prediction	Predicts mutagenicity	www.vegahub.eu
Lazar	Using structural fragment analysis, Lazar makes prediction of toxicological endpoints such as mutagenicity, human liver toxicity, rodent and hamster carcinogenicity, etc.	www.lazar.in-silico.ch/predict
HazardExpert	HazardExpert predicts toxicity based on bioavailability parameters and toxic fragments covering mutagenicity, teratogenicity, carcinogenicity, membrane irritation, immunotoxicity, and neurotoxicity endpoints	www.compudrug.com/hazardexpertpro
MultiCASE/MC4PC	Based on hierarchical statistical, MC4PC analyzes compares active (mutagenic) and inactive (nonmutagenic) molecules and identifies biophores	www.multicase.com/case-ultra-models

Name	Description	URL
Leadscope model applier (LSMA)	Pretrained models of Leadscope model applier (LSMA) predicts genetic toxicity	www.leadscope.com/model_appliers/
ChemSilico	ChemSilico gives online calculation of various biological and physiochemical parameters for mutagenicity prediction	http://www.labhoo.com/labhoosite.asp?SID=68327280
SciQSAR (formerly MDL-QSAR) software	Using E-state descriptors and nonparametric discriminant, SciQSAR predicts Ames mutagenicity	www.pharmaceuticalonline.com/doc/sciqsar-2d-0001
Sarah nexus	By using a unique machine-learning approach, a statistical model is built for Ames mutagenicity	www.lhasalimited.org
Sarah and Derek Nexus	The combination provides toxicological evaluation from intuitive interface	www.lhasalimited.org
AMBIT	Based on various precalibrated toxicity models, generates a toxicity report	www.ambit.sourceforge.net/
OpenTox	Predicts mutagenicity and provides a transparent reasoning behind each prediction	www.opentox.net/
eTOX	Predicts mutagenicity and carcinogenicity of human relevance	www.etoxproject.eu/
ToxBoxes	Based on machine learning, mutagenicity prediction is done using fragment-based Advanced Algorithm Builder (AAB)	www.acdlabs.com/
MDL QSAR	On the basis of new properties, new compound library is generated, which predicts toxicity endpoints	www.mdl-qsar1.software.informer.com/
Scaffold Hunter	Open source tool for scaffold analysis of chemicals	www.scaffoldhunter.sourceforge.net/
Bioclipse	Predicts mutagenicity, having modules for structural alerts, similarity searches, and QSARs	www.bioclipse.net/decision-support
PreADMET	Calculates descriptors and neuronal network for toxicity prediction system	www.preadmet.bmdrc.kr

[a]As on July 2017.

5.3.1 Molecular Descriptors and Fingerprints

Each molecule contains the chemical information, which cannot be extracted by experiments. Molecular descriptors are developed, which help in establishing a quantitative relationship of a compound between the biological activity and its structural and experimental properties. Molecular descriptors can be defined based on its nature such as topological, constitutional, geometrical, quantum–chemical, electronic, etc. and their dimensions [24,42,43]. Fingerprints are the binary bits strings form of a molecular structure. Molecular fingerprints can be used as the descriptors for developing QSAR models. It is also utilized in classification and similarity searching.

5.3.2 Substructures or Structural Alerts

SAs are the substructures of a molecule, which are linked with an adverse consequences or one of the toxicological endpoints. The SAs for mutagenicity are the molecular functional or substructural groups that are linked to the mutagenic activity of the chemical compounds. Because DNA is modified by the actions of a mutagen, such groups are also valid for carcinogenicity. SAs used with QSAR models are proved to be a potential way for predicting various endpoints of toxicity including mutagenicity. DEREK and Toxtree are the softwares that utilize SAs for the prediction of mutagenicity [44]. The ToxCast (Toxicity Forecaster) project uses quantitative high-throughput screening to predict chemical toxicity of several biological pathways that follow national toxicology program guidelines [45].

5.3.3 Scaffold Analysis

Scaffold represents the core of a compound or a compound without its functional group. The prediction of mutagenicity involving structural alert has a drawback that it neither deals nor evaluates the core of a chemical. Most of these approaches mainly emphasize on the functional groups or side chain analysis, whereas scaffolds or cores are neglected. In case, if the mutagenicity is caused by the specific scaffold, then scaffold analysis is necessary. One such tool is scaffold hunters that extracts important compound scaffold from a chemical by removing all side chains except linking or exocyclic double bonds [46]. Rings are removed from the parent compound to get the child scaffolds. Scaffold trees are constructed and mutagenicity values are given to each scaffold tree [47].

5.4 Understanding Mechanism and Interaction of Mutagens

Systematic study of chemical mutagens can provide essential mechanistic understanding of the process that trigger the formation of DNA damage and mutations. When mutagens react with DNA, they covalently/noncovalently bound to it to form DNA adducts. Computational techniques provide a way to study such interaction using multiscale approaches. The most common among them is quantum mechanics (QM) methods and MD simulations. QM and molecular mechanics approaches were used to study three-dimensional DNA model (B-DNA (Dickerson model)) for the identification of molecules forming preferred noncovalent DNA interaction through major/minor groove-binding, DNA intercalation, or both [48–50]. These investigations evaluate the ability of a chemical/drug to bind between two adjacent DNA base pairs and calculate the binding energy based on electrostatic, hydrogen bonding, and Van der Waals interactions. These works provide insights on predicting genotoxicity of noncovalent DNA binding molecules. MD simulations help in studying the modification of DNA via chemicals. MD simulations are used to elucidate the conformational changes and fluctuations of nucleic acid (DNA, RNA) and proteins [51]. It is reported from an MD simulation study that alkylating adducts cause methylation of deoxyguanine and pyridyloxobutyl (POB)-DNA [52]. These modifications are linked to lung cancer. Computational techniques are now mostly used to scrutinize the structure and thermodynamics of biological molecules and their complexes.

These are few selected emerging techniques and tools that are currently available and applied successfully for genotoxicity and mutagenicity prediction. Several computational online tools, online QSAR models, various toxicology databases, metabolite prediction, and ADME prediction are accessible both commercially and through open sources. Each predictive method has their own advantages and disadvantages in analysis of chemicals, data quality, rules and guidelines used for model development, reliability and correlations, accessibility, and predictive power. Development of more accurate and integrated computational approaches will provide high sensitivity and specificity of chemical mutagenicity prediction and will minimize their genotoxic risks.

6. SUMMARY

In summary, this chapter provides the information necessary to establish the practical use of in silico toxicology approaches for the safety assessment of chemicals, APIs, and drug impurities. It is out of question that all the available in silico approaches add value to identify and predict chemically induced mutagenicity and genotoxicity prediction. The potential of such techniques can be understood by the fact that some of the in silico techniques are now being proposed by ICH guidelines for various chemical screening strategy. In silico predictive tools provide time-sensitive results and not only just label a chemical as mutagen or nonmutagen but also highlights which part of the molecule potentially inducing mutagenesis and/or directly comes in contact with DNA. The initial phase of prediction for safety evaluation is shifting from the experimental testing toward in silico knowledge and information-based tools. However, this transition completely depends on the reliability and success of the applied in silico tools. Despite the advantages provided by computational approaches, extensive calibration and improvement is required to extract precise information about various toxicological endpoints to replace biological assays in the near future.

ACKNOWLEDGMENTS

Funding from the Council of Scientific and Industrial Research (CSIR), New Delhi, and CSIR-Indian Institute of Toxicology Research, Lucknow, is gratefully acknowledged.

REFERENCES

[1] Guideline IHT. Guidance on specific aspects of regulatory genotoxicity tests for pharmaceuticals S2A.
[2] Committee IS. Genotoxicity: a standard battery for genotoxicity testing of pharmaceuticals. ICH harmonised tripartite guideline. 1997. p. 1–6.
[3] Greene N. Computer systems for the prediction of toxicity: an update. Adv Drug Deliv Rev 2002;54(3):417–31.
[4] Cunny H, Hodgson E. Toxicity testing. A textbook of modern toxicology. 2004. p. 353.
[5] Zeiger E. History and rationale of genetic toxicity testing: an impersonal, and sometimes personal, view. Environ Mol Mutagen 2004;44(5):363–71.
[6] Benigni R, Battistelli CL, Bossa C, Colafranceschi M, Tcheremenskaia O. Mutagenicity, carcinogenicity, and other end points. Comput Toxicol 2013;II:67–98.
[7] Benigni R, Bossa C. Mechanisms of chemical carcinogenicity and mutagenicity: a review with implications for predictive toxicology. Chem Rev 2011;111(4):2507–36.
[8] Tennant RW. Mutagens and carcinogens. McGraw-Hill Education; 2014. [updated 2017/7/12]. AccessScience. Available from: https://www.accessscience.com:443/content/mutagens-and-carcinogens/441100.
[9] Miller EC, Miller JA. Mechanisms of chemical carcinogenesis. Cancer 1981;47(S5):1055–64.
[10] Miller EC, Miller JA. Searches for ultimate chemical carcinogens and their reactions with cellular macromolecules. Cancer 1981;47(10):2327–45.
[11] Ames BN, Durston WE, Yamasaki E, Lee FD. Carcinogens are mutagens: a simple test system combining liver homogenates for activation and bacteria for detection. Proc Natl Acad Sci 1973;70(8):2281–5.
[12] Snyder RD, Smith MD. Computational prediction of genotoxicity: room for improvement. Drug Discov Today 2005;10(16):1119–24.
[13] Claxton LD, de Umbuzeiro GA, DeMarini DM. The Salmonella mutagenicity assay: the stethoscope of genetic toxicology for the 21st century. Environ Health Perspect 2010;118(11):1515.
[14] Assessment and control of DNA reactive (mutagenic) impurities in pharmaceuticals to limit potential carcinogenic risk. In: Guideline IHT, editor. International Conference on Harmonization of Technical Requirements for Registration of pharmaceuticals for Human Use (ICH), Geneva. 2014.

[15] Barber C, Amberg A, Custer L, Dobo KL, Glowienke S, Van Gompel J, et al. Establishing best practise in the application of expert review of mutagenicity under ICH M7. Regul Toxicol Pharmacol 2015;73(1):367–77.

[16] Modi S, Li J, Malcomber S, Moore C, Scott A, White A, et al. Integrated in silico approaches for the prediction of Ames test mutagenicity. J Comput Aided Mol Design 2012;26(9):1017–33.

[17] Singh S, Handa T, Narayanam M, Sahu A, Junwal M, Shah RP. A critical review on the use of modern sophisticated hyphenated tools in the characterization of impurities and degradation products. J Pharm Biomed Anal 2012;69:148–73.

[18] Muster W, Breidenbach A, Fischer H, Kirchner S, Müller L, Pähler A. Computational toxicology in drug development. Drug Discov Today 2008;13(7):303–10.

[19] Lapenna S, Gatnik MF, Worth AP. Review of QSAR models and software tools for predicting acute and chronic systemic toxicity. Luxembourg: Publications Office of the European Union; 2010.

[20] Lionta E, Spyrou G, Vassilatis DK, Cournia Z. Structure-based virtual screening for drug discovery: principles, applications and recent advances. Curr Top Med Chem 2014;14(16):1923–38.

[21] Ekins S, Nikolsky Y, Nikolskaya T. Techniques: application of systems biology to absorption, distribution, metabolism, excretion and toxicity. Trends Pharmacol Sci 2005;26(4):202–9.

[22] Parthasarathi R, Elango M, Padmanabhan J, Subramanian V, Roy D, Sarkar U, et al. Application of quantum chemical descriptors in computational medicinal chemistry and chemoinformatics. 2006.

[23] Hansch C, Hoekman D, Gao H, Comparative QSAR. Toward a deeper understanding of chemicobiological interactions. Chem Rev 1996;96(3):1045–76.

[24] Hansch C, Hoekman D, Leo A, Weininger D, Selassie CD. Chem-bioinformatics: comparative QSAR at the interface between chemistry and biology. Chem Rev 2002;102(3):783–812.

[25] Hansch C, Maloney PP, Fujita T, Muir RM. Correlation of biological activity of phenoxyacetic acids with Hammett substituent constants and partition coefficients. Nature 1962;194(4824):178–80.

[26] Cronin MT, Jaworska JS, Walker JD, Comber MH, Watts CD, Worth AP. Use of QSARs in international decision-making frameworks to predict health effects of chemical substances. Environ Health Perspect 2003;111(10):1391.

[27] Cronin MTD. In silico tools for toxicity prediction. In: New horizons in predictive toxicology: current status and application. The Royal Society of Chemistry; 2012. p. 9–25.

[28] Knudsen TB, Keller DA, Sander M, Carney EW, Doerrer NG, Eaton DL, et al. FutureTox II: in vitro data and in silico models for predictive toxicology. Toxicol Sci 2015;143(2):256–67.

[29] Raies AB, Bajic VB. In silico toxicology: computational methods for the prediction of chemical toxicity. Wiley Interdisc Rev Computa Mol Sci 2016;6(2):147–72.

[30] Wijeyesakere SJ, Richardson RJ. In silico chemical–protein docking and molecular dynamics. Comput Sys Pharmacol Toxicol 2017:174–90.

[31] Holt PA, Chaires JB, Trent JO. Molecular docking of intercalators and groove-binders to nucleic acids using Autodock and Surflex. J Chem Inf Model 2008;48(8):1602–15.

[32] Warshel A, Levitt M. Theoretical studies of enzymic reactions: dielectric, electrostatic and steric stabilization of the carbonium ion in the reaction of lysozyme. J Mol Biol 1976;103(2):227–49.

[33] Pearlman DA, Case DA, Caldwell JW, Ross WS, Cheatham TE, DeBolt S, et al. AMBER, a package of computer programs for applying molecular mechanics, normal mode analysis, molecular dynamics and free energy calculations to simulate the structural and energetic properties of molecules. Comput Phys Commun 1995;91(1–3):1–41.

[34] Brooks BR, Brooks CL, MacKerell AD, Nilsson L, Petrella RJ, Roux B, et al. CHARMM: the biomolecular simulation program. J Comput Chem 2009;30(10):1545–614.

[35] Hess B, Kutzner C, Van Der Spoel D, Lindahl E. GROMACS 4: algorithms for highly efficient, load-balanced, and scalable molecular simulation. J Chem Theory Comput 2008;4(3):435–47.

[36] Phillips JC, Braun R, Wang W, Gumbart J, Tajkhorshid E, Villa E, et al. Scalable molecular dynamics with NAMD. J Comput Chem 2005;26(16):1781–802.

[37] Giarratano JC, Riley G. Expert systems. PWS publishing Co.; 1998.

[38] Segall MD, Barber C. Addressing toxicity risk when designing and selecting compounds in early drug discovery. Drug Discov Today 2014;19(5):688–93.

[39] Dearfield KL, Gollapudi BB, Bemis JC, Benz RD, Douglas GR, Elespuru RK, et al. Next generation testing strategy for assessment of genomic damage: a conceptual framework and considerations. Environ Mol Mutagen 2016;58:264–83.

[40] Greene N, Dobo KL, Kenyon MO, Cheung J, Munzner J, Sobol Z, et al. A practical application of two in silico systems for identification of potentially mutagenic impurities. Regul Toxicol Pharmacol 2015;72(2):335–49.

[41] Sutter A, Amberg A, Boyer S, Brigo A, Contrera JF, Custer LL, et al. Use of in silico systems and expert knowledge for structure-based assessment of potentially mutagenic impurities. Regul Toxicol Pharmacol 2013;67(1):39–52.

[42] Cherkasov A, Muratov EN, Fourches D, Varnek A, Baskin II, Cronin M, et al. QSAR Modeling: where have you been? Where are you going to? J Med Chem 2014;57(12):4977–5010.

[43] Randić M. Generalized molecular descriptors. J Math Chem 1991;7(1):155–68.

[44] Ridings J, Barratt M, Cary R, Earnshaw C, Eggington C, Ellis M, et al. Computer prediction of possible toxic action from chemical structure: an update on the DEREK system. Toxicology 1996;106(1):267–79.

[45] Kavlock R, Dix D. Computational toxicology as implemented by the US EPA: providing high throughput decision support tools for screening and assessing chemical exposure, hazard and risk. J Toxicol Environ Health B 2010;13(2–4):197–217.

[46] Wetzel S, Klein K, Renner S, Rauh D, Oprea TI, Mutzel P, et al. Interactive exploration of chemical space with Scaffold Hunter. Nat Chem Biol 2009;5(8):581–3.

[47] Hsu K-H, Su B-H, Tu Y-S, Lin OA, Tseng YJ. Mutagenicity in a molecule: identification of core structural features of mutagenicity using a scaffold analysis. PLoS One 2016;11(2):e0148900.

[48] Arockiasamy DL, Radhika S, Parthasarathi R, Nair BU. Synthesis and DNA-binding studies of two ruthenium (II) complexes of an intercalating ligand. Eur J Med Chem 2009;44(5):2044–51.

[49] Uma Maheswari P, Rajendiran V, Palaniandavar M, Parthasarathi R, Subramanian V. Enantiopreferential DNA Binding:[{(5, 6-dmp) 2Ru} 2 (μ-bpm)]$^{4+}$ induces a B-to-Z conformational change on DNA. Bull Chem Soc Japan 2005;78(5):835–44.

[50] Vijayalakshmi R, Kanthimathi M, Parthasarathi R, Nair BU. Interaction of chromium (III) complex of chiral binaphthyl tetradentate ligand with DNA. Bioorg Med Chem 2006;14(10):3300–6.

[51] Kuhn H, Demidov VV, Coull JM, Fiandaca MJ, Gildea BD, Frank-Kamenetskii MD. Hybridization of DNA and PNA molecular beacons to single-stranded and double-stranded DNA targets. J Am Chem Soc 2002;124(6):1097–103.

[52] Kotandeniya D, Murphy D, Yan S, Park S, Seneviratne U, Koopmeiners JS, et al. Kinetics of O 6-pyridyloxobutyl-2′-deoxyguanosine repair by human O 6-alkylguanine DNA alkyltransferase. Biochemistry 2013;52(23):4075–88.

CHAPTER NINE

Overview of Nonclinical Aspects for Investigational New Drugs Submission: Regulatory Perspectives

Ranjeet Prasad Dash[1], Manish Nivsarkar[2]

[1]Auburn University, Auburn, AL, United States; [2]B.V. Patel Pharmaceutical Education and Research Development (PERD) Centre, Ahmedabad, India

LIST OF ABBREVIATIONS

ADME Absorption distribution metabolism and excretion
ANVISA Agência nacional de vigilância sanitária, Brazil
BCRP Breast cancer resistance protein
BSEP Bile salt extrusion protein
$\mathbf{CYP_{450}}$ Cytochrome P_{450}
EMA European Medical Agency
HTS High-throughput screening
ICH International Council on Harmonization
IND Investigational new drug
MATE Multidrug and toxin extrusion protein
MRP Multidrug resistance protein
NCE New chemical entity
NDA New drug application
NMR Nuclear magnetic resonance
NOAEL No observed adverse effect level
P-gp Permeability glycoprotein
PMDA Pharmaceuticals and Medical Devices Agency, Japan
US FDA United States Food and Drug Administration

1. INTRODUCTION

Various diseases and clinical complications have been identified so far and are found to be affecting different target organs. To cater the needs of the patients, researchers all over the world are pouring their untiring efforts in developing new molecules to alleviate the pain and the associated symptoms as well as the associated complications [1]. Research studies are conducted

Mutagenicity: Assays and Applications
ISBN 978-0-12-809252-1
http://dx.doi.org/10.1016/B978-0-12-809252-1.00009-2

both in academia as well as pharmaceutical industries, which include understanding the disease pathology and identifying the suitable drug candidate [2]. In other words, the process can be described as the selection of a target and identifying the lead. The lead may be from the natural source or a synthetic candidate or a biological molecule. The lead candidate has to go through various drug development phases (Fig. 9.1) till it reaches the patients, which include both nonclinical and clinical developments (Fig. 9.2) [3]. The whole process of developing a drug is a costly affair that incurs approximately $1 billion and a time period of 10–15 years [4,5]. The drug approval process involves stringent verifications by the regulatory authorities such as US FDA, ANVISA, EMA, PMDA, etc. keeping in view the final implications on the patients' health. Fig. 9.3 shows the drug (new combinations and monotherapy) approval rate by US FDA since 2010 [6]. Apart from synthetic drugs,

Figure 9.1 Schematic representation of drug development phases. *IND*, investigational new drug; *NCE*, new chemical entity; *NDA*, new drug application.

Figure 9.2 Nonclinical and clinical aspects of drug development process. *HTS*, high-throughput screening; *NCE*, new chemical entity.

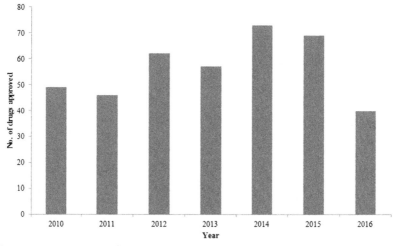

Figure 9.3 Drugs (monotherapy and combinations) approved by US FDA since 2010.

various phytopharmaceuticals as well as biologics have also been approved by the regulatory authorities for the management of various diseases [7]. The steps involved in drug discovery and development process include (1) target identification, (2) target validation, (3) hit series identification, (4) assay development, (5) hit to lead identification, (6) lead optimization, (7) investigational new drug (IND) submission, (8) clinical studies, and (9) new drug application (NDA) filing [1]. This review will give a highlight about the various steps involved in drug discovery and development with primary focus on the data required for IND submission. Further, we will also discuss about the regulatory requirements for IND submission according to the International Council on Harmonization (ICH) and US FDA guidelines. This article will be helpful for experimental design as required by the regulatory bodies.

2. TARGET IDENTIFICATION

The two primary reasons for a drug failure are either its compromised efficacy and/or safety, which have a direct connection with the target identification, and its validation. A target may be a protein, gene, and/or RNA [8]. Druggability is the primary criterion that needs to be met for a discovery compound and that must be basically safe, efficacious, and complies with the clinical and manufacturing requirements. In other words, a druggable target might be accessible to the drug to elicit the pharmacological response. Thus, proper target identification and its validation are essential to establish the dose–response relationship and have a mechanistic insight about the drug's pharmacological action [8].

3. TARGET VALIDATION

The next step after target identification is the target validation that is done using various in vitro and in vivo techniques with an objective to mimic the actual target in the patient [9]. The target validation approach is multidimensional that includes gene and protein expression studies, analysis of molecular signaling pathway, drug–receptor interactions, cell-based assays, safety and efficacy studies as well as pharmacokinetic studies [9].

4. HIT SERIES IDENTIFICATION

Subsequent to the target validation, hit series identification stage starts, which includes the screening of the compounds for various targets. Hit series identification involves multiple approaches such as high-throughput screening, focused screening, fragment screening, structural aided drug design, virtual screening, physiological screening, and nuclear magnetic resonance (NMR) screening [10].

5. ASSAY DEVELOPMENT

The assay development is a crucial stage and is customized based on the target to be worked out. This stage includes establishing the method for screening the compound with significant level of confidence. Assay development primarily involves the use of mammalian cell lines or primary cell culture models or developing specific biochemical assays [11]. The cell-based assays are preferred to monitor the drug–receptor interaction, whereas the biochemical assays give an insight about the drug–enzyme or drug–protein interaction [12,13]. Although, a variety of assay methods are available, the selection of the method depends on the biology of the drug target protein, the equipment infrastructure in the host laboratory, the experience of the scientists in that laboratory, whether an inhibitor or activator molecule is sought, and the scale of the compound [1].

6. HIT TO LEAD IDENTIFICATION

This stage involves refining the hit series and developing more potent compounds based on the structure–activity relationship with an objective to achieve optimum efficacy and bioavailability [14]. The structure of the compound is modified based on the observed effect using various molecular modeling techniques such as X-ray crystallography and NMR. This type of investigation will also often give rise to the discovery of new binding sites on the target protein [14].

7. LEAD OPTIMIZATION

The objective of this stage is to maintain the desirable properties of the lead compound and minimize the associated drawbacks [15]. This step involves investigation of the lead compound in various animal species and using cell-based assays as mentioned by the regulatory guidelines to declare the compound as a preclinical candidate [16,17]. Some of the crucial experiments that need to be performed at this stage include evaluation of the compound for genotoxicity using Ames test and in vivo models of general behavior such as the Irwin's test [18]. Experiments such as high-dose pharmacology, pharmacokinetic–pharmacodynamic studies, dose linearity and repeat dosing pharmacokinetic studies, effect of food and gender on pharmacokinetic profile, drug-induced metabolism, and metabolic profiling need to be carried out by the end of this stage. Furthermore, the chemical stability and salt selection has to be finalized in this stage [19]. This stage is followed by IND submission to the regulatory authorities and obtaining an approval for clinical studies [1].

8. INVESTIGATIONAL NEW DRUG SUBMISSION

The IND submission is the crucial stage where a sponsor compiles and submits all the chemical, manufacturing, control, stability, in vitro efficacy and preclinical safety, efficacy, and pharmacokinetic data to a regulatory authority to obtain the approval for clinical investigations [20]. The regulatory authority to which the data need to be submitted is decided by the sponsor based on the intent of future business and sale [17]. The primary objectives of IND are to ensure the safety and rights of subjects in all phases of investigation and to assure the quality of scientific evaluation in phase 2 and 3 [20,21]. The research studies that require IND include the following:

1. The research that involves a "drug-article intended for use in the diagnosis, cure, mitigation, treatment, or prevention of disease."
2. The research is a "clinical investigation experiment in which a drug is administered or dispensed to, or used involving, one or more human subjects."

In this section, we will discuss the requirements pertaining to nonclinical studies as per ICH and US FDA guidelines.

8.1 M4S(R2) Guidelines

The preclinical data to be submitted for IND need to be arranged in the specific format as mentioned in "The Common Technical Document for the Registration of Pharmaceuticals for Human Use: Safety – M4S(R2)." The data should be presented under two heads such as (1) nonclinical overview and (2) nonclinical written and tabulated summaries. The key points of the M4S(R2) technical guideline is discussed in the subsequent section. By the end of this section, the readers will have an insight about the nonclinical studies to be performed for IND submission [22].

1. Nonclinical overview: The nonclinical overview should be presented in the following sequence:
   a. Overview of the nonclinical testing strategy
   b. Pharmacology
   c. Pharmacokinetics
   d. Toxicology
   e. Integrated overview and conclusions
   f. List of literature references

   All the critical aspects such as pharmacodynamic effects, the mode of action, and potential side effects should be evaluated and concerns regarding any experiment need to be addressed [22]. A clear explanation of the analytical methods used and their application to pharmacokinetic, toxicokinetic, and metabolite(s) identification studies is needed to correlate the findings and the derived parameters. Certain value-added parameters such as impact of the disease states, changes in physiology, antiproduct antibodies, and cross-species consideration of toxicokinetic data, if evaluated, should be added to the nonclinical data [22]. Variability in data needs to be addressed with appropriate justifications. The data regarding multispecies comparison of metabolism and systemic exposure shall be used for allometric scaling to predict the human pharmacokinetic parameters [22]. However, the allometric scaling should be done using radiolabeled pharmacokinetic study in animals that will give a more accurate estimation of the human exposure [23]. This will also help in estimating the mass balance for the amount of drug administered and amount distributed, metabolized, and excreted [24]. However, animal to human scaling should consider (1) animal species, (2) numbers of animals, (3) routes of administration, (4) dose level, (5) duration of treatment or of the study, and (6) systemic exposures [22].

   The onset, severity, and duration of the toxic effects, their dose-dependency and degree of reversibility (or irreversibility), and species- or gender-related differences should be evaluated and important features discussed, particularly with regard to pharmacodynamics, toxic signs, causes of death, pathologic findings, genotoxic activity, carcinogenic potential, and reproductive toxicity [22].

2. Nonclinical written and tabulated summaries: This section contains the following:
   a. Introduction
   b. Written summary of pharmacology
   c. Tabulated summary of pharmacology

   d. Written summary of pharmacokinetics

   e. Tabulated summary of pharmacokinetics

   f. Written summary of toxicology

   g. Tabulated summary of toxicology

The data need to be generated and presented in individual sections, which are discussed in this section.

1. Introduction: This section shall include a briefing about the chemical class and its therapeutic application.

2. Summary of pharmacology: The primary and secondary pharmacodynamics, safety pharmacology, and drug interaction data should be provided in this section. It is required to correlate and compare the pharmacology data of the drug (selectivity, safety, potency) with that of other drugs belonging to the same class. Secondary pharmacodynamics data sometimes help in understanding and predicting the potential adverse effect(s) in humans and can be combined with the safety pharmacology data [22].

3. Summary of pharmacokinetics: This section includes method of analysis, absorption, distribution, metabolism, excretion, and pharmacokinetic interactions. The content of each subsection are discussed below:

   a. Absorption: Various studies pertaining to drug absorption includes $Caco_2$ permeability, single-dose absorption study in mice and rat, dose escalation study in mice and rat, effect of food and gender in mice and rats, absorption studies in beagle dogs and rhesus monkeys (if facility is available) [22].

   b. Distribution: This includes single-dose tissue distribution, in vitro plasma protein binding in animal and human, blood plasma partitioning assay, and placental transfer studies [22].

   c. Metabolism: The experiments conducted to understand the metabolism of the investigational drug primarily include in vitro studies such as metabolic stability in liver microsomes, cytochrome P450 reaction phenotyping in human recombinant enzymes, metabolic stability in hepatocytes, CYP inhibitions in human liver microsomes, time-dependent CYP inhibition in human liver microsomes, and identification of metabolite(s). Metabolite identification has to be done from in vivo studies by administering high dose of the investigation compound to identify the specific metabolite(s) as observed from in vitro studies [22].

   d. Excretion: The excretion study is done to monitor the elimination of the parent or its metabolite(s) in the urine, feces, and bile and predict the primary route of drug elimination [22].

    e. Drug interactions: This attempt is made to understand the CYP-mediated drug–drug interactions using human liver microsomes and CYP-specific substrates against CYP 1A2, 2C8, 2C9, 2C19, 2D6, and 3A4/5, primarily. The role of transporters such as P-gp, BCRP, BSEP, MRP, MATE needs to be evaluated, which will give an insight about the drug–drug interaction in the absorption and disposition process [22,25].

    f. Other pharmacokinetic studies: These include estimating the aqueous solubility, log P and log D, stability of the drug in simulated gastric and intestinal fluid [22].

4. Summary of toxicology: The toxicity data should include single-dose toxicity, repeat-dose toxicity, genotoxicity, carcinogenicity, reproductive and developmental toxicity, studies in juvenile animals, and local tolerance study [22].

    a. Single-dose toxicity: The data should be summarized in order of species and route of administration.

    b. Repeat dose toxicity: This section should describe the methodology followed and highlight the important findings such as nature and severity of target organ toxicity, dose (exposure)–response relationships, no observed adverse effect levels (NOAELs), etc. [22].

    c. Genotoxicity: The data generated using in vitro nonmammalian and mammalian cell system, in vivo mammalian system should be provided. These findings should also be supported with toxicokinetic study data [22].

    d. Carcinogenicity: The carcinogenicity studies should be supported with toxicokinetic data describing the basis of studies chosen and rationale for high-dose selection. Apart from that, the carcinogenicity study data should include the findings of long-term studies, short- or medium-term studies (in order by species) including range-finding studies that cannot appropriately be included under repeat-dose toxicity or pharmacokinetics [22].

    e. Reproductive toxicity: The reproductive toxicity study data should summarize the findings pertaining to fertility and early embryonic development, embryo–fetal development, prenatal and postnatal development, including maternal functions. Juvenile animal toxicity data can be included, if available [22].

All the pharmacodynamics, pharmacokinetics, and safety pharmacology data should be represented in the order of species, route, and duration of treatment [22]. Species should be ordered as follows: mouse, rat, hamster, other rodent, rabbit, dog, nonhuman primate, other nonrodent mammal, non-mammals. Routes of administration should be ordered as follows: the intended route for human use, oral, intravenous, intramuscular, intraperitoneal, subcutaneous, inhalation, topical, other routes [22].

8.2 Metabolites in Safety Testing for Investigational New Drug

One of the major concerns that need to be addressed in nonclinical drug development is the safety aspect of the metabolite(s). The US FDA guidance on safety testing of metabolites suggests the evaluation of the pharmacokinetic and toxicity profile of the metabolite if it is formed at greater than 10% of parent drug systemic exposure at steady state [26]. However, there are few debates regarding the experimental design. The first concern is regarding the safety evaluation of disproportionate metabolites (minor metabolites in animals that turned out to be major in human) [27]. But this could be explained by the fact that nonclinical studies in animals are often conducted at a higher dose level than human when normalized by weight; these major human metabolites(s) may still have been acceptably evaluated in the animals with dosing of the parent drug, and no further toxicology studies on the metabolites are warranted [28]. However, not all disproportionate metabolites are of concern, including most glutathione conjugates, most glucuronide metabolites, and metabolites with an additional hydroxyl group [29].

The ICH M3(R2) guidelines have laid down the following salient points for conducting the safety assessment of the metabolites:

- The exposure of the metabolite in human should be significantly greater. In other words, the animal exposure should be at least 50% of the exposure seen in humans at the marketed dose [27].
- Metabolite should be identified in one species subjected to different toxicity studies such as general toxicity, carcinogenicity study, and embryo–fetal development study [27].
- A better understanding can be achieved from the single-dose radiolabeled ADME study in humans. Nonclinical evaluation needs to be performed if the metabolite exceeds more than 10% of the total drug following multiple dose studies in humans [27].
- Metabolite exposure shall be done at maximum tolerable dose in animals and at the no-adverse effect level dose in humans [27].
- In case, any safety pharmacology concerns for the metabolites is observed during phase 1, which has not been observed in nonclinical studies, additional safety pharmacology studies of these human metabolites can be considered to better understand the mechanism [27].

- Standard testing approach shall be followed for prodrugs [27].
- The anticancer drug testing can avail the waiver for the metabolite in safety testing experiment [26].

8.3 Nonclinical Data Required for Exploratory Investigational New Drug Studies

"Exploratory IND studies" encompasses evaluation of the human exposure at lower dose level that is not likely to have therapeutic and diagnostic effect. These studies are designed to determine the mode of action of the drug in humans as established from the nonclinical studies (pertaining to pharmacokinetic and pharmacodynamics studies) [30]. The nonclinical data required for exploratory IND submission are categorized into pharmacokinetics and pharmacodynamics studies:

Pharmacokinetics studies: Exploratory IND pharmacokinetic studies are executed at microdose levels (1/100th of the dose of a test substance). The microdose studies can be conducted in a single mammalian species with justification from in vitro metabolism data and by comparative data on in vitro pharmacodynamic effects [30]. The route for administration needs to be the same as the intended clinical route. Animals in the study are monitored for 14 days where an interim necropsy is done on day 2 to monitor the vital signs and pathological conditions. Allometric scaling needs to be done to predict the human dose based on the pharmacokinetic/pharmacodynamics data [30].

Pharmacodynamic studies: The next aspect is to address the pharmacodynamic effects. This encompasses the execution of 2-week repeat dose toxicology study in a sensitive species followed by toxicokinetic study [30]. This experiment will help in predicting the safe dose to initiate the clinical trial. The study can be preferably done in rats ($N = 4$ per treatment) and supported with some dog studies. Gender effect has to be evaluated. This is followed by dose escalation study in rats to establish the NOAEL. The route of administration shall be the same in all the tested species as intended for clinical trials as well as the end-point parameters [30].

8.4 Specific Concerns With Investigational New Drug Submission

There are numerous aspects that need attention regarding submission of IND such as studies involving endogenous compounds, live organisms, cosmetics, and dietary supplement. This section will emphasize the need for IND studies with detailed explanations [30].

1. Endogenous compounds: IND requirements are applicable for the studies involving endogenous compounds intended for provocation and challenge such as histamine, bradykinin, angiotensin, etc., for evoking a physiological response [30].
2. Live organisms: IND is required for the studies involving the use of live organism (virus, bacteria, or fungi), which is to be administered to the animal/subject to study the disease pathology. Under this condition, the organism is considered as a biological product [30].
3. Cosmetics: An IND is required for cosmetics if the ingredient is being studied for use to affect the structure or function of the body or to prevent, treat, mitigate, cure, or diagnose a disease [30].
4. Dietary supplement: IND submission is also required for dietary supplement intended for use in the diagnosis, cure, mitigation, treatment, or prevention of disease [30].

9. CONCLUSIONS

Nonclinical investigation is of prime importance in drug development process. The regulatory authorities have laid some specifications and requirements pertaining to the conduct and evaluation of nonclinical studies. The specifications have been designed keeping in view the safety concern during clinical investigation. Furthermore, the significance of nonclinical studies is considered to be most important for the development of anticancer drugs as the final evaluation is done in patients. Thus the data reliability and accuracy of nonclinical data has to be double verified to ensure better safety and efficacy. Researchers are suggested to understand the regulatory requirements pertaining to the pharmacology, toxicology, and pharmacokinetic studies to ensure better data integrity and faster drug development process as any missed out aspect will require further investigation and delay the drug development process. Some of the crucial aspects that need to be addressed are the analytical method and safety testing of metabolite(s). The method should be rugged enough to support the study. Frequent change in the analytical method should be avoided. The major concern with the safety testing of metabolite(s) is the availability of pure metabolite standards that need to be addressed. Thus, all these aspects need a follow-up discussion on industry practice using innovative approaches to evaluate the safety and efficacy of the drugs in nonclinical setup. Individual investigators shall seek the advice of regulatory personnel prior to the study initiation to scheme out an appropriate experimental design.

REFERENCES

[1] Hughes JP, Rees S, Kalindjian SB, Philpott KL. Principles of early drug discovery. Br J Pharmacol 2011;162(6):1239–49.

[2] Ohlmeyer M, Zhou MM. Integration of small-molecule discovery in academic biomedical research. Mt Sinai J Med 2010;77(4):350–7.

[3] Booth B, Zemmel R. Prospects for productivity. Nature Rev Drug Discov 2004;3:451–6.

[4] DiMasi JA, Hansen RW, Grabowski HG. The price of innovation: new estimates of drug development costs. J Health Econ 2003;22:151–85.

[5] Adams C, Brantner VV. Estimating the cost of new drug development: is it really $802 million? Health Aff 2006;2:420–8.

[6] https://www.centerwatch.com/drug-information/fda-approved-drugs/year/2010.

[7] Dash RP, Nivsarkar M. Pharmacokinetics of phytopharmaceuticals: a peek into contingencies and impediments in herbal drug development. In: Tsay HS, Shyur LF, Agrawal DC, Wu YC, Wang SW, editors. Medicinal plants-recent advances in research and development. Singapore: Springer; 2016. p. 297–308.

[8] Chan JNY, Nislow C, Emili A. Recent advances and method development for drug target identification. Trends Pharmacol Sci 2010;82–8.

[9] Wang S, Sim TB, Kim YS, Chang YT. Tools for target identification and validation. Curr Opin Chem Biol 2004;8(4):371–7.

[10] Keseru˝ GM, Makara GM. Hit discovery and hit-to-lead approaches. Drug Discov Today 2006;11(15–16):741–8.

[11] Dunne A, Jowett M, Rees S. Use of primary cells in high throughput screens. Meth Mol Biol 2009;565:239–57.

[12] Michelini E, Cevenini L, Mezzanotte L, Coppa A, Roda A. Cell based assays: fuelling drug discovery. Anal Biochem 2010;397:1–10.

[13] Moore K, Rees S. Cell-based versus isolated target screening: how lucky do you feel? J Biomol Scr 2001;6:69–74.

[14] Bleicher KH, Bohm HJ, Muller K, Alanine AI. Hit and lead generation: beyond high throughput screening. Nat Rev Drug Discov 2003;2(5):369–78.

[15] Lalonde RL, Kowalski KG, Hutmacher MM, Ewy W, Nichols DJ, Milligan PA, et al. Model-based drug development. Clin Pharmacol Ther 2007;82(1):21–32.

[16] DiMasi JA. The value of improving the productivity of the drug development process: faster times and better decisions. Pharmacoeconomics 2002;20(Suppl. 3):1–10.

[17] Earl J. What makes a good forecaster? Nat Rev Drug Discov January 2003;2(1):83.

[18] Hunt CA, Guzy S, Weiner DL. A forecasting approach to accelerate drug development. Stat Med 1998;17(15–16):1725–40.

[19] Andrade EL, Bento AF, Cavalli J, Oliveira SK, Freitas CS, Marcon, et al. Non-clinical studies required for new drug development – Part I: early in silico and in vitro studies, new target discovery and validation, proof of principles and robustness of animal studies. Braz J Med Biol Res 2016;49(11):e5644.

[20] Holbein ME. Understanding FDA regulatory requirements for investigational new drug applications for sponsor-investigators. J Investig Med 2009;57(6):688–94.

[21] Guidance for clinical investigators, sponsors, and IRBs. Rockville (MD, USA): U.S. Department of Health and Human Services, US FDA; 2013.

[22] The common technical document for the registration of pharmaceuticals for human use: safety – M4S(R2). December 2002.

[23] Sharma V, McNeill JH. To scale or not to scale: the principles of dose extrapolation. Br J Pharmacol 2009;157(6):907–21.

[24] Kalgutkar AS, Tugnait M, Zhu T, Kimoto E, Miao Z, Mascitti V, et al. Preclinical species and human disposition of PF-04971729, a selective inhibitor of the sodium-dependent glucose cotransporter 2 and clinical candidate for the treatment of type 2 diabetes mellitus. Drug Metab Dispos 2011;39(9):1609–19.

[25] Guidance for Industry. Drug interaction studies-study design, data analysis, implications for dosing, and labelling recommendations. Rockville, MD: U.S. Department of Health and Human Services, US FDA; 2012. USA (Draft Guidance).

[26] Guidance for industry safety testing of drug metabolites. Rockville (MD, USA): U.S. Department of Health and Human Services, US FDA; 2008.

[27] Gao H, Jacobs A, White RE, Booth BP, Obach RS. Meeting report: metabolites in safety testing (MIST) symposium - safety assessment of human metabolites: what's really necessary to ascertain exposure coverage in safety tests? AAPS J 2013;15(4):970–3.

[28] Baillie TA, Cayen MN, Fouda H, Gerson RJ, Green JD, Grossman SJ, et al. Drug metabolites in safety testing. Toxicol Appl Pharmacol 2002;182:188–96.

[29] Smith DA, Obach RS. Metabolites and safety: what are the concerns, and how should we address them? Chem Res Toxicol 2006;19:1570–9.

[30] Guidance for industry, investigators, and reviewers exploratory IND studies. Rockville (MD, USA): U.S. Department of Health and Human Services, US FDA; 2006.

CHAPTER TEN

Mutagenicity Testing: Regulatory Guidelines and Current Needs

Rajesh Sundar, Mukul R. Jain, Darshan Valani
Zydus Research Centre, Cadila Healthcare Limited, Ahmedabad, India

1. BACKGROUND

The aim of this chapter is to bring together researchers and students to understand the relationship between basic science and regulatory research. Accurately performed, validated, and sensitive battery of mutagenicity assays are basic requirements for regulatory submissions, although specific requirements may vary considerably between countries and for the type of product evaluated.

Mutagenicity refers to the induction of permanent transmissible changes in the amount or structure of the genetic material of cells or organisms. These changes may involve a single gene or gene segment, a block of genes or chromosomes. The genetic change is referred to as a mutation, and the agent causing the change is referred to as a mutagen.

Mutagenicity data are used worldwide in regulatory decision-making. There are two major types of regulatory decisions made by agencies such as the Environmental Protection Agency (EPA) and the Food and Drug Administration (FDA): (1) the approval and registration of pesticides, pharmaceuticals, medical devices, and medical use products and (2) the setting of standards for acceptable exposure levels in air, water, and food. Mutagenicity data are utilized for both of these regulatory decisions. The current default assumption for regulatory decisions is that chemicals that are shown to be genotoxic in standard tests are, in fact, capable of causing mutations in humans (in somatic and/or germ cells) and that they contribute to adverse health outcomes through a "genotoxic/mutagenic" mode of action (MOA). The new EPA Guidelines for Carcinogen Risk Assessment (Guidelines for Carcinogen Risk Assessment, US EPA, 2005, EPA Publication No. EPA/630/P-03/001F) emphasizes the use of MOA information in risk assessment and provides a framework to help identify a possible mutagenic and/or nonmutagenic MOA for potential adverse effects. An analysis of

Mutagenicity: Assays and Applications
ISBN 978-0-12-809252-1
http://dx.doi.org/10.1016/B978-0-12-809252-1.00010-9

the available mutagenicity data is now, more than ever, a key component to consider in the derivation of an MOA for characterizing observed adverse health outcomes such as cancer. A two-step strategy for evaluating mutagenicity data is followed for optimal use in regulatory decision-making. The strategy includes integration of all available information and provides, first, for a weight-of-evidence (WOE) analysis as to whether a chemical is a mutagen, and second, whether an adverse health outcome is mediated through a mutagenic MOA.

The purpose of mutagenicity testing is to identify substances that can cause genetic alterations in somatic and/or germ cells and use this information in regulatory decisions. Compared to most other types of toxicity, genetic alterations may result in effects that are manifested only after long periods of exposure. However, the deleterious effect could be caused by DNA damage that occurs in a single cell at low exposures. Rather than destroying that cell, the genetic alteration can result in a phenotype that not only persists, but can be amplified, as the cell divides, creating an expanding group of abnormal cells within a tissue or organ. Genetic alterations in somatic cells may cause cancer if they affect the function of specific genes (e.g., tumor suppressor genes, protooncogenes, and/or DNA damage response genes). Mutations in somatic and germ cells are also involved in a variety of other (noncancer) genetic diseases. Accumulation of DNA damage in somatic cells has been related to some degenerative conditions, such as accelerated aging, immune dysfunction, cardiovascular and neurodegenerative diseases [1–3]. In germ cells, DNA damage is associated with spontaneous abortions, infertility, malformation, or heritable damage in the offspring and/or subsequent generations resulting in genetic diseases, such as Down syndrome, hemophilia, and cystic fibrosis [4].

2. MUTAGENICITY ENDPOINTS

Two types of studies considered include the following:

1. Those measuring direct, irreversible damage to the DNA that is transmissible to the next cell generation (i.e., mutagenicity)
2. Those measuring early, potentially reversible effects to DNA or on mechanisms involved in the preservation of the integrity of the genome (genotoxicity).

Mutagenicity, a subset of genotoxicity, includes DNA damage, which may be reversed by DNA repair or other known cellular processes, therefore, may or may not result in permanent alterations in the structure or information content in a surviving cell or its progeny. Thus, tests include evaluate induced damage to DNA (but not direct evidence of mutation) through effects such as unscheduled DNA synthesis (UDS), DNA strand breaks (e.g., comet assay), and DNA adduct formation, e.g., primary DNA damage tests.

Mutagenicity results in events that alter the DNA and/or chromosomal number or structure that are irreversible, therefore, capable of being passed to subsequent cell generations if they are not lethal to the cell in which they occur. Mutations include the following: (1) changes in a single base pair involving partial, single, or multiple genes; or chromosomes; (2) breaks in chromosomes that result in the stable (transmissible) deletion, duplication, or rearrangement of chromosome segments; (3) a change (gain or loss) in chromosome number (e.g., aneuploidy) resulting in cells that do not have an exact multiple of the haploid number; and (4) mitotic recombination. Positive results in mutagenicity tests can be caused by test substances that do not act directly on DNA. Examples are aneuploidy caused by topoisomerase inhibitors or gene mutations caused by metabolic inhibition of nucleotide synthesis. An extensive literature is available on these mechanisms which uses follow-up testing as a part of risk assessment strategies that are beyond the scope of this chapter.

No-observed-adverse-effect-level (NOAEL) cannot be obtained from mutagenicity studies due to study design and different MOA. It is generally assumed that even a small dose of mutagenic chemicals may have potential adverse effects. In general, the advice given by risk assessors in Europe is to keep exposure to such compounds at the lowest possible level responding to the ALARA principle ("as low as reasonably achievable").

For several decades, the mutagenicity data have been applied for hazard characterization and generally not for risk assessment. As a result, genetic toxicology studies usually focus on a dichotomous (positive/negative) determination. Recently, however, there has been a significant amount of work in quantitative evaluation of mutagenicity endpoints [5,6]. Specifically, determination of an estimated point of departure (PoD) and other statistically derived descriptions of the dose (concentration)–response clearly demonstrated the existence of doses that are not discernable from the background, and the indicated that mutagenicity data can be viewed similarly as other toxicology data with no-effect-levels, NOAELs, and so on. Currently, the genotoxicity tests are unable to determine these standard toxicity parameters. However, modifications in the study design such as additional dose levels with closer spacing and analyzing more samples across dose levels and time points with application of relatively high-throughput technologies may provide adequate statistical power to generate these values. This is especially true for mutagenic effects caused by well-accepted threshold mechanisms, e.g., aneugenicity.

3. REGULATORY FRAMEWORK

Based on the type of product to be tested and country to be registered, mutagenicity guidelines may vary, and it is well covered in OECD, ICH, Schedule Y, ISO, and other regional guidelines. ICH guidance, such as S2(R1) and M7, are followed for drugs and related impurities across the United States and other European Union (EU) countries.

In Australia, based on production size, mutagenicity data for industrial chemicals are managed by the National Industrial Chemicals Notification and Assessment Scheme (NICNAS), Department of Health under the Industrial Chemicals (Notifications and Assessment) Act 1989. According to the NICNAS Handbook [7], mutagenicity data must be included as part of the standard notification application. In Canada, industrial chemicals are regulated under the Canadian Environmental Protection Act 1999 (CEPA, 1999). In China, under the "China REACH," companies are required to submit a notification to the Chemical Registration Center (CRC) of Ministry of Environmental Protection [8] for a new chemical, i.e., a chemical not listed on the Inventory of Existing Chemical Substances Produced or Imported. In the EU countries, industrial chemicals are regulated under the Registration, Evaluation, Authorization, and Restriction of Chemicals (REACH), which came into force on June 1, 2007, and is managed by the European Chemicals Agency (ECHA). In Japan, general industrial chemical products are regulated under Japanese Chemical Substances Control Law (CSCL), which was initially implemented in 1974. In the US, industrial chemicals are regulated under the Toxic Substances Control Act (TSCA) of 1976 and the US EPA is responsible for their implementation. TSCA enables EPA to handle chemical reporting, testing requirements, and limit the use of chemical substances and/or mixtures [9]. However, in India, currently, there is no specific legislation pertaining to registration of general chemical substances, preparation of a national inventory, restriction of hazardous substances, or detailed classification and labeling criteria.

The Committee on Mutagenicity of Chemicals in Food, and Consumer Products, and the Environment (COM) provides advice to the UK

Government on the mutagenic hazard of chemicals. The COM has recently published a strategy for genotoxicity testing of chemicals starting from the position where no genotoxicity data are available, such as in the development of new chemical agents [10].

Fig. 10.1 shows an overview strategy for testing chemical substances for genotoxicity, whereas Fig. 10.2 shows strategy for assessment and testing of chemicals with inadequate genotoxicity data.

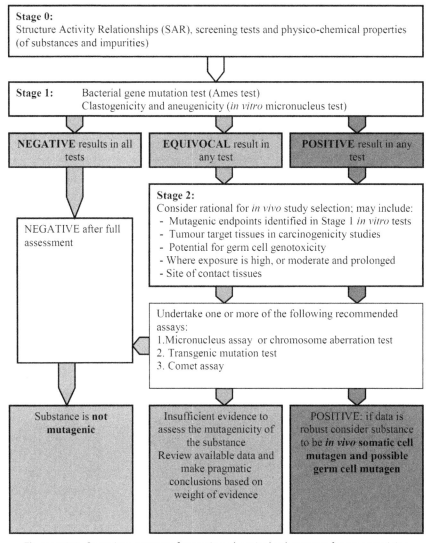

Figure 10.1 Overview strategy for testing chemical substances for genotoxicity.

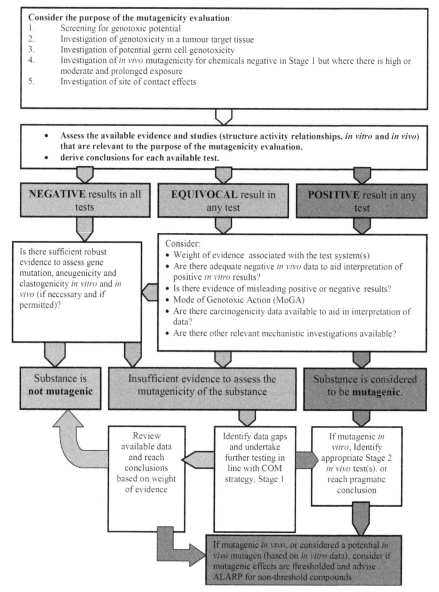

Figure 10.2 Strategy for assessment and testing of chemicals with inadequate geno-toxicity data.

4. REGULATORY STUDIES FOR MUTAGENICITY TESTING

Tests with internationally accepted guidelines (e.g., OECD Guidelines for the Testing of Chemicals and Regulation (EC) No 440/2008) are preferred where possible. However, not all test methods have an internationally accepted test guideline that specifically characterizes the testing protocol. When a method without an OECD or EU test guideline is used, thoroughly evaluated protocols should be followed. In many cases, subject matter experts such as International Workshop on Genotoxicity Testing (IWGT) have been formed to drive scientific method and test guideline development [9].

Both in vitro and in vivo mutagenicity studies are available for germ cells and somatic cells. Usually, in vitro studies are employed first. The Ames test, an in vitro gene mutation study in bacteria, is the basic test for mutagenicity. Positive result within in vitro tests, is followed by in vivo mutagenicity studies for further confirmation.

For an adequate evaluation of the mutagenic potential, usually three endpoints are assessed:
- Gene mutation,
- Structural chromosome aberrations, and
- Numerical chromosome aberrations.

The commonly used in vitro and in vivo studies are summarized as follows:

4.1 In Vitro Studies
- Bacterial reverse mutation test—Ames test
- Mammalian chromosome aberration test
- Mammalian cell gene mutation test
- UDS in mammalian cells

4.2 In Vivo Studies
- Rodent dominant lethal mutation test
- Mouse heritable translocation assay
- Mouse-specific locus test
- Sister chromatid exchange (SCE) analysis in spermatogonia
- UDS test in testicular cells
- Transgenic rodent somatic and germ cell gene mutation assay
- Mammalian erythrocyte micronucleus test

- Mammalian bone marrow chromosome aberration test
- Liver UDS
- Transgenic rodent somatic and germ cell gene mutation assays
- Mouse spot test
- Mammalian bone marrow SCE

5. REGULATORY REQUIREMENTS FOR REGISTRATION

5.1 Pesticides

Approximately 1.8 billion people engage in agriculture worldwide and most use pesticides to protect the food and commercial products produced by them. Others use pesticides occupationally for public health programs, and in commercial applications, whereas many others use pesticides for lawn and garden applications in their home. Pesticides are defined as "chemical substances used to prevent, destroy, repel, or mitigate any pest ranging from insects (i.e., insecticides), rodents (i.e., rodenticides), and weeds (herbicides) to microorganisms (i.e., algicides, fungicides, or bactericides)."

In the US, before a pesticide can be marketed or sold, the Federal Insecticide, Fungicide and Rodenticide Act (FIFRA) requires the EPA to ensure that the pesticide would not harm health or the environment with reasonable certainty, even when you used according to label instructions. The EPA, in turn, assesses carcinogenic risk of the pesticides by relying primarily on animal genotoxicity studies and/or short-term mutagenicity assays, as genotoxicity and mutagenicity play a major role in carcinogenesis. However, sometimes a pesticide that is determined to be noncarcinogenic based on genotoxicity/mutagenicity assays (thus labeled by the US EPA as "Not Likely to Be Carcinogenic in Humans" or " Evidence of Non-Carcinogenicity for Humans"), has been found to show cancer incidences in epidemiologic studies. Some nonmutagenic human carcinogens passes through the existing screening procedures, although other mechanisms of screening are being explored, through epigenetic cellular control measures within the cell including DNA methylation, covalent posttranslational modifications of historic proteins, and RNA-mediated gene silencing.

For technical pesticides, in addition to Ames test, a battery of in vitro and in vivo tests with specific endpoints, namely, mutation, chromosomal aberration (chromatids can be seen separately), and DNA are to be submitted as per EPA specifications. For pesticide formulations, mutagenicity data are not generally required. All studies should follow the OECD guidelines. Pesticides with known mutagenic potential to germ cells are categorized according to Globally Harmonized System (GHS) classification.

GHS hazard class germ cell mutagenicity is primarily concerned with chemicals that may cause mutations in germ cells of humans and can be transmitted to the progeny. However, in vivo genotoxicity to somatic cells are also considered.

GHS classification criteria for germ cell mutagenicity are summarized in Table 10.1.

5.2 Pharmaceuticals

ICH guidance for pharmaceuticals on genotoxicity testing intended for human use S2(R1) intends to optimizes the standard battery for prediction of potential risks, and provides guidance on interpretation of results, with the ultimate goal of improving risk characterization for carcinogenic effects that have their basis in changes in the genetic material. ICH describes standards for follow-up testing and interpretation of positive results in vitro and in vivo in the standard genetic toxicology battery. The focus of this guidance

Table 10.1 Globally Harmonized System classification criteria for germ cell mutagenicity

Category	Criteria
Category 1A	Chemicals known to induce or regarded as if they induce heritable mutations in human germ cells. Known to induce heritable mutations—positive evidence from human epidemiological studies. Mixtures containing ≥0.1% of such a category 1A mutagen.
Category 1B	Chemicals known to induce or regarded as if they induce heritable mutations in human germ cells. Regard as if they induce heritable mutations—positive results from in vivo heritable germ cell or somatic cell mammalian mutagenicity tests, or positive results showing mutagenic effects in the germ cells of humans without demonstration of transmission to progeny. Mixtures containing ≥0.1% of such a category 1B mutagen.
Category 2	Chemicals that may induce heritable mutations in human germ cells. Positive evidence obtained from in vivo somatic cell mutagenicity or somatic cell genotoxicity tests in mammals and in some cases with support from in vitro experiments. Mixtures containing ≥1% of such a category 2 mutagen.

is testing of new "small molecule" drug substances and does not apply to biologics or vaccines.

The following two options are considered equally suitable for mutagenicity testing of drugs.

5.2.1 Option 1

- A test for gene mutation in bacteria.
- A cytogenetic test for chromosomal damage (the in vitro metaphase chromosome aberration test or in vitro micronucleus test), or an in vitro mouse lymphoma *Tk* gene mutation assay.
- An in vivo test, generally a test for chromosomal damage using rodent hematopoietic cells, either for micronuclei or for chromosomal aberrations in metaphase cells.

5.2.2 Option 2

- A test for gene mutation in bacteria.
- An in vivo assessment with two different tissues, usually an assay for micronuclei using rodent hematopoietic cells and a second in vivo assay. Typically this would be a DNA strand breakage assay in liver, unless otherwise justified.

In cases where compounds are toxic to bacteria (e.g., antibiotics), the bacterial reverse mutation (Ames) test should still be carried out, just as cytotoxic compounds are tested in mammalian cells because mutagenicity can occur at lower, less toxic concentrations. In such cases, any one of the in vitro mammalian cell assays should also be done, i.e., Option 1 should be followed.

5.2.3 Quantitative Structure–Activity Relationships Modeling of Pharmaceuticals

The safety assessment process of drugs requires information on their mutagenic potential. The experimental determination of mutagenicity of a large number of chemicals is tedious and time- and cost-intensive, thus compelling for alternative methods. There are few local and global quantitative structure–activity relationships (QSAR) models available for discriminating low and high mutagenic compounds and predicting their mutagenic activity in a quantitative manner in *Salmonella typhimurium* (TA) bacterial strains. Relevant structural features of diverse chemicals that are responsible and influence the mutagenic activity have been identified. The applicability domains of the developed models have been defined. The developed models can be used as tools for screening new drugs or drug impurities for their mutagenicity assessment for regulatory purpose.

Structurally alerting compounds are usually detectable in the standard test battery because the majority of "structural alerts" are defined in relation to bacterial mutagenicity. A few chemical classes are known to be more easily detected in mammalian cell chromosome damage assays than bacterial mutation assays. Thus a negative result in either test battery with a compound that has a structural alert is usually considered sufficient assurance of the lack of genotoxicity. However, for compounds bearing certain specific structural alerts, modification to standard protocols can be appropriate. The choice of additional test(s) or protocol modification(s) depends on the chemical nature, the known reactivity, and any metabolism data of the structurally alerting compound in question. QSAR methods usually followed are mentioned in Table 10.2.

Table 10.2 Quantitative structure–activity relationships methods

Method	Sensitivity identification of mutagens or rodent carcinogens	Specificity identification of nonmutagens or rodent noncarcinogens	Comments/references
DEREK	No data reported	No data reported	Agreement with Ames positive 65% (416 compounds) [11]
TOPKAT	No data reported	No data reported	Agreement with Ames positive 73% (416 compounds) [11]
MDL QSAR	81%	76%	3338 compounds tested in bacterial mutagenicity tests [12]
MultiCASE (MC4PC)	71% (bacterial)	88% (bacterial)	1485 compounds, bacterial
	63% (mouse lymphoma)	74% (mouse lymphoma)	328 compounds for mouse lymphoma
	44% (clastogenicity in vitro)	92% (clastogenicity in vitro)	556 compounds for clastogenicity [13]
	53% (clastogenicity)	75% (clastogenicity)	679 compounds [14]
Toxtree (version 1.50)	74% (rodent carcinogenicity)	64% (rodent carcinogenicity)	878 chemicals with carcinogenicity data
	85% (bacterial mutagenicity)	72% (bacterial mutagenicity)	698 chemicals with mutagenicity data [15]

5.2.4 DNA Reactive Drug Impurities

The toxicological assessment of DNA reactive impurities is important in the regulatory framework for pharmaceuticals. ICH M7 guidance [16] mentions the assessment and control of DNA reactive (mutagenic) impurities in pharmaceuticals to limit potential carcinogenic risk. A Threshold of Toxicological Concern (TTC) concept was developed to define an acceptable intake for any unstudied chemical that poses a negligible risk of carcinogenicity or other toxic effects. These risk levels represent a small theoretical increase in risk when compared to overall human lifetime incidence of developing any type of cancer, which is greater than 1 in 3. It is noted that established cancer risk assessments are based on lifetime exposures. Less-than-lifetime exposures both during development and marketing can have higher acceptable intake of impurities and still maintain comparable risk levels. The use of a numerical cancer risk value (1 in 100,000) and its translation into risk-based doses (TTC) is a highly hypothetical concept that should not be regarded as a realistic indication of the actual risk. Nevertheless, the TTC concept provides an estimate of safe exposures for any mutagenic compound. In addition, in cases where a mutagenic compound is a noncarcinogen in a rodent bioassay, no predicted increase in cancer risk could be expected. Based on all the above considerations, any exposure to an impurity that is later identified as a mutagen is not necessarily associated with an increased cancer risk for patients already exposed to the impurity. There may be cases when an impurity is also a metabolite of the drug substance. In such cases, the risk assessment that addresses mutagenicity of the metabolite can qualify the impurity. Earlier, some structural groups have been identified to be of such high potency that intakes even below the TTC would theoretically be associated with a potential for a significant carcinogenic risk. This group consists of high potency mutagenic carcinogens, includes aflatoxin-like-, N-nitroso-, and alkyl-azoxy compounds.

Examples of clinical use scenarios with different treatment durations for applying acceptable intakes scenario include the following:

Treatment duration <1 month (acceptable intake—120 µg/day), e.g., drugs used in emergency procedures.

Treatment duration >1–12 months (acceptable intake—20 µg/day), e.g., antiinfective therapy with maximum up to 12 months treatment (HCV), parenteral nutrients, prophylactic flu drugs (~5 months), peptic ulcer (these are acute use but with long half-lives).

Treatment duration of >1–10 years (acceptable intake—10 μg/day), e.g., stage of disease with short life expectancy (severe Alzheimer), nongenotoxic anticancer treatment being used in a patient population with long-term survival (breast cancer, chronic myelogenous leukemia), drugs specifically labeled for less than 10 years of use, drugs administered intermittently to treat acute recurring symptoms (chronic herpes, gout attacks, substance dependence such as smoking cessation), and macular degeneration.

Treatment duration >10 years to lifetime (acceptable intake—1.5 μg/day), e.g., chronic use indications with high likelihood for lifetime use across broader age range (hypertension, dyslipidemia, asthma, hormone therapy (e.g., growth hormone, thyroid hormone, parathyroid hormone), lipodystrophy, schizophrenia, depression, psoriasis, atopic dermatitis, COPD, cystic fibrosis, and seasonal and perennial allergic rhinitis.

Application of promising computational methods (e.g., QSAR, Structure–Activity Relationships (SARs) and/or expert systems) for the evaluation of genotoxicity is needed, especially when very limited information on impurities is available. To gain an overview of how computational methods are used internationally in the regulatory assessment of pharmaceutical impurities, the current regulatory documents were reviewed. The software recommended in the guidelines (e.g., MCASE, MC4PC, DEREK for Windows) or used practically by various regulatory agencies (e.g., US FDA, US and Danish Environmental Protection Agencies), as well as other existing programs were analyzed. Both statistic based and knowledge-based (expert system) tools were analyzed. The overall conclusions on the available in silico tools for genotoxicity and carcinogenicity prediction are quite optimistic, and the regulatory application of QSAR methods is constantly growing. For regulatory purposes, it is recommended that predictions of genotoxicity/carcinogenicity should be based on a battery of models, combining high-sensitivity models (low rate of false negatives) with high-specificity ones (low rate of false positives) and in vitro assays in an integrated manner.

5.2.5 Biotechnology-Derived Products

ICH S6 (R1) guidance is intended primarily for products derived from characterized cells through the use of a variety of expression systems including bacteria, yeast, insect, plant, and mammalian cells. The intended indications may include in vivo diagnostic, therapeutic, or prophylactic uses. The active substances include proteins and peptides, and their derivatives and products. They could be derived from cell cultures or produced using recombinant DNA technology including production by transgenic plants and animals. Examples include but are not limited to cytokines, plasminogen activators, recombinant plasma factors, growth factors, fusion proteins, enzymes, receptors, hormones, monoclonal antibodies, recombinant DNA protein vaccines, chemically synthesized peptides, plasma-derived products, endogenous proteins extracted from human tissue, and oligonucleotide drugs. Active substance does not include antibiotics, allergenic extracts, heparin, vitamins, cellular blood components, conventional bacterial or viral vaccines, DNA vaccines, or cellular and gene therapies.

The range and type of mutagenicity studies routinely conducted for pharmaceuticals are not applicable to biotechnology-derived pharmaceuticals and therefore are not needed. Moreover, the administration of large quantities of peptides/proteins may yield uninterpretable results. These substances are not expected to interact directly with DNA or other chromosomal material due to its biological nature and higher molecular weight. Studies in available and relevant systems, including newly developed systems, should be performed in those cases where there is cause for concern about the product (because of the presence of an organic linker molecule in a conjugated protein product). The use of standard genotoxicity studies for assessing the genotoxic potential of process contaminants is not considered appropriate.

5.2.6 Concerns of Anticancer, Photogenotoxic, and Liposomal Drugs

According to ICH S9 [17], genotoxicity studies are not considered essential to support clinical trials for therapeutics intended to treat patients with advanced cancer. Genotoxicity studies should be performed to support marketing. The principles outlined in ICH S6 should be followed for biopharmaceuticals that are used to treat cancer. If the in vitro assays are positive, an in vivo assay might not be warranted.

Photochemical activation by nonionizing radiation (in the range 290–700 nm) may result in photochemical reactions, which can include photogenotoxicity. When exposure is sufficient to radiation and prolonged, it may result in photocarcinogenicity. A number of in vitro photogenotoxicity tests have been developed based mainly on the existing "dark" versions of these assays. Positive results in in vitro tests may trigger in vivo testing such as photocomet and photomicronucleus in animal models dermally exposed to nonionizing radiation and treated with the test chemical (either dermally or orally).

The concordance between positive phototoxicity and photogenotoxicity is 57%. Chemicals known to be strong photogenotoxins were predicted to be positive in photoreactivity assays. A number of chemicals were negative for photogenotoxicity despite being positive for photochemical reactivity and in vitro phototoxicity. There are no in vivo photogenotoxicity assays currently recommended by regulatory agencies. However, some developmental work on in vivo skin and eye photocomet and photomicronucleus assays in both rats and mice are available. Photogenotoxicity testing need not be undertaken routinely as part of a photosafety assessment and would only be required when the molar extinction coefficient was >1000 L/mol cm.

Drugs with liposomal encapsulation mask genotoxicity of a chemotherapeutic agent in regulatory toxicology assessments. For example, the mutagenicity of the free drug, doxorubicin hydrochloride with its nano-encapsulated form, doxorubicin encapsulated liposome expressed a significantly lower mutant frequency in the Ames assay and was nongenotoxic in the in vitro micronucleus assay. Further investigation of the systems' cytotoxicity and their interaction with the bacterial cell envelope suggests that the modification of the test parameters and release of the encapsulated drug prior to the Ames test have comparable mutagenic potential to a free drug.

5.3 Medical Devices

Although measures of a medical product's biocompatibility have largely been reported in terms of irritation, sensitization, and systemic toxicity, there is growing concern that devices, their components, or material extracts may also exert genotoxic effects. Thus, any attempt to assess the safety of a device intended for intimate body contact or permanent implantation would be incomplete without testing for the presence of toxins that exert an effect on the genetic material of cells. In its set of harmonized standards for the biological evaluation of medical devices, the International Organization for Standardization (ISO) has outlined the need for such genotoxicity testing in ISO 10993-3: "Tests for Genotoxicity, Carcinogenicity, and Reproductive Toxicity."

Considerations include accommodation of the evaluation of fluid extracts, as most biomaterials are insoluble. The United States Pharmacopeia (USP) has established standard preparation methods for materials testing that can be used for genotoxicity testing, and ISO 10993-12 "Sample Preparation and Reference Materials" describes standard methods for the preparation of extracts of device materials. Selection of the appropriate extraction vehicle may vary with the test system of choice. For example, bacterial test systems are frequently exposed to 0.9% sodium chloride solution along with ethanol or dimethyl sulfoxide extract. Because in vitro mammalian test systems require media that can support cell growth, the nutrient medium used for culture is often employed as the extraction vehicle. In vivo test models frequently employ the standard aqueous and nonaqueous USP extraction fluids that are capable of extracting both water-soluble (polar) and lipid-soluble (nonpolar) chemicals.

Classical in vitro and in vivo tests can be used to evaluate the genotoxicity of medical device materials. In all cases, adverse or equivocal findings warrant further investigations. Confirmation testing by dose–response relationship is the standard course of action. In addition, a presumptive positive finding in an in vitro assay can be confirmed by conducting an alternative in vivo model.

Acceptable results from a battery of genotoxicity tests will not only go a long way toward ensuring the safety of a proposed biomaterial; such data can, in some cases, justify not pursuing in vivo carcinogenicity studies, particularly if there is existing information about the lack of genotoxicity of the material in question.

5.4 Food Additives Ingredients

The FDA recommends the use of a battery of short-term mutagenicity tests for all when the cumulative estimated dietary intake exceeds 1.5 μg per person per day, which corresponds to 0.5 parts per billion (ppb) in the total diet. The recommended tests directly measure gene mutations and/or chromosomal effects. The Agency uses such data, in the absence of long-term animal feeding studies, to determine whether or not a chemical should be considered to be a possible carcinogen. Such data may also indicate whether a chemical may have adverse heritable effects. When long-term animal feeding studies are available for the evaluation of carcinogenicity, genetic toxicity data may assist in the interpretation of the results of such studies.

The European Food Safety Authority (EFSA) focuses on genotoxicity testing of food additives and other food ingredients to minimize the health risk for consumers through the primary prevention of the exposure to genotoxic substances.

The genetic toxicity test endpoints recommended for food ingredients whose cumulative estimated daily intake exceeds 50 ppb in the diet (150 μg per person per day) are gene mutation, structural and numerical (aneuploidy) chromosomal alterations.

The EFSA emphasizes the consideration of all three endpoints for an adequate genotoxicity assessment of chemical substances, as all of them have detrimental health effects. No single test system can detect all three endpoints, therefore a battery of complementary tests is necessary.

A stepwise approach is recommended for the generation and evaluation of data on genotoxic potential, beginning with a basic battery of in vitro tests, comprising a bacterial reverse mutation assay, and an in vitro micronucleus assay. Tests should be considered based on whether specific features of the test substance might require substitution of one or more of the recommended in vitro tests by other in vitro or in vivo tests in the basic battery. In the event of negative in vitro results, it can be concluded that the substance has no genotoxic potential. In case of inconclusive, contradictory or equivocal results, it may be appropriate to conduct further in vitro tests. In case of positive in vitro results, review of the available relevant data on the test substance and, where necessary, an appropriate in vivo study to assess whether the genotoxic potential observed in vitro is also expressed in vivo is recommended. Suitable in vivo tests are the mammalian erythrocyte micronucleus test, transgenic rodent assay, and comet assay. The approach to in vivo testing should be stepwise. If the first in vivo test is positive, no further testing is necessary, and the substance should be considered as an in vivo genotoxin. If the test is negative, it may be possible to conclude that the substance is not an in vivo genotoxin. However, in some cases, a second in vivo test may be necessary (e.g., if the first test is negative but more than one endpoint in the in vitro tests are positive, an in vivo test on a second endpoint may be necessary). The combination of assessing different endpoints in different tissues in the same animal in vivo should also be considered.

5.5 Industrial Chemicals

Mutagenicity is an important toxicological endpoint that requires thorough evaluation during the industrial chemical registration process. Regulatory requirements for mutagenicity assessment in registration of industrial chemicals vary in geographic regions (and in some cases by intended application). The mutagenicity testing requirements for registration of industrial chemicals from representative geographic regions (in alphabetical order), i.e., Australia, Brazil, Canada, China, EU, India, Japan, South Korea, Taiwan, and US are reviewed. We further discuss the challenges faced by the industry to meet global regulations, e.g., different testing requirements among geographic regions, different strategies in follow-up tests to in vitro positive findings, NOAEL in genetic toxicity testing, and human relevance of mutagenicity.

Industrial chemicals are regulated under REACH, which came into force on June 1, 2007, and is managed by the ECHA. Mutagenicity data requirements are tailored to the tonnage band (i.e., production volume).

Industrial chemicals are regulated under the TSCA of 1976, and the US EPA is responsible for implementation. Specifically, the TSCA enables the EPA to administer chemical reporting, testing requirements, and may limit the use of chemical substances and/or mixtures. Within Section 5 of TSCA, the New Chemicals program facilitates a chemical usage in commerce by evaluating potential risk to human health and environment. Currently, in India there is no specific legislation pertaining to registration of general chemical substances, preparation of national inventory, restriction of hazardous substances, or detailed classification and labeling criteria.

Mutagenicity data requirements for registration of industrial chemicals vary considerably based on production/usage volumes, intended application, and geographic region, which poses challenges to the industry during the registration process. For instance, for a substance intended to be market in the EU at 10–100 ton/year, if the Ames test, in vitro mammalian cell clastogenicity/aneugenicty test, and in vitro mammalian cell gene mutation test are all negative, further in vivo testing is generally not recommended by the REACH and as the substance is considered nongenotoxic.

However, this data package will not satisfy the requirements of China or Korea at the same tonnage level. In these countries, an in vivo test is mandatory these countries believe that the in vitro battery does not reveal any mutagenic potential. Furthermore, some regulatory agencies require permission or approval of a testing plan if animals are to be utilized; however, as mentioned above, animal-based testing may be required in other situations, hence there may not be a need for the approval of animal testing. Another notable geographic difference is the testing requirement for following positive in vitro findings. For instance, at lower tonnage levels in the EU REACH, one well-conducted relevant in vivo test is generally considered sufficient to clarify the in vivo mutagenicity of the test material. In China, however, to evaluate if the test chemical is an in vivo mutagen, two in vivo tests are required to follow up one positive in vitro test, even though this is not specified in the China guidance. Thus, it is evident that, a globally harmonized regulation is warranted, even though it is not likely to be established in the near future.

5.6 Cosmetics

In the EU, the Regulation (EC) No. 1223/2009 has prohibited animal testing for cosmetics from 2013. Three in vitro mutagenicity assays are recommended for the basic level testing of cosmetic substances: (1) the Ames test; (2) an in vitro mammalian cell gene mutation test; and (3) an in vitro mammalian cell micronucleus test or chromosomal aberration test. If all the three tests are negative, the substance is considered as a nonmutagen. If a positive result is seen in any of the tests, the substance is considered as a mutagen.

5.7 Nanomaterials

The mutagenic properties of nanomaterials are often closely linked to oxidative damage to DNA and proteins, caused by oxidative stress resulting from hyper production of reactive oxygen species (ROS) and reactive nitrogen species (RNS). Nanomaterial-induced oxidative stress is perhaps the most broadly developed and accepted mechanism for the potential toxic activity of NPs (nanoparticles). NP-mediated ROS and RNS production mechanisms can be classified into three groups: intrinsic production, production by interaction with cell targets, and production mediated by the inflammatory reaction. All these groups share responsibility for most of the primary (direct or indirect) or secondary genotoxic effects of the nanomaterials. Carbon nanomaterials, such carbon nanotubes, graphene, and fullerenes are some of the most promising nanomaterials. Although carbon nanomaterials have been reported to possess genotoxic potential, it is imperative to analyze the data on the genotoxicity of carbon nanomaterials in vivo and in vitro and check the validity and predictability of different assays [18].

Nanomaterials are currently regulated by the REACH in EU. Owing to the size and specific physical–chemical properties, manufactured nanomaterials may require different or additional testing beyond the standard tests used for other chemicals and the REACH guidance for nanomaterials is being revised. Some EU member states have issued their own regulations requiring companies to notify nanomaterials to their national products register or nano register. The OECD has released a series of publication on the safety of manufactured nanomaterials. Some consensus on mutagenicity testing has been reached in expert meetings, e.g., the Ames test is not a recommended test for the investigation of the genotoxicity of nanomaterials [19], although definitive recommendations are still in the process of getting framed.

6. PREDICTIVE MUTAGENICITY TESTING

Predictivity of in vitro genotoxicity studies in relation to rodent carcinogenicity studies are provided in Table 10.3.

7. LIMITATIONS AND CHALLENGES OF MUTAGENICITY ASSESSMENT

In addition to the recommended genotoxicity tests, the inclusion of an appropriate in vivo genotoxicity test following a positive result in the in vitro assay has been emphasized. Although, the micronucleus assay has been conventionally used [21,22,23], other tests such as those based

Table 10.3 Predictivity of in vitro genotoxicity studies in relation to rodent carcinogenicity studies

Method	Sensitivity[b]	Sensitivity[c]	Comments/references
Ames	59%	74%	541 chemicals [20]
Ames	52%	72%	3711 chemicals including tests with *Salmonella* and *Escherichia* [13]
Micronucleus (in vitro)	79%	31%	89 chemicals [20]
Micronucleus (in vitro)	88%	23%	182 chemicals [13]
Chromosomal aberrations (in vitro)	66%	45%	352 chemicals [20]
Chromosomal aberrations (in vitro)	55%	63%	1391 chemicals [13]
Mouse lymphoma assay	73%	39%	245 chemicals [20]
Mouse lymphoma assay	71%	44%	827 chemicals [13,20]
Ames + micronucleus[a] combined	94%	12%	372 chemicals, positive results in at least one test [20]
Ames + mouse lymphoma[a] combined	89%	32%	436 chemicals [20]
Ames + mouse lymphoma + chromosomal aberrations combined	84%	23%	202 chemicals [20]
Ames + mouse lymphoma + micronucleus[a] combined	91%	5%	54 chemicals [20]

[a]Positive results in at least one test.
[b]Accurate prediction of rodent carcinogenicity.
[c]Accurate prediction of rodent noncarcinogenicity.

on DNA adduct formation [24], in vivo chromosomal aberration [25], transgenic mutation [26,27], and comet assay [28–30] have also been routinely employed in the past as in vivo follow-up assay when equivocal or inconclusive results are obtained in the in vitro assay. Kawaguchi et al. [31] reported an identical sensitivity of the micronucleus test and the comet assay in detecting the studied mutagens however, they believe that the power of the comet assay to detect a low level of genotoxic potential can be superior to that of micronucleus test, if DNA resynthesis inhibitors are included in the comet assay. In another study a comparison between traditional micronucleus test and comet assay was done. The results revealed that transgenic assay detected about 50% of the 67 known carcinogens that were missed in the traditional micronucleus test, UDS about 20%, whereas the comet assay detected up to 90% suggesting the superiority of comet assay. The comet assay was also observed to have a high specificity in terms of producing negative results for noncarcinogen with an accuracy of up to 78% [32]. When compared to other genotoxicity assessment methods, the comet assay is flexible, requiring only a small amount of sample and cells [30]. It is more adaptable for the detection of various forms of DNA damage including double strand breaks (DSBs), cross DNA damage, and apoptosis [29,33]. Nevertheless, there is a need for the standardization of the methodologies of the comet assay to make it acceptable to all international regulatory agencies. Similarly, more studies are required to populate and validate comet assay data, especially those obtained with organs other than the liver [34,35].

7.1 Concurrent Negative and Positive Controls

In addition to establishing laboratory competence, negative and positive historical control data are important for assessing the acceptability of individual experiments, and the interpretation of test data. In particular, it is necessary to determine whether specific responses fall within or outside the distribution of the negative control. With the 3R principles in mind, the recommendations for positive controls differ for in vitro and among various in vivo tests.

Negative control groups are important for providing a contemporaneous control group for use in comparisons with the treated groups. This group can also be used to assess whether the experiment is of acceptable quality in comparison to a set of historical control groups.

Negative controls usually consist of solvent- or vehicle-treated (i.e., without test chemical) cells or animals. They should be incorporated into each in vitro and in vivo test and handled in the same way as the treatment groups. It should be noted that when choosing a solvent or vehicle, the decision should be based on obtaining maximum solubility of the test material without interacting with the test chemical and/or test system.

To reduce unnecessary animal usage for in vivo tests, if consistent inter-animal variability and frequencies of cells with genotoxicity are demonstrated by historical negative control data at each sampling time for the testing laboratory, only a single sampling for the negative control may be necessary. Where a single sampling is used for negative controls, it should be the first sampling time used in the study.

The inclusion of concurrent positive controls (reference controls/well-known genotoxic substances) is designed to demonstrate the effectiveness of a particular genetic toxicology test on the day it is performed. Each positive control should be used at a concentration or dose expected to reliably and reproducibly result in a detectable increase over background to demonstrate the ability of the test system to efficiently detect DNA damage, gene mutations, and/or chromosomal aberrations depending on the test, and in the case of in vitro tests, the effectiveness of the exogenous metabolic activation system. Therefore, positive control responses (of both direct-acting substances and substances requiring metabolic activation) should be observed at concentrations or doses that produce weak or moderate effects that will be detected when the test system is optimized, but not so dramatic that positive responses will be seen in suboptimal test systems, and immediately reveal the identity of the coded samples to the scorer (i.e., for tests using coded samples).

7.2 Historical Control Distribution and Control Charts

Historical control data (both negative and positive) should be compiled separately for each genetic toxicology test type; each species, strain, tissue, cell type, metabolic condition, treatment and sampling time, route of exposure; as well as for each solvent or vehicle within each laboratory. All control data of each individual genetic toxicology test, strain, etc. during a certain time period (e.g., five years) or from the last tests performed (e.g., the last 10 or 20 tests) should initially be accumulated to create the historical control data set. The laboratory should not only establish the historical negative (untreated, vehicle) and positive control range, but also define the distribution (e.g., Poisson distribution 95% control limits) as this information will be used for data interpretation. This set should be updated regularly. Any changes to the experimental protocol should be considered in terms of their impact on the resulting data remaining consistent with the laboratory's existing historical control database. Only major changes in experimental conditions should result in the establishment of a new historical control database where expert judgment determines that it differs from the previous distribution. Further recommendations on how to build and use the historical data (i.e., criteria for inclusion and exclusion of data in historical data and the acceptability criteria for a given experiment) can be found in the literature [36].

7.3 Data Interpretation and Criteria in Mutagenicity Testing

Extensive research has been carried out on how to interpret the findings, and several new concepts that are included in the revised/new tests. Prior to considering whether a particular experiment is positive or negative, it is important to ascertain whether that experiment is properly conducted. Therefore, the revised tests clarify the acceptance criteria for each assay. In addition, guidance was developed to provide recommendations for defining a biologically relevant positive result. Previous tests indicated that positive responses should be biologically relevant, but did not provide a means to determine biological relevance. The revised/new tests include three equal considerations when assessing whether a response is positive or negative. First, the test chemical response should be assessed as to whether there is a statistically significant increase from the concurrent negative controls. Second, the response should be concentration/dose related. Finally, a new concept, that utilizes the historical negative control distributions, is introduced to provide for assessing biological relevance. For the mouse lymphoma assay (MLA) (TG 490), the use of the global evaluation factor (GEF) to define the biological relevance of the response is introduced (see below). It should be noted, however, that tests not revised in the current round of revisions (TG 471, TG 485, and TG 486) are not affected by this new approach.

7.4 Individual Test Acceptability Criteria

The revised tests clarify recommendations for individual assay acceptability as follows:
- The concurrent negative control is considered acceptable for addition to the laboratory historical negative control database and/or is consistent with published norms (depending on the assay).
- Concurrent positive controls induce responses that are compatible with those generated in the laboratory's historical positive control database, and produce a statistically significant increase compared with the concurrent negative control.

- For in vitro assays, all experimental conditions (based on the recommended treatment times and including the absence and presence of metabolic activation) are tested unless one resulted in clear positive results.
- An adequate numbers of animals/cells are treated and carried through the experiment or scored (as appropriate for the individual test).
- An adequate number of doses/concentrations covering the appropriate dose/concentration range are analyzable.
- The criteria for the selection of top dose/concentration are consistent with those described in the individual TGs. Apart from the above criteria, MLA-specific acceptability criteria have been defined based on the IWGT MLA expert workgroup's [37–40] data evaluation for several negative control data parameters. Consistent with the general approach to establishing acceptability criteria for the revised genetic toxicology TGs, these recommendations are based on distributions of a very large number of experiments from laboratories proficient in the conduct of the MLA. There are also MLA-specific criteria for positive controls that assure good recovery of both small and large colony mutants. The specific recommendations (i.e., acceptable ranges for the main parameters) for the MLA are given in detail in TG 490.

7.5 Criteria for a Positive/Negative Result

If a genetic toxicity test is performed according to the specific test, and all acceptability criteria are fulfilled (as outlined above), the data can be evaluated as to whether the response is positive or negative. The tests required to recognize that chemicals determined to be positive should demonstrate biologically relevant increases which are concentration/dose related. As with the acceptability criteria, the assessment of biological relevance takes the distribution of the negative control data into consideration (e.g., Poisson 95% control limits).

For both in vitro and in vivo assays (with the exception of the MLA) a response is considered a clear positive in a specific test if it meets all the criteria in at least one experimental condition: (i) at least one of the data points exhibits a statistically significant increase compared to the concurrent negative control; (ii) the increase is concentration or dose related at least at one sampling time, when evaluated with an appropriate trend test; and (iii) the result is outside the distribution of the historical negative control data (e.g., Poisson-based 95% control limits). A test chemical is considered clearly negative if, none of the above criteria for a positive result are met in all experimental conditions examined.

7.6 Role of Statistics in Mutagenicity Assays

As a part of the genotoxicity assay revisions, an extensive evaluation has been undertaken to analyze how the selection of specific parameters impact the overall ability of the various tests to detect induced genetic damage. In particular, this analysis better defined an appropriate approach for using spontaneous background frequencies both for individual experiment acceptability and data interpretation, and for understanding the impact of assay-specific background frequencies on the statistical power of the assay. This analysis was used to develop the new recommendations for the number of cells to be treated for the in vitro gene mutation assays, and the number of cells to be scored for the cytogenetic tests (both in vitro and in vivo). A discussion of this analysis can be found in OECD documents [19].

Recommendations were included in the latest revisions to the genotoxicity assays to discourage overreliance on P-values associated with the statistical significance of differences found by pair-wise comparisons. Statistical significance based on a particular P-value is relevant, but is only one of the criteria to be used to decide whether to categorize a result as positive or negative. For example, the confidence intervals around the means for the controls and the treated cultures/animals should also be evaluated and compared within an individual experiment.

One of the goals for the genotoxicity assays revision was to include recommendations that would insure that test results deemed to be positive would be based on biologically relevant responses. Initially, it was proposed that in the revised OECD genetic toxicology guidelines, studies should be designed to detect a doubling (i.e., twofold increase) in the treated group responses over the negative control level. However, subsequent discussions focused on the fact that the sample sizes needed to detect a doubling will depend on the background level; e.g., a doubling from 1% to 2% is a smaller absolute change than one from 3% to 6%. Furthermore, it was recognized that defining the level of response required to achieve biological relevance, requires an appreciation of the nature of the endpoint, consideration of the background (negative control) incidence, and consideration of an absolute or relative difference versus negative control. These considerations are different for each of the assays and have been taken into account in the new recommendations found in the individual TGs.

The in vitro mutagenicity assays were evaluated, and in some cases revised, to increase the power of the various assays to detect biologically significant increases. For the in vitro gene mutation studies, where the cell is the experimental unit, power calculations showed that designs with relatively small numbers of cells per culture had low power to detect biologically relevant differences. For the cytogenetic tests, an acceptable level of statistical power (conventionally 80%) to detect two- to threefold changes would only be achievable if the number of cells scored were increased appreciably in some tests. For revisions to the recommendations for the in vitro cytogenetic tests, both the ideal number of scored cells and to the technical practicalities of actually scoring that number of cells, particularly for the chromosome aberration test were considered.

For the MLA, the IWGT MLA expert workgroup recommendation for determination of a biologically relevant positive result does not rely on statistically significant increases compared to the concurrent negative control, but on the use of a predefined induced mutant frequency (i.e., increase in MF above concurrent control) designated the GEF, which is based on the analysis of the distribution of the negative control MF data from participating laboratories [40]. For the agar version of the MLA, the GEF is 90×10^{-6}, and for the microwell version of the MLA, the GEF is 126×10^{-6}. Responses determined to be positive should also demonstrate a dose response (which can be assessed using a trend test) [41,42].

7.7 Weight of the Evidence and Interpretation of Data

The OECD tests provide recommendations on how to conduct the various genetic toxicology tests. However, in some cases, regulatory organizations have recommended modifications to specific guidelines appropriate for specific product types, ICH [43].

The OECD tests do not make any specific recommendations as to which tests to be used in a test battery. Regulatory agencies publish their own recommendations, which should be consulted prior to initiating testing. Generally, the recommended genetic toxicology test batteries include tests to detect both gene mutations and structural as well as numerical chromosomal damage in both in vitro and in vivo tests; however, more recently, in some jurisdictions the emphasis has been on using only in vitro (and no, or fewer well chosen, in vivo) tests.

Some publications provide basic information on using genetic toxicology information for regulatory decisions [44]. Expert work groups have also discussed appropriate follow-up testing strategies for chemicals which have been found to be positive in vitro tests and/or in vivo tests [45–48].

It is important to emphasize that the results from the different assays should be evaluated in line with the applicable regulatory test strategy. The amount of data available for a WOE evaluation will vary enormously, particularly among different product categories. Data-rich packages prepared for drug or pesticide regulations may permit analyses that would be impossible for substances involving other uses for which less data are available.

8. CURRENT NEEDS IN MUTAGENICITY TESTING

For several years, regulatory testing methods for mutagenicity have been standardized where the tests being utilized examined mutations and structural and numerical chromosomal damages. This has served the genetic toxicity community well when most of the substances being tested were amenable to such assays. The outcome is usually a binary (yes/no) evaluation of test results, and in many instances, the information is only used to determine whether a chemical has carcinogenic potential or not. Over the same time period, mechanisms and MOAs that elucidate a wider range of genomic damage involved in many adverse health outcomes have been recognized. In addition, a paradigm shift in applied genetic toxicology is moving the field toward a more quantitative dose–response analysis and PoD determination with a focus on risks to exposed humans. This is directing emphasis on genomic damage that is likely to induce changes associated with a variety of adverse health outcomes. This paradigm shift is moving the testing emphasis for mutagenic risk from a hazard identification evaluation to a more comprehensive risk assessment approach that provides more insightful information for decision makers regarding the potential risk of mutagenicity to exposed humans. To enable this broader context for examining genetic damage, a next generation testing strategy needs to take into account a broader, more flexible approach to testing, and ultimately modeling, of genomic damage as it relates to human exposure. This is consistent with the larger risk assessment context being used in regulatory decision-making.

The historical use of genomic damage data as a yes/no gateway for possible cancer risk has been too narrowly focused in risk assessment. The recent advances in assaying for and understanding genomic damage, including eventually epigenetic alterations, evidently add a greater wealth of information for determining potential risk to humans. Regulatory bodies need to embrace this paradigm shift from hazard identification to quantitative analysis and to incorporate the wider range of genomic damage in their assessments of risk to humans. Experts continue to report a lot of new findings that greatly contribute to regulatory science, and discuss the evaluation of the risk of mutagens/genotoxicants correctly based on the newly available data and their interpretation. Various screening assays are mentioned in Table 10.4.

8.1 GreenScreen HC Assay

GADD45a (growth arrest and DNA damage-inducible, alpha) gene is one of many genes that show increased transcript levels following stressful growth conditions and treatment with DNA-damaging agents. It plays a central role in DNA stress, therefore, damage, and repair cascades within the mammalian cells. Mutagens, clastogens, and aneugens cause increased expression of the mammalian GADD45a gene [55]. This feature has been exploited in the development of a green fluorescent protein (GFP) reporter–based assay (GreenScreen) for the gene using human lymphoblastoid cell line TK6 [52]. TK6 cells have a wild-type p53 tumor suppressor gene [56]. This attribute makes this cell line very useful for assessing the genotoxic response properly.

The assay protocols, described in detail by Hastwell et al. [52], are summarized here. Assay uses 96 well plate formats, in which four compounds are each tested in two series of nine twofold dilutions. One series is tested using the TK6 cell line expressing the GADD45a–GFP reporter (test cells), and another is tested using TK6 cells in which the reporter has a non-expressed GFP gene (control cells). Assay can be done in presence and absence of metabolic activation system. Data from the control cells identify compounds that are themselves fluorescent, or induce cellular autofluorescence, and thus allow for normalization with the test-cell data. Data from S9-treated samples are collected using flow cytometry [53,57], and data from samples not treated with S9 can also be collected using fluorescence/absorbance spectrometry in addition to flow cytometry as both give comparable results [53]. The GreenScreen assay is reported to exhibit sensitivity and high specificity for various classes of genotoxins compared with other in vitro mammalian cell assays [51,52,58]. International multilaboratory "ring trials" have demonstrated transferability of assay versions both with and without S9 metabolic activation [59].

Table 10.4 Various screening assays

Method	Sensitivity[a]	Sensitivity[b]	Comments/References
HepG2 (cystatin, p53, Nrf2) Luciferase reporter	85% (17/20 ECVAM list) 74% (bacterial mutagenicity) 45% (clastogenicity)	81% (34/42 ECVAM list) 80% (bacterial mutagenicity) 83% (clastogenicity)	62 ECVAM listed chemicals, 192 additional chemicals [49]
Vitotox (bacterial SOS reporter assay for mutagenicity) RadarScreen (RAD54 reporter assay in yeast for clastogenicity)	70% bacterial mutagenicity (14/20 ECVAM list) 86% (bacterial mutagenicity) 70% clastogenicity (14/20 ECVAM list) 77% (clastogenicity)	93% (39/42 ECVAM list) 94% (bacterial mutagenicity) 83% clastogenicity (35/42 ECVAM list) 74% (clastogenicity)	62 ECVAM listed chemicals, 192 additional chemicals [50]
GADD45a-GFP	18/20 ECVAM list of mammalian cell mutagens (90%) 95% for genotoxic carcinogens 63% (regulatory battery of Ames, CA/MNvit, or CA/MLA) 94% of genotoxicants 30% (in vitro genotoxicants) 30% (rodent carcinogens)	22/23 ECVAM list of mammalian cell nonmutagens (96%) 100% noncarcinogens 100% (regulatory battery of Ames, CA/MNT, or CA/MLA) 83% nongenotoxicants 97% (in vitro genotoxicants) 88% (rodent carcinogens)	[51] 75 compounds studied [52] Validation data for 56 compounds requiring metabolic activation [53] 57 in vitro genotoxiants, 50 chemicals with rodent carcinogenicity data [54]

[a]Accurate prediction of positive responses compared to comparator data set, given in *parenthesis*.
[b]Accurate prediction of negative responses compared to comparator data set, given in *parenthesis*.

8.2 Pig-A Gene Mutation Assay

A cytoplasmic membrane–bound glycolipid structure, termed glycosyl phosphatidylinositol (GPI) anchors that links various protein markers to the surface of several types of mammalian cells, including hematopoietic cells [60]. An enzyme required at the initial step for the synthesis of GPI is

governed by the endogenous phosphatidylinositol glycan complementa-
tion group A gene (Pig-A) [61,62]. In addition to Pig-A, many genes are
required for GPI synthesis (e.g., Pig-B and Pig-C). As the Pig-A gene is
located on the X chromosome [63,64] single mutation is sufficient to cause
alteration in cell phenotype. Numerous protein markers bind with GPI
on cell surface of various tissues [65,66], any mutations in the Pig-A gene
could result in the lack of GPI synthesis in turn resulting in deficiency in
GPI-anchored proteins. Rapid flow cytometric methods for measuring the
frequency of rodent and nonhuman primate peripheral blood cells deficient
in GPI-anchored proteins (presumed Pig-A mutant cells) have been developed. One common approach is to evaluate the expression of proteins (in
rats, usually CD59) anchored by GPI to the surface of peripheral red blood
cells (RBCs).

The detailed procedure is mentioned by Miura et al. [60]. Briefly, the
animals are dosed for single or multiple treatment regimens with test item.
At the end of treatment, peripheral blood is collected in anticoagulant and
diluted in PBS and the cells are labeled with anti–rat CD45 antibody. After
incubation for 1 h in the dark at room temperature, the cells are washed,
resuspended in PBS, and assayed for CD59-negative RBCs or reticulocytes
using flow cytometer. The change in CD59-negative RBCs or reticulocytes
in treatment group as compared to those of control group is analyzed as
endpoint.

The Pig-A assay is reported to be a promising tool for evaluating in vivo
mutagenicity, hence the recent draft guideline for impurity testing, ICH
M7 (Step 2 version), includes Pig-A assay to assess the in vivo relevance of
in vitro mutagens detected with positive bacterial mutagenicity [67].

8.3 γH2AX by Flow Assay

DNA DSBs are the toxic lesions that can drive genetic instability [68].
Histone variant H2AX is a key component of DNA damage response. It
becomes rapidly phosphorylated at the carboxyl terminus to form the so-
called γH2AX at DSB sites [69]. The measurement of serine139–phosphorylated histone H2AX provides a biomarker of DNA DSBs and thus identifies
potential genotoxic activity [70]. The γH2AX facilitates the repair of DNA
DSBs as an integral component in the DNA damage response machinery of
mammalian cells [69]. Flow cytometric measurement of γH2AX is a novel
approach for detection of genotoxic potential of a compound. The assessment of γH2AX over conventional assays like MLA and chromosome aberration assay is reported to be more advantageous [71,72]. For example, the

genotoxic signatures in the form of DNA damage response and cell cycle information are readily elucidated, obtained at low amount of compound with rapid and high-throughput data acquisition [70].

The detailed assay protocol reported by D.J. Smart et al. [70] includes mouse lymphoma L5178Y cells (tk+/−). This assay can be done in presence and absence of S9. Relative cell counts (RCC; % control) using Coulter counter provide an index of cytotoxicity.

To evaluate γH2AX, cells are lysed to separate the nuclei, which are resuspended in PBS containing antihuman γH2AX–FITC antibody and 7-AAD (7-aminoactinomycin D) to form a suspension of single nuclei. Nuclear, i.e., γH2AX and genomic DNA are labeled with green and red fluorescence, and forward/side light scatter from 104 nuclei are measured using a flow cytometer with standard emission filters. The difference in fluorescence values are expressed as relative fold change in γH2AX fluorescence over control in correlation with relative cell count as a measure of cytotoxicity in terms of % control. The data are used to determine whether a compound induced a positive, negative, or equivocal response in the assay.

Many genotoxins and nongenotoxins are reported to be identified by γH2AX flow assay, including a wide range of chemicals, viz, alkylating agents, aromatic amines, amides, nitroso, epoxides, hydroperoxides, N–nitrosos, with both non- and structurally alerting chemicals (based on SAR using DEREK for Windows). Thus the flow cytometric assay of γH2AX offers broad range genotoxicity assay [70]. It has shown high sensitivity to detect DNA reactive genotoxic compounds, although the utility for metabolic activation dependent compounds is limited and needs to be improved. Newer approaches for in vitro improvement in sensitivity (i.e., toxicogenomics) may be combined with this automated assay to obtain improved efficiency in genotoxicity testing [73].

8.4 HUPKI Mouse Model

HUPKI mouse model considered as the human p53 knock-in (HUPKI) mouse is a relevant and useful model for investigating chemically induced mutations. The mice carry a human wild-type TP53 DNA sequence (from exon 4–9), which replaces both copies of the murine TP53 sequence, is expressed at physiological levels and functions as normal p53 [74]. The spontaneous tumor responses are similar to those of mice with murine p53 [75]. AFB1 exposure resulted in enhanced formation of HCC in treated HUPKI mice compared to wild-type mice. However, the typical codon

249 TP53 mutations observed in humans were not observed in either wild-type mice or in HUPKI mice [76]. Embryonic HUPKI cells can be cultured, provide an in vitro systems in which chemically induced mutation spectra can be examined [77]. The embryo fibroblasts readily undergo immortalization in culture, generating cells in which TP53 is dysfunctional, enabling selection from their growth characteristics [74]. Mutation spectra from treated cultures are compared with mutations from spontaneously immortalized cultures to determine the specificity of any mutations. In HUPKI cells, the AA-induced mutation spectrum comprised primarily AT→TA transversions (57%), and this compares with 78% AT→TA transversions in human urothelial cancer from the Balkans [78]. BaP generated predominantly GC→TA transversions (49%), GC→CG transversions (22%), and GC→AT (19%) transitions in comparison with 30% GC→TA transversions, 12% GC→CG transversions, and 26% GC→AT transitions in lung cancers from smokers. The embryonic stem cells cultured from the HUPKI mouse do not always reflect human TP53 mutation response, but as an in vitro model they do appear to offer some advantages over other models.

8.5 Next Generation Sequencing

"Next generation sequencing" is the term currently used to describe high-throughput, whole genome sequencing. Members agreed that "next generation sequencing" would soon start to provide better insight into the evaluation and interpretation of chemically induced mutation spectra. "Next generation sequencing" methodology reads DNA templates randomly, enabling a picture of the entire genome to be generated. The more detailed pattern of mutations will provide greater resolution of mutation spectra and improve mechanistic insights [79]. Whole genome sequencing will identify genetic variants, including single nucleotide polymorphisms, small insertions and deletions, and structural and genomic variants (>1000 bp) across the entire DNA sequence and not just in specific genes such as TP53 from single gene sequencing studies. This methodology is combined with advanced bioinformatics methods and database searching to enable detailed analyses of cancer genomes. Mathematical models are being developed, which facilitate the process of extracting mutational signatures from the complex data sets generated from the whole genome sequencing of tumor DNA. It is possible to identify a variety of mutational patterns (in 96-element signatures) and to quantify the contribution of each signature for each tumor [80]. "Next generation sequencing" is used to accurately identify the entire mutation spectra of cancers in an attempt to examine the mutational changes, which lead to cancer. It has been used to investigate gene expression signatures in a wide variety of cancers, including a review of over 7000 cancers from 30 different sites when more than 20 distinct mutational spectra were observed [81]. It is envisaged that correlating these changes with mutational signatures from known chemical exposures in defined systems will advance the understanding of cancer etiology. Mutation signatures using current, single gene, approaches may, on a case-by-case basis, provide useful mechanistic insight into genotoxic MOA and contribute to WOE genotoxicity assessments.

REFERENCES

[1] De Flora S, Izzotti A. Mutagenesis and cardiovascular diseases: molecular mechanisms, risk factors, and protective factors. Mutat Res Fundam Mol Mech Mutagen 2007;621(1):5–17.

[2] Frank SA. Evolution in health and medicine Sackler colloquium: somatic evolutionary genomics: mutations during development cause highly variable genetic mosaicism with risk of cancer and neurodegeneration. Proc Natl Acad Sci USA 2010;(Suppl. 1):1725–30.

[3] Slatter MA, Gennery AR. Primary immunodeficiencies associated with DNA-repair disorders. Exp Rev Mol Med 2010.

[4] Yauk CL, Aardema MJ, van Benthem J, Bishop JB, Dearfield KL, DeMarini DM, Dubrova YE, Honma M, Lupski JR, Marchetti F, Meistrich ML. Approaches for identifying germ cell mutagens: report of the 2013 IWGT workshop on germ cell assays. Mutat Res Genet Toxicol Environ Mutagen 2015;783:36–54.

[5] Gollapudi BB, Johnson GE, Hernandez LG, Pottenger LH, Dearfield KL, Jeffrey AM, Julien E, Kim JH, Lovell DP, Macgregor JT, Moore MM. Quantitative approaches for assessing dose–response relationships in genetic toxicology studies. Environ Mol Mutagen 2013;54(1):8–18.

[6] Johnson GE, Soeteman-Hernández LG, Gollapudi BB, Bodger OG, Dearfield KL, Heflich RH, Hixon JG, Lovell DP, MacGregor JT, Pottenger LH, Thompson CM. Derivation of point of departure (PoD) estimates in genetic toxicology studies and their potential applications in risk assessment. Environ Mol Mutagen 2014;55(8): 609–23.

[7] National Industrial Chemicals Notification, Assessment Scheme (NICNAS) Handbook. Department of Health; 2014.

[8] Guidance Documents for New Chemical Substance Notification in China. Ministry of Environmental Protection; 2010.

[9] Ji Z, Ball N, LeBaron M. Global regulatory requirements for mutagenicity assessment in the registration of industrial chemicals. Environ Mol Mutagen 2017;58(5):345–53.

[10] Committee on Mutagenicity of Chemicals in Food Consumer Products and the Environment (COM). Guidance on a strategy for genotoxicity testing of chemical substances. 2011.

[11] Cariello NF, Wilson JD, Britt BH, Wedd DJ, Burlinson B, Gombar V. Comparison of the computer programs DEREK and TOPKAT to predict bacterial mutagenicity. Deductive estimate of risk from existing knowledge. Toxicity prediction by Komputer assisted technology. Mutagenesis 2002;4:321–9.

[12] Contrera JF, Mathews EJ, Kruhlak N, Benz RD. In silico screening of chemicals for bacterial mutagenicity using electropological E-state indices and MDL QSAR software. Regul Toxicol Pharmacol 2005;43:313–23.

[13] Matthews EJ, Kruhlak N, Cimino M, Benz RD, Contrera JF. An analysis of genetic toxicology, reproductive and developmental toxicity and carcinogenicity data I: identification of carcinogens using surrogate end points. Regul Toxicol Pharmacol 2006;44:83–96.

[14] Roithfuss A, Steger-Hartmann T, Heinrich N, Wichard J. Conspirational prediction of the chromosome-damaging potential of chemicals. Chem Res Toxicol 2006;19:1313–9.

[15] Benigni R, Bossa C. Structure alerts for carcinogenicity, and the Salmonella assay system: a novel insight through the chemical relational databases technology. Mutat Res 2008:248–61.

[16] ICH M7 – assessment and control of DNA reactive (mutagenic) impurities in pharmaceuticals to limit potential carcinogenic risk, guidance for industry. May 2015.

[17] ICH S9 (step – 4): Nonclinical evaluation for anticancer pharmaceuticals. October 2009.

[18] Wallin H, Jacobsen NR, White PA, Gingerich J, Møller P, Loft S, Vogel U. Mutagenicity of carbon nanomaterials. J Biomed Nanotechnol 2011;7(1):29.

[19] OECD Migration Outlook 2014. Paris: OECD; 2014.

[20] Kirkland D, Aardema M, Henderson L, Müller L. Evaluation of the ability of a battery of three in vitro genotoxicity tests to discriminate rodent carcinogens and non-carcinogens I. Sensitivity, specificity and relative predictivity. Mutat Res 2005;584(1–2):1–256.

[21] Heddle JA, Hite M, Kirkhart B, Mavournin K, MacGregor JT, Newell GW, Salamone MF. The induction of micronuclei as a measure of genotoxicity: a report of the US Environmental Protection Agency Gene-Tox Program. Mutat Res Rev Genet Toxicol 1983;123(1):61–118.

[22] Heddle JA, Cimino MC, Hayashi M, Romagna F, Shelby MD, Tucker JD, Vanparys P, MacGregor JT. Micronuclei as an index of cytogenetic damage: past, present, and future. Environ Mol Mutagen 1991;18(4):277–91.

[23] Kirsch-Volders M, Plas G, Elhajouji A, Lukamowicz M, Gonzalez L, Vande Loock K, Decordier I. The in vitro MN assay in 2011: origin and fate, biological significance, protocols, high throughput methodologies and toxicological relevance. Arch Toxicol 2011;85(8):873–99.

[24] Dybing E, et al. Genotoxicity studies with paracetamol. Mutat Res Genet Toxicol 1984;138(1):21–32.

[25] You Z, Brezzell MD, Das SK, Espadas-Torre MC, Hooberman BH, Sinsheimer JE. Ortho-Substituent effects on the in vitro and in vivo genotoxicity of benzidine derivatives. Mutat Res Genet Toxicol 1993;319(1):19–30.

[26] Heddle JA, Dean S, Nohmi T, Boerrigter M, Casciano D, Douglas GR, Glickman BW, Gorelick NJ, Mirsalis JC, Martus HJ, Skopek TR. In vivo transgenic mutation assays. Environ Mol Mutagen 2000;35(3):253–9.

[27] Lambert IB, Singer TM, Boucher SE, Douglas GR. Detailed review of transgenic rodent mutation assays. Mutat Res Rev Mutat Res 2005;590(1):1–280.

[28] Kumaravel TS, Jha AN. Reliable Comet assay measurements for detecting DNA damage induced by ionising radiation and chemicals. Mutat Res Genet Toxicol Environ Mutagen 2006;605(1):7–16.

[29] Olive PL, Banáth JP. The comet assay: a method to measure DNA damage in individual cells. Nat Protocols 2006;1(1):23.

[30] Tice RR, Agurell E, Anderson D, Burlinson B, Hartmann A, Kobayashi H, Miyamae Y, Rojas E, Ryu JC, Sasaki YF. Single cell gel/comet assay: guidelines for in vitro and in vivo genetic toxicology testing. Environ Mol Mutagen 2000;35(3):206–21.

[31] Kawaguchi S, Nakamura T, Yamamoto A, Honda G, Sasaki YF. Is the Comet assay a sensitive procedure for detecting genotoxicity? J Nucleic Acids 2010.

[32] Kirkland D, Speit G. Evaluation of the ability of a battery of three in vitro genotoxicity tests to discriminate rodent carcinogens and non-carcinogens: III. Appropriate follow-up testing in vivo. Mutat Res Genet Toxicol Environ Mutagen 2008;654(2):114–32.

[33] Speit G, Hartmann A. The comet assay (single-cell gel test). A sensitive genotoxicity test for the detection of DNA damage and repair. In: DNA repair protocols: eukaryotic systems. 1999. p. 203–12.

[34] Burlinson B, Tice RR, Speit G, Agurell E, Brendler-Schwaab SY, Collins AR, Escobar P, Honma M, Kumaravel TS, Nakajima M, Sasaki YF. Fourth International Workgroup on Genotoxicity testing: results of the in vivo Comet assay workgroup. Mutat Res Genet Toxicol Environ Mutagen 2007;627(1):31–5.

[35] Lovell DP, Omori T. Statistical issues in the use of the comet assay. Mutagenesis 2008;23(3):171–82.

[36] Hayashi M, Dearfield K, Kasper P, Lovell D, Martus HJ, Thybaud V. Compilation and use of genetic toxicity historical control data. Mutation Res Genet Toxicol Environ Mutagen 2011;723(2):87–90.

[37] Moore MM, Honma M, Clements J, Awogi T, Bolcsfoldi G, Cole J, Gollapudi B, Harrington-Brock K, Mitchell A, Muster W, Myhr B. Mouse lymphoma thymidine kinase locus gene mutation assay: international workshop on genotoxicity test procedures workgroup report. Environ Mol Mutagen 2000;35(3):185–90.

[38] Moore MM, Honma M, Clements J, Harrington-Brock K, Awogi T, Bolcsfoldi G, Cifone M, Collard D, Fellows M, Flanders K, Gollapudi B. Mouse lymphoma thymidine kinase gene mutation assay: follow-up International Workshop on Genotoxicity Test Procedures, New Orleans, Louisiana, April 2000. Environ Mol Mutagen 2002;40(4):292–9.

[39] Moore MM, Honma M, Clements J, Bolcsfoldi G, Cifone M, Delongchamp R, Fellows M, Gollapudi B, Jenkinson P, Kirby P, Kirchner S. Mouse lymphoma thymidine kinase gene mutation assay: International Workshop on Genotoxicity Tests Workgroup Report—Plymouth, UK 2002. Mutat Res Genet Toxicol Environ Mutagen 2003;540(2):127–40.

[40] Moore MM, Honma M, Clements J, Bolcsfoldi G, Burlinson B, Cifone M, Clarke J, Delongchamp R, Durward R, Fellows M, Gollapudi B. Mouse lymphoma thymidine kinase gene mutation assay: follow-up Meeting of the International Workshop on Genotoxicity Testing—Aberdeen, Scotland, 2003—assay acceptance criteria, positive controls, and data evaluation. Environ Mol Mutagen 2006;47(1):1–5.

[41] Kim BS, Cho MH, Kim HJ. Statistical analysis of in vivo rodent micronucleus assay. Mutat Res Genet Toxicol Environ Mutagen 2000;469(2):233–41.

[42] Lovell DP, Anderson D, Albanese R, Amphlett GE, Clare G, Ferguson R, Richold M, Papworth DG, Savage JR. Statistical analysis of in vivo cytogenetic assays. Statist Eval Mutagen Test Data 1989:184–232.

[43] ICH S6 (R1) – Preclinical safety evaluation of biotechnology-derived pharmaceuticals. June 2011.

[44] Dearfield KL, Moore MM. Use of genetic toxicology information for risk assessment. Environ Mol Mutagen 2005;46(4):236–45.

[45] Dearfield KL, Thybaud V, Cimino MC, Custer L, Czich A, Harvey JS, Hester S, Kim JH, Kirkland D, Levy DD, Lorge E. Follow-up actions from positive results of in vitro genetic toxicity testing. Environ Mol Mutagen 2011;52(3):177–204.

[46] Thybaud V, Aardema M, Clements J, Dearfield K, Galloway S, Hayashi M, Jacobson-Kram D, Kirkland D, MacGregor JT, Marzin D, Ohyama W. Strategy for genotoxicity testing: hazard identification and risk assessment in relation to in vitro testing. Mutat Res Genet Toxicol Environ Mutagen 2007;627(1):41–58.

[47] Thybaud V, MacGregor JT, Müller L, Crebelli R, Dearfield K, Douglas G, Farmer PB, Gocke E, Hayashi M, Lovell DP, Lutz WK. Strategies in case of positive in vivo results in genotoxicity testing. Mutat Res Genet Toxicol Environ Mutagen 2011;723(2):121–8.

[48] Tweats DJ, Blakey D, Heflich RH, Jacobs A, Jacobsen SD, Morita T, Nohmi T, O'donovan MR, Sasaki YF, Sofuni T, Tice R. Report of the IWGT working group on strategies and interpretation of regulatory in vivo tests: I. Increases in micronucleated bone marrow cells in rodents that do not indicate genotoxic hazards. Mutat Res Genet Toxicol Environ Mutagen 2007;627(1):78–91.

[49] Westerink WM, Stevenson JC, Horbach GJ, Schoonen WG. The development of RAD51C, Cystatin A, p53 and Nrf2 luciferase-reporter assays in metabolically competent HepG2 cells for the assessment of mechanism-based genotoxicity and of oxidative stress in the early research phase of drug development. Mutat Res 2010;696(1): 21–40.

[50] Westerink WM, Stevenson JC, Lauwers A, Griffioen G, Horbach GJ, Schoonen WG. Evaluation of the Vitotox and RadarScreen assays for the rapid assessment of genotoxicity in the early research phase of drug development. Mutat Res 2009;676(1–2): 113–30.

[51] Birrell L, Cahill P, Hughes C, Tate M, Walmsley RM. GADD45a-GFP GreenScreen HC assay results for the ECVAM recommended lists of genotoxic and non-genotoxic chemicals for assessment of new genotoxicity tests. Mutat Res Genet Toxicol Environ Mutagen 2010;695(1):87–95.

[52] Hastwell PW, Chai LL, Roberts KJ, Webster TW, Harvey JS, Rees RW, Walmsley RM. High-specificity and high-sensitivity genotoxicity assessment in a human cell line: validation of the GreenScreen HC GADD45a-GFP genotoxicity assay. Mutat Res Genet Toxicol Environ Mutagen 2006;607(2):160–75.

[53] Jagger C, Tate M, Cahill PA, Hughes C, Knight AW, Billinton N, Walmsley RM. Assessment of the genotoxicity of S9-generated metabolites using the GreenScreen HC GADD45a–GFP assay. Mutagenesis 2008;24(1):35–50.

[54] Olaharski A, Albertini S, Kirchner S, Platz S, Uppal H, Lin H, Kolaja K. Evaluation of the GreenScreen GADD45alpha-GFP indicator assay with non-proprietary and proprietary compounds. Mutat Res 2009;672(1):10–6.

[55] Ellinger-Ziegelbauer H, Aubrecht J, Kleinjans JC, Ahr HJ. Application of toxicogenomics to study mechanisms of genotoxicity and carcinogenicity. Toxicol Lett 2009;186(1):36–44.

[56] Hernández LG, van Steeg H, Luijten M, van Benthem J. Mechanisms of non-genotoxic carcinogens and importance of a weight of evidence approach. Mutat Res Rev in Mutat Res 2009;682(2):94–109.

[57] Knight AW, Birrell L, Walmsley RM. Development and validation of a higher throughput screening approach to genotoxicity testing using the GADD45a-GFP GreenScreen HC assay. J Biomol Screen 2009:16–30.

[58] Hastwell PW, Webster TW, Tate M, Billinton N, Lynch AM, Harvey JS, Rees RW, Walmsley RM. Analysis of 75 marketed pharmaceuticals using the GADD45a-GFP 'GreenScreen HC' genotoxicity assay. Mutagenesis 2009;24(5):455–63.

[59] Billinton N, Hastwell PW, Beerens D, Birrell L, Ellis P, Maskell S, Webster TW, Windebank S, Woestenborghs F, Lynch AM, Scott AD. Interlaboratory assessment of the GreenScreen HC GADD45a-GFP genotoxicity screening assay: an enabling study for independent validation as an alternative method. Mutat Res Genet Toxicol Environ Mutagen 2008;653(1):23–33.

[60] Miura D, Dobrovolsky VN, Kasahara Y, Katsuura Y, Heflich RH. Development of an in vivo gene mutation assay using the endogenous Pig-A gene: I. Flow cytometric detection of CD59-negative peripheral red blood cells and CD48-negative spleen T-cells from the rat. Environ Mol Mutagen 2008;49(8):614–21.

[61] Brodsky RA, Hu R. PIG-A mutations in paroxysmal nocturnal hemoglobinuria and in normal hematopoiesis. Leuk Lymphoma 2006;47(7):1215–21.

[62] Nishimura JI, Murakami Y, Kinoshita T. Paroxysmal nocturnal hemoglobinuria: an acquired genetic disease. Am J Hematol 1999;62(3):175–82.

[63] Kawagoe K, Takeda J, Endo Y, Kinoshita T. Molecular cloning of murine pig-a, a gene for GPI-anchor biosynthesis, and demonstration of interspecies conservation of its structure, function, and genetic locus. Genomics 1994;23(3):566–74.

[64] Takeda J, Miyata T, Kawagoe K, Iida Y, Endo Y, Fujita T, Takahashi M, Kitani T, Kinoshita T. Deficiency of the GPI anchor caused by a somatic mutation of the PIG-A gene in paroxysmal nocturnal hemoglobinuria. Cell 1993;73(4):703–11.

[65] Cross GA. Glycolipid anchoring of plasma membrane proteins. Annu Rev Cell Biol 1990;1:1–39.

[66] Low MG. The glycosyl-phosphatidylinositol anchor of membrane proteins. Biochim Biophys Acta (BBA) – Rev Biomembr 1989;988(3):427–54.

[67] ICH M7 (Step 3): Assessment and control of DNA reactive (mutagenic) impurities in pharmaceuticals to limit potential carcinogenic risk. April 2013.

[68] Chapman JR, Taylor MR, Boulton SJ. Playing the end game: DNA double-strand break repair pathway choice. Mol Cell 2012;47(4):497–510.

[69] Bonner WM, Redon CE, Dickey JS, Nakamura AJ, Sedelnikova OA, Solier S, Pommier Y. γH2AX and cancer. Nat Rev Cancer 2008;8(12):957–67.

[70] Smart DJ, Ahmedi KP, Harvey JS, Lynch AM. Genotoxicity screening via the γH2AX by flow assay. Mutat Res Fundam Mol Mech Mutagen 2011;715(1):25–31.

[71] Huang X, Halicka HD, Traganos F, Tanaka T, Kurose A, Darzynkiewicz Z. Cytometric assessment of DNA damage in relation to cell cycle phase and apoptosis. Cell Prolif 2005;38(4):223–43.

[72] Smart DJ. Genotoxicity of topoisomerase II inhibitors: an anti-infective perspective. Toxicology 2008;254(3):192–8.

[73] Tsamou M, Jennen DG, Claessen SM, Magkoufopoulou C, Kleinjans JC, van Delft JH. Performance of in vitro γH2AX assay in HepG2 cells to predict in vivo genotoxicity. Mutagenesis 2012;27(6):645–52.

[74] Luo JL, Yang Q, Tong WM, Hergenhahn M, Wang ZQ, Hollstein M. Knock-in mice with a chimeric human/murine p53 gene develop normally and show wild-type p53 responses to DNA damaging agents: a new biomedical research tool. Oncogene 2001;20(3).

[75] Kucab JE, Phillips DH, Arlt VM. Linking environmental carcinogen exposure to TP53 mutations in human tumours using the human TP53 knock-in (Hupki) mouse model. FEBS J 2010;277(12):2567–83.

[76] Tong WM, Lee MK, Galendo D, Wang ZQ, Sabapathy K. Aflatoxin-B exposure does not lead to p53 mutations but results in enhanced liver cancer of Hupki (human p53 knock-in) mice. Int J Cancer 2006;119(4):745–9.

[77] Olivier M, Weninger A, Ardin M, Huskova H, Castells X, Vallée MP, McKay J, Nedelko T, Muehlbauer KR, Marusawa H, Alexander J. Modelling mutational landscapes of human cancers in vitro. Sci Rep 2014;4.

[78] Hollstein M, Moriya M, Grollman AP, Olivier M. Analysis of TP53 mutation spectra reveals the fingerprint of the potent environmental carcinogen, aristolochic acid. Mutat Res Rev Mutat Res 2013;753(1):41–9.

[79] Alexandrov LB, Stratton MR. Mutational signatures: the patterns of somatic mutations hidden in cancer genomes. Curr Opin Genet Dev 2014;24:52–60.

[80] Helleday T, Eshtad S, Nik-Zainal S. Mechanisms underlying mutational signatures in human cancers. Nat Rev Genet 2014;15(9):585–98.

[81] Alexandrov LB, Nik-Zainal S, Wedge DC, Aparicio SA, Behjati S, Biankin AV, Bignell GR, Bolli N, Borg A, Børresen-Dale AL, Boyault S. Signatures of mutational processes in human cancer. Nature 2013;7463:415–21.

FURTHER READING

[1] Guidelines for the notification and testing of new substances: chemicals and Polymers, environment Canada. 2005.

[2] Chapter R.7a: Endpoint specific guidance, Version 5.0. Guidance on information requirements and chemical safety assessment. European Chemicals Agency; 2016.

[3] Submission method of test data on physico-chemical properties, hazard. Ministerial Decree (Enforcement Regulation) Ministry of Environment; 2015.

[4] EFSA Scientific Committee. Scientific opinion on genotoxicity testing strategies applicable to food and feed safety assessment. EFSA J 2011;9:2379.

CHAPTER ELEVEN

Detecting Mutations In Vivo

Vasily N. Dobrovolsky, Robert H. Heflich
National Center for Toxicological Research, US Food and Drug Administration, Jefferson, AR, United States

1. INTRODUCTION

Germline mutations are involved in hereditary genetic disorders, while somatic cell mutations cause acquired genetic disorders, premature tissue and organ senescence; some somatic mutations are responsible for the onset and progression of cancer. Because of these severe impacts on human health, mutation has been a subject of concern to regulatory authorities. Methods to detect de novo (e.g., treatment-induced) phenotypically expressed mutations were described for lower phylogeny organisms (e.g., in *Drosophila* species) almost a century ago [1,2]. Yet a detailed understanding of mutagenesis processes and the development of methods for detecting mutation in animal cell cultures and in mammalian species in vivo (including humans) only occurred following reports describing DNA structure and function in the 1950s. The discoveries that happened in the following two to three decades clarified mechanisms of DNA replication, helped in understanding the roles of specific genes in various biochemical pathways, and suggested ways of using the phenotypes associated with specific endogenous genes to measure mutation. A new wave of models for detecting in vivo mutation arrived with the invention of methods for manipulating the genomes of laboratory animals, i.e., adding artificially designed reporter transgenes and inactivating endogenous genes via targeted homologous recombination.

The most recently developed approaches to detecting and characterizing mutation are a direct result of quantum improvements in DNA sequencing technology and the relative affordability of instrumentation for high-throughput analysis of large cell samples. This review describes several in vivo gene mutation assays that have gained various degrees of regulatory acceptance. Included is the *Pig-a* assay, a novel assay that is in the advanced stages of validation and test guideline development. Details as to which model or assay is most appropriate in a specific situation can be found in guidance documents issued by consensus groups (e.g., Test Guidelines

Mutagenicity: Assays and Applications
ISBN 978-0-12-809252-1
http://dx.doi.org/10.1016/B978-0-12-809252-1.00011-0

(TGs) published by the Organisation for Economic Co-operation and Development (OECD); http://www.oecd.org/) and various regulatory agencies.

For clarity, we define mutation as a stable change to primary DNA sequence that can be propagated with cell division. All nonheritable genomic DNA changes (e.g., micronuclei, single- and double-stranded DNA breaks, altered bases, adducts to bases and the phosphate backbone, nonheritable aneuploidies, and gross chromosomal abnormalities) will be referred to as DNA damage.

2. REGULATORY APPLICATIONS OF MUTATION DATA

Mutation may be the driver of evolution, adaptability, and species diversity, but at the level of an individual organism, mutation in somatic cells has at best neutral or at worst extremely deleterious consequences. Up until the late 1960s, mammalian in vivo mutation (usually, germline mutation) was evaluated in large-scale studies on mice exposed to ionizing radiation or chemicals [3]. In vivo studies utilizing the mouse-specific locus test (MSLT) required complicated breeding schemes and could involve thousands of animals per experiment [4]. The MSLT and the similarly resource-intensive mouse spot test (OECD TG 484) proved impractical as routine safety tests for detecting potential mutagens (even for detecting germline mutagens). Mutation testing changed dramatically when Bruce Ames and colleagues demonstrated that many known carcinogens are mutagens in prokaryotic models [5,6], implying a mechanistic link between somatic cell mutation and cancer and providing a method for safety testing. Subsequently, the mutations that actually cause cancer were identified in key genes involved in cell cycle control and DNA repair/maintenance; and regulatory agencies adopted the concept that chemicals should be tested for mutagenicity as a means to identify potential carcinogens.

While the bacterial test for mutagenicity (the Ames test) was inexpensive, and easy to perform and interpret, it was recognized that bacterial assays lack the ability to detect a number of mutations that are associated with human health effects, including structural and numerical chromosomal aberrations (ChrAbs) such as partial and complete chromosome losses/gains (clastogenic and aneugenic events) and loss of heterozygosity (LOH; e.g., due to large deletions or mitotic recombination). Further development of the bacterial and complimentary eukaryotic cell–based assays also was influenced by understanding that cell culture models may lack the metabolic capabilities for converting many xenobiotics into ultimate mutagens. The approach to screening for potential carcinogens has evolved into using a battery of tests that includes assays for the detection of mutation in bacterial cells and in mammalian cells, with and without the addition of some sort of metabolic activation system (typically a rat liver homogenate).

Different regulatory agencies have their own requirements as to the battery that should be used [7], yet most batteries are based on similar principles. Besides in vitro tests, agencies often require in vivo testing for genotoxicity in nonclinical trials (for better accounting for such factors as the absorption, distribution, metabolism, and excretion that are characteristic of animal models). For instance, agencies that regulate drugs intended for human use require both in vitro and in vivo testing before allowing investigational new drugs (INDs) into human clinical studies. The US Food and Drug Administration (FDA) requires submitting the results of nonclinical safety evaluations of INDs in accordance with a guidance that was developed by the Safety implementation Working Group of the International Council on Harmonization of Technical Requirements for Registration of Pharmaceuticals for Human Use (ICH), ICH S2R1 (accessible via the FDA website at http://www.fda.gov/downloads/drugs/ guidancecomplianceregulatoryinformation/guidances/ucm074931.pdf). ICH S2R1 was also adopted by regulatory bodies in the European Union and Japan.

Paradoxically, S2R1 has no absolute requirement for testing INDs for mutagenicity in biological systems other than in bacteria. For in vivo assessments, the guidance suggests measuring micronucleus (MN) formation in erythrocytes (derived from peripheral blood or bone marrow) or ChrAbs induction in nucleated cells of bone marrow. The earlier ICH guidance (S2A and S2B, now superseded by S2R1) mentioned several cell culture assays for assessment of mutation in the endogenous *Hprt*, *Gpt*, and *Tk* genes (in vivo analogs of these in vitro assays also have been developed; they are discussed in subsequent sections of this review). But the recommended assay was the mouse lymphoma assay (MLA) in L5178Y $Tk^{+/-}$ cells [8], mostly because it was capable of measuring gene and chromosomal mutation. At the same time, the old guidance considered as equally acceptable measuring metaphase ChrAbs in cultures of mammalian cells (including in L5178 $Tk^{+/-}$ cells). Thus, detection of mutation in mammalian cells, in vitro or in vivo, was not mandatory in the old ICH guidance, and it remains so in the new guidance.

Classical approaches to cytogenetic testing (i.e., testing for MN and ChrAbs) involve examination of cells on slides using laborious visual microscopic techniques which results in rather limited statistical power; besides, the overall process is subjected to experimenter bias. One reason that gene mutation assays have not been emphasized in S2R1 is improvements in methodology and instrumentation that have increased dramatically the throughput and statistical power of cytogenetic tests (e.g., using flow cytometry and computerized image analysis). However, most of what is detected in these assays is not stable heritable mutation but is best characterized as DNA damage. The MN endpoint is not directly implicated in carcinogenesis. In addition, the micronuclei that are scored in nucleated cells often lack centromeres and thus are nonheritable events; in erythrocytes, MNs are nonheritable even if they have centromeres. Similarly, the types of aberrations that are typically measured in ChrAb assays in nucleated cells are nonheritable and result in cell death. The specific types of stable ChrAbs that are heritable and that could contribute to cancer require special techniques to score and are not evaluated in standard screening assays.

The in vivo Comet assay is recommended by S2R1 and other guidances for detecting potential mutagens. Like the MN and ChrAb assays, the Comet assay also detects DNA damage, but it can detect the damage in almost any animal tissue (including solid tissues). For most screening applications, the convenience of performing the Comet and/or cytogenetic assays outweighs the benefits of performing a mutation assay, even though the mutation assay provides a stronger link between the measured endpoint and cancer development. S2R1 acknowledges that the recommended in vivo DNA damage detection tests have limitations and that the results of these tests may be subjected to alternative interpretations (i.e., they have the potential of detecting "misleading positive results" that may not indicate true genotoxicity). To mitigate against these problems, the guidance recommends taking into account available toxicological and hematological findings while evaluating the results of genotoxicity tests.

An important feature of the cytogenetic tests and the Comet assay is that they are truly short-term in vivo genotoxicity tests. They are typically conducted with acute treatments that approach the maximum tolerated dose. In fact, their sensitivity may suffer when the tests are integrated into subchronic or chronic testing designs where doses are often lower (but perhaps more relevant to human exposures). Thus, they take only days to conduct. On the other hand, the requirement for phenotypic expression in mutation assays makes them long-term procedures, taking weeks to months to complete. Unlike the Comet and cytogenetic tests, the sensitivity of most gene mutation assays profits from subchronic or chronic dosing designs. Heritable mutation is a permanent event, and it is possible to measure cumulative mutant frequencies induced in long exposures.

In vivo mutation assays have theoretical advantages for predicting disease, and they often benefit from integration into repeat dose studies. However, the lower costs, the established predictive power for cancer, as well as decades of improvements to methodology and data interpretation have made the cytogenetic and Comet assays primary tools for assessing in vivo genetic risk. Even with these practical considerations, assays that measure in vivo gene mutation are recommended in many guidance documents as "secondary" tests. For instance, they are recommended where germ cell mutation is a particular concern. Also they are recommended as a follow-up for positive findings in sensitive in vitro mutation assays to determine relevance of the in vitro results to in vivo exposures (a process commonly referred to as derisking).

Regulatory guidance documents strive to be as unambiguous and comprehensive as possible, and the difficulty in achieving consensus on their content means that they change infrequently, and only to reflect major scientific advances. Yet, to take advantage of rapid progress in technology and science, most guidances contain wording indicating that alternative approaches for proving safety can be used if they satisfy "the requirements of the applicable statues and regulations" (quote from ICH S2R1). Thus, novel models for safety assessment are often adopted before formal acceptance by a guidance document. Perhaps in vivo mutation detection in the *Pig-a* gene (described in a subsequent section of this review) will be one such model.

3. DETECTION OF MUTATION IN THE *HPRT* GENE

Hypoxanthine guanine phosphoribosyltransferase (*Hprt*) is an endogenous gene residing on the X-chromosome in mammalian species. The product of *Hprt* (the Hprt enzyme) is involved in the purine nucleotide salvage pathway. In mammals, the *Hprt* gene is present in a single functional copy: in male cells there is only one copy of the X-chromosome, in female cells there are two X-chromosomes but one of them is permanently inactivated. Under normal circumstances, random inactivation (also called lyonization) of one of the X-chromosomes happens early in female embryonic development. After the initial inactivation, the transcriptional shutdown pattern is maintained in mitotically dividing cells, i.e., the replicated copies generated from the inactivated X-chromosome will remain inactive in daughter cells. A single mutation in the functional copy of the *Hprt* gene can completely abolish Hprt enzymatic activity in the cell. A rare human heritable recessive genetic disorder, Lesch–Nyhan syndrome, is due to *Hprt* gene mutation [9].

Hprt can metabolize a toxic purine analog, 6-thioguanine (6TG), into a nucleoside. The toxic nucleoside analog is further phosphorylated into mono-, di-, and triphosphate forms. These metabolites interfere with normal purine nucleotide metabolism, incorporating into DNA and stalling replication. Cycling cells that are wild type for the endogenous *Hprt* gene are sensitive to the presence of 6TG; mutant cells deficient in Hprt function divide and form clones in the presence of 6TG. This property is exploited in in vivo models where *Hprt* serves as a reporter for detecting somatic mutation.

The Hprt assay was first described in the 1970s for measuring mutation in humans (reviewed in Ref. [10]) and in the late 1980s for measuring mutation in rodents [11]. In most in vivo models, the target cell population used for detecting *Hprt* mutation is peripheral blood T-cells [12,13], although the assay can be performed with other cell types [14,15]. T-cells can be expanded ex vivo using a medium supplemented with appropriate growth factors and mitogens. For small laboratory animals (e.g., mice and rats), T-cells are derived from spleens (up to ¼ of spleen mononuclear cells are positive for T-cell receptor, although not all T-cells are capable of ex vivo proliferation); for large animals and humans, a sufficient number of T-cells can be derived from 5 to 10 mL of peripheral blood. A typical protocol for performing the T-cell *Hprt* assay is shown in Fig. 11.1.

The *Hprt* assay is not a short-term assay; it requires a period of time for fixation of the initial damage to the *Hprt* gene into a stable mutation and for expression of the *Hprt*-deficient phenotype, usually 4–8 weeks for adult rodents. Potentially, the *Hprt* assay can be used in conjunction with repeated-dose 28-day, 90-day, and 2-year toxicology and carcinogenicity studies. The relatively low daily doses of test compounds used in long-term studies may not induce measurable levels of DNA damage, but, cumulatively, the doses may be high enough to induce increases in *Hprt* mutation. The strength of the *Hprt* assay is that it can be performed on males and females of most laboratory rodents used in toxicology assessments, regardless of their genetic background, as well as on less common laboratory mammalian species (e.g., nonhuman primates [16,17]) and on humans.

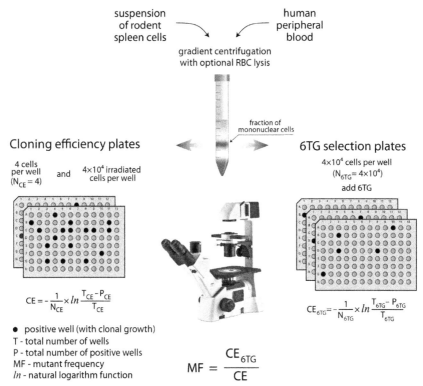

Cloning efficiency plates

$$CE = -\frac{1}{N_{CE}} \times ln\, \frac{T_{CE} - P_{CE}}{T_{CE}}$$

6TG selection plates

$$CE_{6TG} = -\frac{1}{N_{6TG}} \times ln\, \frac{T_{6TG} - P_{6TG}}{T_{6TG}}$$

- positive well (with clonal growth)
T - total number of wells
P - total number of positive wells
MF - mutant frequency
ln - natural logarithm function

$$MF = \frac{CE_{6TG}}{CE}$$

Figure 11.1 T-cell *Hprt* mutation detection assay. Spleen cells (from rodents) and peripheral blood (from large animals and humans) are amenable for the *Hprt* assay. First, the sample containing the target cell population is depleted of red blood cells (RBCs) by centrifugation on a ficoll density gradient and/or selective RBC lysis in a hypotonic ammonium chloride buffer. Mononuclear cells form a distinct band above the ficoll phase. The RBC-free population of mononuclear cells containing T-lymphocytes is primed overnight in the presence of growth factors and mitogens and then seeded into 96-well plates to establish two types of cultures: at low cell concentration (e.g., 4 cells per well) in growth medium promoting expansion of T-lymphocytes (cloning efficiency plates) and at high cell concentration (e.g., at 40,000 cells per well; selection plates) promoting expansion of *Hprt* mutant T-lymphocytes (i.e., growth medium supplemented with 6TG). In addition, the low cell concentration cultures are supplemented with mitotically arrested cells (usually the same RBC-depleted cells but exposed to 80–90 Gy of ionizing radiation; 40,000 cells per well); so that both cloning efficiency and selection plates have a similar number of cells per well. After 12–14 days in culture, the plates are scored with a tissue culture microscope or using a viability dye and plate reader. The number of wells with clonal growth is counted on each plate, and cloning efficiencies (CE) of cells in the absence and in the presence of 6TG (CE_{6TG}) are calculated assuming a Poisson distribution. Finally, the frequency of *Hprt* mutant cells (MF) in the sample is calculated as a ratio [CE_{6TG}]/[CE].

Although the *Hprt* assay has been conducted in several laboratories, it has not been widely used or recommended for safety evaluations. The assay was mentioned in the old ICH S2A guidance, not as part of a standard battery, but as a follow-up test for evaluating potentially "biologically irrelevant" positive results determined in in vitro tests, e.g., in the MLA. ICH S2A went on to note that in vivo tests measuring mutation in endogenous genes (including *Hprt*) were not sufficiently validated for routine use. Besides not being validated, three properties of the assay limit its use for routine mutagenicity assessments: the assay predominantly detects mutation induced in bone marrow, the assay is technically challenging to perform (laborious aseptic techniques are required for handling tissues and cell cultures); and the assay requires weeks to months for the in-life animal portion and in vitro cell culturing. Also, the single-copy X-linked *Hprt* gene cannot detect LOH mutations.

4. DETECTION OF MUTATION IN THE *TK1* AND *APRT* GENES

To partially address the last limitation of the *Hprt* assay, two models were designed using genetically engineered knockout mice. The autosomal genes used as reporters in these mouse models, the adenine phosphoribosyltransferase (*Aprt*) and cytosolic thymidine kinase (*Tk1*) genes, are also involved in nucleotide salvage. In heterozygous animals, for either of these genes (in $Aprt^{+/-}$ or $Tk1^{+/-}$ animals), the functional copy of the gene can serve as a reporter for detection of mutation. The selective expansion of Aprt- or Tk1-deficient mutant cells can be achieved using toxic analogs of natural substrates metabolized by these enzymes, 8-azaadenine (8-AA) and 2,6-diaminopurine (DAP) or bromodeoxyuridine (BrdU), respectively. Mice naturally heterozygous for autosomal *Aprt* or *Tk1* have not been identified, so they were generated through targeted inactivation of one copy of the gene in mouse embryonic stem cells using homologous recombination [18–20].

The *Aprt* and *Tk1* mutation detection assays are performed in 96-well plates similarly to the *Hprt* assay, with the exception of using different selecting agents and seeding a different number of cells per well for selection and expansion of mutants [12]. The mouse *Tk1* mutation detection assay was conceived as an in vivo analog of the MLA in L5178Y $Tk^{+/-}$ cells, and as such, the in vivo model is capable of detecting gene mutation and LOH-type mutation caused by large deletions and recombination.

The *Aprt*$^{+/-}$ mouse model has similar properties. The usefulness of the heterozygous mouse models for mutation studies is that each has two reporter genes suitable for detection of mutation, an autosomal *Aprt* or *Tk1* gene and an X-linked *Hprt* gene. Interestingly, the background mutant frequencies for the autosomal genes of heterozygous animals are about 10-fold higher than the background mutant frequency for the *Hprt* gene in the same animals. That by itself implies that the autosomal targets can detect a wider spectrum of mutational events than the X-linked target.

The *Tk1*$^{+/-}$ and *Aprt*$^{+/-}$ mouse models were used intensively for characterizing the mutagenic properties of drugs and in studies of DNA repair, but not often in regulatory studies. The models are not considered sufficiently validated. As is true of the *Hprt* model, they require aseptic techniques, extended ex vivo culturing, and are limited to detecting mutations in hematopoietic tissue. These models are not referred to in the new ICH S2R1 guidance.

5. TRANSGENIC MODELS FOR DETECTING IN VIVO MUTATION

Originally designed in the late 1980s–1990s, the transgenic rodent (TGR) mutation assay is capable of detecting mutation in vivo. Actually, several TGR models have been developed for mutagenicity testing [21–25], the most commonly used models are Muta Mouse, Big Blue (mouse and rat models), and *gpt* delta (mouse and rat models). They are similar in the sense that the transgenes in these models are recoverable bacterial vectors (bacteriophage λ genomes or linearized plasmids) present in several copies and integrated in tandem arrays into rodent genomes. Each copy of the vector contains a gene or genes that serve as reporters of mutation in vivo. Upon recovery from the animals, the vectors are introduced into appropriate bacterial hosts in which the mutant vectors produce a specific mutant phenotype plaque or colony (e.g., in the presence of chromogenic substrate, selecting agents and/or antibiotics, and at an appropriate temperature). Fig. 11.2 shows the general approach for detecting mutation in the various TGR assays.

For the purposes of S2R1, TGR mutation assays are considered as sufficiently validated, and they are specifically recommended for safety testing. A standardized test guideline for using various TGR assays was published in 2013 (OECD TG 488). While TG 488 is informative and detailed in terms of treatment schedule, choice of tissues, sampling timing, and data

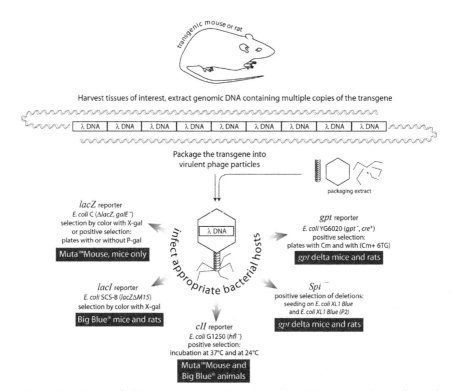

Figure 11.2 Basics of TGR mutation detection assays. The TGR assay can be performed on any tissue or organ from various transgenic rats and mice. The targets for detection of mutation are bacterial reporter genes, *lacZ*, *lacI*, *cII*, or *gpt*, present in the λ-transgene of genetically engineered animals; the *Spi⁻* model detects up to 10 kb deletions affecting the *red* and *gam* bacterial genes present in λ-transgene of *gpt* delta animals. High-molecular weight genomic DNA is extracted from the desired tissues, and individual copies of the λ-transgene are excised from genomic DNA and packaged into virulent phage particles. In the next step, depending upon the transgene, appropriate bacterial hosts are infected with the newly packaged phage and seeded onto agar plates containing selecting agents or chromophores for forming plaques or colonies at appropriate temperatures. For the *lacZ* plasmid-based model, instead of packaging, genomic DNA is digested with a suitable restriction endonuclease, the excised copies of the linear plasmid transgene are purified from the rest of digested DNA using immunomagnetic separation. The plasmid then is circularized by ligation, electroporated into *Escherichia coli* C host (Δ*lacZ*, *galE⁻*) and seeded on plates with appropriate antibiotics and selecting/chromophore agents.

interpretation of a typical study, it refers to published manuscripts for technical details on conducting each particular assay. There are significant differences between the assays as far as the use of bacterial host strains, prokaryotic culture conditions, and the overall steps that are involved.

In contrast to assays based on the endogenous *Hprt*, *Aprt*, and *Tk1* genes, TGR assays do not require establishing primary cell cultures; the assays require only DNA from the TGR animals. Thus, the assays can be performed on any tissue of interest assuming that a reasonable quantity and quality of genomic DNA can be extracted from the tissue [26]. This feature is especially useful for detecting mutation in liver, the major tissue for xenobiotic metabolism, and in male germ cells. As is true for all in vivo gene mutation detection assays, TGR assays are not short term. They require a certain period after exposure to fix treatment–induced DNA damage into stable mutations, a period that is determined by the proliferation kinetics of different tissues. Like other true mutation detection assays, TGR assays detect mutations that accumulate with each treatment, so they benefit from multiple treatments and, in theory, can be integrated into standard repeated-dose toxicology and carcinogenicity studies (provided that the genetic background of TGR animals is acceptable, which often is not).

Nevertheless, TGR assays are considered impractical for routine regulatory safety assessments: the assays are performed on specialized transgenic animals, which limit their integration potential; the assays are labor intensive and expensive to conduct. Most TGR models cannot detect medium-size (several Kb) deletions and LOH-type mutations. The transgenes are of bacterial origin, heavily methylated, and not transcribed in vivo, and thus are not subject to transcription-coupled DNA repair. Because of this they may have higher mutation rate than endogenous, transcribed genes.

6. DETECTION OF MUTATION IN THE ENDOGENOUS *PIG-A* GENE

A model based on the detection of mutation in the endogenous X-linked *Pig-a* gene is currently being developed for regulatory safety assessments. *Pig-a* stands for phosphatidylinositol glycan class–A gene; residing on the X-chromosome, the gene, like *Hprt*, is present as a single functional copy in male and female cells. The product of the *Pig-a* gene is the catalytic subunit of a protein complex involved in an early step of glycosylphosphatidylinositol (GPI) biosynthesis. GPI serves as an anchor for tethering a number of proteins (markers) to the exterior leaflet of the cytoplasmic membrane of almost all eukaryotic cells. Besides *Pig-a*, about two dozen other (autosomal) genes are involved in GPI synthesis. The synthesis of markers that are anchored by GPI is independent of the synthesis of anchors. The markers are conjugated with the core GPI in the lumen of the endoplasmic reticulum, after which the anchor–marker assembly is exported to the surface of the cell [27]. A single inactivating mutation in the *Pig-a* gene results in disruption of GPI synthesis; thus, a *Pig-a* mutant cell acquires a distinct GPI- and (GPI-anchored) marker-deficient surface phenotype (see Fig. 11.3).

There are two ways to identify *Pig-a* mutant-phenotype cells. The first is appropriate for primary cells capable of ex vivo clonal expansion (e.g., T-cells derived from spleen or peripheral blood). The steps involved in the *Pig-a* mutant T-cell cloning assay are similar to those described for the *Hprt* assay, but instead of 6TG another selective agent is used, the bacterial toxin proaerolysin from *Aeromonas hydrophila* [13,28]. In culture, proaerolysin is converted into an active form, aerolysin, by proteases present at the surface of most mammalian cells. Aerolysin has a specific affinity for the GPI-marker assembly; once bound, molecules of mature toxin create a pore-forming oligomer complex around the GPI anchor. The pores cause lysis of GPI-proficient cells. GPI-deficient cells are resistant to aerolysin and can form clones in the presence of proaerolysin in culture medium. Molecular analysis indicates that proaerolysin-resistant clones of primary human and rodent lymphocytes have *Pig-a* mutations. Although multiple genes are involved in the synthesis of GPI, only *Pig-a* is on the X-chromosome. The likelihood of mutation in the single copy of the *Pig-a* gene is orders of magnitude higher than simultaneous mutations in the two copies of one of the autosomal genes required for GPI synthesis.

The other approach for detecting GPI-anchored marker-deficient phenotype cells (potential *Pig-a* mutants) is based on immunofluorescent labeling and flow cytometric analysis of cell samples. This particular approach was developed as an extension of a diagnostic protocol for characterization of a human acquired (nonheritable) genetic disease, paroxysmal nocturnal hemoglobinuria (PNH), which is caused by mutation in the *PIG-A* gene of bone marrow hematopoietic precursors. Fluorescent antibodies against GPI-anchored markers make marker-positive cells fluoresce, while marker-deficient cells will not fluoresce [29]. The nonfluorescent (mutant-phenotype) cells can be efficiently enumerated in samples of 10,00,000+ target cells using a high-throughput flow cytometer; tissue culturing is not needed for scoring the mutant phenotype. Immunocytometric detection of marker-deficient *Pig-a* mutant cells can be performed on cells that do not have genetic material or cannot clonally expand ex vivo, e.g., red blood cells (RBCs) or terminally differentiated granulocytes from peripheral blood, respectively. Mammalian erythrocytes purge the nucleus and cytoplasmic organelles before being released into circulation from bone marrow; but, if an RBC has matured from a GPI-deficient precursor, it retains a GPI-deficient and GPI-anchored marker-deficient phenotype.

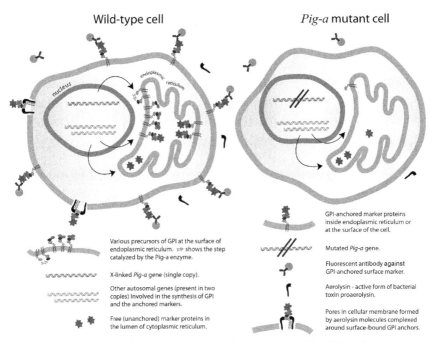

Figure 11.3 Detection of in vivo mutation in the *Pig-a* gene. The GPI anchor is synthesized at the membrane of the endoplasmic reticulum. The Pig-a enzyme is involved in the first step of GPI synthesis, the transfer of acetyl glucosamine onto phosphatidylinositol. This reaction occurs on the cytoplasmic side of the membrane; later, GPI synthesis flips into the lumen (in-depth description of GPI synthesis can be found in Ref. [27]). Protein markers attach to fully synthesized anchors in the lumen. The completed GPI-marker assemblies are exported from the endoplasmic reticulum to the surface of the cell. In *Pig-a* mutant cells, the synthesis of GPI is disrupted; as a result, the cells become deficient in GPI anchors and GPI-anchored surface markers. The marker proteins are synthesized in *Pig-a* mutant cells, but they remain trapped inside the endoplasmic reticulum.

The bacterial toxin, aerolysin, has affinity for GPI core. Molecules of the toxin form leaky pores in the cytoplasmic membrane around the surface-bound GPI, ultimately killing wild-type cells. *Pig-a* mutant cells that lack GPI anchors at the surface expand ex vivo in the presence of aerolysin in culture medium. The frequency of in vivo–derived *Pig-a* mutant T-cells can be determined using limiting dilution cultures like in the *Hprt* assay.

In addition, wild-type cells can be labeled with fluorescent antibodies against surface-bound GPI-anchored markers. *Pig-a* mutant cells will not label with such antibodies. The fraction of nonfluorescent *Pig-a* mutants can be determined efficiently by flow cytometry. With appropriate instrumentation, mutant-phenotype cells can be sorted out for analysis of *Pig-a* mutations by sequencing. In mammalian species, RBCs developing from *Pig-a* mutant progenitor cells expel nuclei and cellular organelles but retain marker-deficient mutant phenotype that can be determined by flow cytometry.

The great advantage of the erythrocyte-based *Pig-a* assay is that small (sub milliliter) volumes of peripheral blood contain sufficient number of RBCs for performing even the most sensitive tests. A basic RBC *Pig-a* assay can be performed using just 3 µL of peripheral blood. Even small laboratory animals (i.e., mice and rats) tolerate repeated samplings of such blood volumes. As a result, longitudinal mutation detection studies can be performed on the same set of animals in extended-time or repeated-dose studies. This is in contrast to the TGR assay, where animals generally are sacrificed to measure mutations. Just like other gene mutation assays, the erythrocyte *Pig-a* assay detects the cumulative effect of treatment [30], which makes it suitable for use in regulatory repeated-dose toxicology and carcinogenicity investigations. *Pig-a* assays can be performed on most laboratory rodents that are commonly used in toxicology or cancer bioassays, which is a major advantage for integration into preexisting studies.

A typical labeling protocol for an in vivo flow-based *Pig-a* assay includes fluorescent antibodies against at least one GPI-anchored marker (e.g., antibodies against CD24, CD48, CD55, and CD59 have been employed in various in vivo assays; see Table 11.1 and references therein). Additional fluorescent antibodies or stains that help in identifying the target cells or help in excluding contaminating (nontarget) cells are frequently used in the assays. An important consideration for designing a successful *Pig-a* assay is that the GPI-anchored marker that provides the best discrimination between wild-type and mutant cells varies with the cell type and species.

In mammals, polychromatic reticulocytes are the youngest erythrocytes present in peripheral blood; as a result, they relay mutations induced in bone marrow precursors as early as 2 weeks after exposure to a potential mutagen. For this reason, multiple *Pig-a* assays have been established that can measure mutation in reticulocytes. In the early versions, relatively small numbers of reticulocytes and mature normochromatic RBCs were interrogated for the frequency of *Pig-a* mutants ($\sim 3 \times 10^5$ and $\sim 1 \times 10^6$, respectively). Those assays had sufficient sensitivity to reliably detect in vivo *Pig-a* mutation induced by powerful mutagens. More advanced versions of the erythrocyte *Pig-a* assay employ immunomagnetic separation techniques that are capable of processing samples containing up to 6×10^6 reticulocytes and up to 2×10^8 total RBCs in a reasonable time frame. Such a dramatic increase in throughput increases the statistical power of the assay and, therefore, its ability

Table 11.1 Flow cytometry–based in vivo *Pig-a* assays designed for various species and cell types

Glycosylphosphatidylinositol - anchored markers	Other essential markers, stains, and steps	References
Rat, peripheral blood RBCs		
CD59	CD45	[41]
CD59	GC, Syto 13	[42]
CD59	GC, Syto 13, CD61, MS	[43]*
CD59	HIS49	[44]
CD59	HIS49, Syto 59	[45]
CD59	GC, CD71, MS, HIS49	[46]*
Rat, peripheral blood (spleen) T-lymphocytes		
CD48	GC, CD3, 7AAD	[41]
CD48	GC, RBC lysis, CD3, 7AAD (MS/sorting)	[32]
Rat, bone marrow erythroid cells		
CD59	Hoechst, CD71 (HIS49) (MS/sorting)	Unpublished
Rat, bone marrow granulocytes		
CD48	RBC lysis, HIS49, CD11b (MS/sorting)	Unpublished
Rat, epididymis spermatozoa		
CD59	Only light scatter	Unpublished
Mouse, peripheral blood RBCs		
CD24	GC, Syto 13	[47]
CD24	GC, Syto 13, MS	[48]*
CD24	Ter119, Syto 59	[45]
CD24	GC, CD71, MS, Ter119	[49]*
Mouse, bone marrow erythroid cells		
CD24	GC, CD71, Ter119 (MS/sorting)	[49]
Human, peripheral blood RBCs		
CD59	Glycophorin	[50]
CD55, CD59	GC, Syto 13, CD61, MS	[51]*

Table 11.1 Flow cytometry–based in vivo *Pig-a* assays designed for various species and cell types—cont'd

Glycosylphosphatidylinositol - anchored markers	Other essential markers, stains, and steps	References
Human, peripheral blood granulocytes		
CD55, CD59	GC, RBC lysis, CD11b (MS/sorting)	[29,52]
Rhesus monkey, peripheral blood RBCs		
CD59	Glycophorin	[16]

Unique gating strategies were developed for each assay. For details consult the original articles. Markers labeled with fluorescent monoclonal antibodies and their purpose in *Pig-a* assays: *CD11b*, granulocyte marker (for positive identification of granulocytes in bone marrow and peripheral blood samples); *CD3*, T-cell receptor (for positive identification of T-cells); *CD45*, pan-leukocyte marker (for gating out contaminating WBCs); *CD59, CD48, CD55, CD24*, GPI-anchored cell surface markers, reporters of *Pig-a* mutation; *CD61*, integrin; platelet marker (for gating out contaminating platelets); *CD71*, transferrin receptor (for positive identification of reticulocytes in peripheral blood and erythroid cells in bone marrow samples); *Glycophorin*, erythroid marker (for positive identification of RBCs in peripheral blood samples); *HIS48*, granulocyte marker (for positive identification of granulocytes in bone marrow samples); *HIS49*, unknown rat erythroid marker (for positive identification of peripheral blood RBCs and bone marrow erythroid precursors); *Ter119*, unknown mouse erythroid marker (for positive identification of RBCs in peripheral blood and erythroid cells in bone marrow samples). Stains and their purpose: *7AAD*, 7-aminoactinomycin D, viability indicator (for gating out dead WBCs); *Hoechst*, Hoechst 33342, DNA-specific stain (for identification of nucleated cells); *Syto 13, Syto 59*, nucleic acid stains (for positive identification of RNA-containing RBCs and gating out DNA-containing contaminating WBCs in peripheral blood).
Steps and their purpose:
GC, gradient centrifugation, depleting peripheral blood off mononuclear cells, depleting spleen cells off contaminating RBCs; purification of granulocyte fraction from peripheral blood; *MS*, magnetic separation (enrichment for mutant cells in high-throughput assays; enrichment for reticulocyte fraction in high-throughput assay; enrichment for mutants for sorting applications); *RBC lysis*, selective lysis of RBCs in hypotonic buffer.
(....), entries in parentheses indicate optional steps.
*, high-throughput RBC protocols.

to identify relatively weak mutagenic responses. The high-throughput immunomagnetic protocols have been employed in multilaboratory trials in which the portability and reproducibility of the methodology were evaluated, using a variety of instrument platforms and various inbred/outbred rodents that are common in toxicological research. A consensus was formed that the erythrocyte *Pig-a* assay can be useful in mutation research and specifically in regulatory studies [31]. An OECD TG for performing the erythrocyte *Pig-a* assay and interpreting the data is currently being prepared.

Until recently there were concerns that the RBC mutant phenotype measured by flow cytometry might be due to events other than mutation in the endogenous *Pig-a* gene (e.g., due to a transient epigenetic event or even due to flow cytometer errors). But the recent demonstration that flow-sorted GPI-anchored marker-deficient rat T-cells contain *Pig-a* mutations [32,33], as well as the wealth of genetic evidence from PNH studies, suggests that a properly designed *Pig-a* mutant-phenotype detection assay detects mutation in the endogenous *Pig-a* gene.

The erythrocyte *Pig-a* mutation detection assay does not detect LOH mutation, and it is limited to analyzing mutations in hematopoietic tissues, at least at the present time. Theoretically, a *Pig-a* assay can be established for any type of cells that can be labeled with antibodies/stains and analyzed on a flow cytometer, i.e., "nonsticky" cells that can be dispersed as a single-cell suspension and those that experienced minimal damage to the repertoire of surface markers during the extraction protocol. Preliminary data reported publically at scientific meetings (as of 2016) suggest that sperm might be a good target for establishing a germline *Pig-a* assay.

7. THE FUTURE OF IN VIVO MUTATION DETECTION

Detection of in vivo mutation is still a complicated process. The methods described in this review rely on reporter genes (whether they are endogenous or bacterial transgenes) and require multiple steps for developing/visualizing a specific phenotype. Even if the phenotype suggests a mutation in the reporter, actual confirmation of mutation may be required to understand the true mutation frequency in the reporter gene (e.g., to differentiate independent and clonally expanded mutations). Also sequencing may be required to confirm the effect of weak mutagens (including mutagens that do not induce a strong response in the reporter gene, but nonetheless affect the spectrum of genetic changes). Sequencing can be incorporated into the models described above, with different degrees of difficulty. But sequencing will not resolve the concern that some reporters are not capable of detecting certain types of mutations; while other reporters may be under different endogenous genetic integrity surveillance and be subject to different DNA repair pathways. All these factors make it difficult to use mutation data from current models in any quantitative manner.

While technologies such as transgenics and flow cytometry have improved our ability to evaluate in vivo mutation for safety assessment, a relatively new technology, massively parallel sequencing, commonly referred to as next generation sequencing (NGS), is poised to transform in vivo mutation testing in the future. Currently NGS is used widely in cancer research, but not in routine safety assessments. The most straightforward use of NGS in mutagenicity studies is in detecting germline mutation transmission [34,35]. Using off-the-shelf chemistry and informatics tools, NGS also can be employed for monitoring somatic mutation. For instance, T-cells and fibroblasts can clonally expand ex vivo. Whole genome sequencing can be performed on several clones expanded from a subject (e.g., a lab animal treated with a mutagen). Comparing genomes determined for each clone individually against the consensus genome sequence (constructed using information determined for all sequenced clones) will reveal mutational load in each clone. Such an approach may be useful for studying mutation in animals for which a reference genome is unavailable (e.g., for hybrid and outbred animals or humans).

Alternative NGS methods can be useful for detecting mutation in any tissue of any species. Multiple protocols for whole genome amplification and sequencing at the level of a single cell have been described [36,37]. Usually the genome of an in vivo–derived cell is amplified in vitro using a high fidelity polymerase, e.g., ϕ29 polymerase. Assuming that sequencing accuracy of a ϕ29-amplified genome is reasonable, then just a few separate cells (rather than a few expanded clones) from any tissue of a subject may be sufficient to evaluate mutational load in the tissue.

Additional possibilities for using NGS in detection of in vivo mutation may open up if the fidelity of individual sequencing reads is improved from the current 10^{-3} error rate (1 error per 1000 nucleotides). This may allow detecting mutations in tissue samples consisting of multiple cells in cases when isolation of individual cells is difficult or unfeasible (e.g., from solid tissues). In this case it is not the whole genome of each cell that will be sequenced, but rather many random DNA fragments from the tissue sample. The multiple fragments will be sufficient for building a consensus sequence of the whole genome, and each individual mutant fragment will signify a true mutation. Mutant frequency in such samples will be expressed as the number of mutations per length of sequenced DNA. Duplex sequencing available for the Illumina platform [38] and novel approaches to chemistry and hardware [39,40] already have sufficient sequencing fidelity for detecting rare mutations in tissue samples.

It is anticipated that with time, NGS methodologies will become more affordable and will generate data with higher fidelity. Thus, we foresee that NGS will provide sensitive and comprehensive information on tissue-specific mutational spectra and on the spatial distribution of mutation across the whole genome, including genes relevant to toxicity and cancer, for agents of concern.

DISCLAIMER

The opinions presented in this chapter do not necessarily reflect the official views or policies of the US Food and Drug Administration.

REFERENCES

[1] Jeffrey EC. Drosophila and the mutation hypothesis. Science 1925;62(1592):3–5.
[2] Kossikov KV. The influence of age and sex of the germ cells on the frequency of mutation in *Drosophila Simulans* and *Drosophila Melanogaster*. Genetics 1937;22(2): 213–24.
[3] Russell WL, Russell LB, Cupp MB. Dependence of mutation frequency on radiation dose rate in female mice. Proc Natl Acad Sci USA 1959;45(1):18–23.
[4] Russell WL, Kelly EM, Hunsicker PR, Bangham JW, Maddux SC, Phipps EL. Specific-locus test shows ethylnitrosourea to be the most potent mutagen in the mouse. Proc Natl Acad Sci USA 1979;76(11):5818–9.
[5] Ames BN. Carcinogens are mutagens: their detection and classification. Environ Health Perspect 1973;6:115–8.
[6] Ames BN, Durston WE, Yamasaki E, Lee FD. Carcinogens are mutagens: a simple test system combining liver homogenates for activation and bacteria for detection. Proc Natl Acad Sci USA 1973;70(8):2281–5.
[7] Cimino MC. Comparative overview of current international strategies and guidelines for genetic toxicology testing for regulatory purposes. Environ Mol Mutagen 2006;47(5):362–90.
[8] Clive D, Johnson KO, Spector JF, Batson AG, Brown MM. Validation and characterization of the L5178Y/TK$^{+/-}$ mouse lymphoma mutagen assay system. Mutat Res 1979;59(1):61–108.
[9] Kelley WN, Wilson JM. Human hypoxanthine-guanine phosphoribosyltransferase: studies of the normal and five mutant forms of the enzyme. Trans Am Clin Climatol Assoc 1983;94:91–9.
[10] Albertini RJ. *HPRT* mutations in humans: biomarkers for mechanistic studies. Mutat Res 2001;489(1):1–16.
[11] Aidoo A, Lyn-Cook LE, Mittelstaedt RA, Heflich RH, Casciano DA. Induction of 6-thioguanine-resistant lymphocytes in Fischer 344 rats following in vivo exposure to N-ethyl-N-nitrosourea and cyclophosphamide. Environ Mol Mutagen 1991;17(3):141–51.
[12] Dobrovolsky VN, Shaddock JG, Heflich RH. Analysis of in vivo mutation in the *Hprt* and *Tk* genes of mouse lymphocytes. Methods Mol Biol 2014;1105:255–70.
[13] Dobrovolsky VN, Shaddock JG, Mittelstaedt RA, Miura D, Heflich RH. Detection of in vivo mutation in the *Hprt* and *Pig-a* genes of rat lymphocytes. Methods Mol Biol 2013;1044:79–95.
[14] Horn PL, Turker MS, Ogburn CE, Disteche CM, Martin GM. A cloning assay for 6-thioguanine resistance provides evidence against certain somatic mutational theories of aging. J Cell Physiol 1984;121(2):309–15.
[15] Turker MS, Monnat Jr RJ, Fukuchi K, Johnston PA, Ogburn CE, Weller RE, et al. A novel class of unstable 6-thioguanine-resistant cells from dog and human kidneys. Cell Biol Toxicol 1988;4(2):211–23.
[16] Dobrovolsky VN, Shaddock JG, Mittelstaedt RA, Manjanatha MG, Miura D, Uchikawa M, et al. Evaluation of *Macaca mulatta* as a model for genotoxicity studies. Mutat Res 2009;673(1):21–8.
[17] Zimmer DM, Aaron CS. In vivo mutagenesis in the cynomolgus monkey: time course of *HPRT* mutant frequency at long time points following ethylnitrosourea exposure. Environ Mol Mutagen 1997;29(2):117–23.

[18] Van Sloun PP, Wijnhoven SW, Kool HJ, Slater R, Weeda G, van Zeeland AA, et al. Determination of spontaneous loss of heterozygosity mutations in *Aprt* heterozygous mice. Nucleic Acids Res 1998;26(21):4888–94.

[19] Stambrook PJ, Shao C, Stockelman M, Boivin G, Engle SJ, Tischfield JA. *APRT*: a versatile in vivo resident reporter of local mutation and loss of heterozygosity. Environ Mol Mutagen 1996;28(4):471–82.

[20] Dobrovolsky VN, Casciano DA, Heflich RH. $Tk^{+/-}$ mouse model for detecting in vivo mutation in an endogenous, autosomal gene. Mutat Res 1999;423(1–2):125–36.

[21] Gossen JA, de Leeuw WJ, Tan CH, Zwarthoff EC, Berends F, Lohman PH, et al. Efficient rescue of integrated shuttle vectors from transgenic mice: a model for studying mutations in vivo. Proc Natl Acad Sci USA 1989;86(20):7971–5.

[22] Nohmi T, Katoh M, Suzuki H, Matsui M, Yamada M, Watanabe M, et al. A new transgenic mouse mutagenesis test system using *Spi-* and 6-thioguanine selections. Environ Mol Mutagen 1996;28(4):465–70.

[23] Kohler SW, Provost GS, Fieck A, Kretz PL, Bullock WO, Sorge JA, et al. Spectra of spontaneous and mutagen-induced mutations in the *lacI* gene in transgenic mice. Proc Natl Acad Sci USA 1991;88(18):7958–62.

[24] Boerrigter ME, Dolle ME, Martus HJ, Gossen JA, Vijg J. Plasmid-based transgenic mouse model for studying in vivo mutations. Nature 1995;377(6550):657–9.

[25] Jakubczak JL, Merlino G, French JE, Muller WJ, Paul B, Adhya S, et al. Analysis of genetic instability during mammary tumor progression using a novel selection-based assay for in vivo mutations in a bacteriophage lambda transgene target. Proc Natl Acad Sci USA 1996;93(17):9073–8.

[26] Lambert IB, Singer TM, Boucher SE, Douglas GR. Detailed review of transgenic rodent mutation assays. Mutat Res 2005;590(1–3):1–280.

[27] Kinoshita T, Fujita M, Maeda Y. Biosynthesis, remodelling and functions of mammalian GPI-anchored proteins: recent progress. J Biochem 2008;144(3):287–94.

[28] Ware RE, Pickens CV, DeCastro CM, Howard TA. Circulating PIG-A mutant T lymphocytes in healthy adults and patients with bone marrow failure syndromes. Exp Hematol 2001;29(12):1403–9.

[29] Araten DJ, Nafa K, Pakdeesuwan K, Luzzatto L. Clonal populations of hematopoietic cells with paroxysmal nocturnal hemoglobinuria genotype and phenotype are present in normal individuals. Proc Natl Acad Sci USA 1999;96(9):5209–14.

[30] Miura D, Dobrovolsky VN, Kimoto T, Kasahara Y, Heflich RH. Accumulation and persistence of *Pig-A* mutant peripheral red blood cells following treatment of rats with single and split doses of N-ethyl-N-nitrosourea. Mutat Res 2009;677(1–2):86–92.

[31] Gollapudi BB, Lynch AM, Heflich RH, Dertinger SD, Dobrovolsky VN, Froetschl R, et al. The in vivo *Pig-a* assay: a report of the International Workshop On Genotoxicity Testing (IWGT) Workgroup. Mutat Res Genet Toxicol Environ Mutagen 2015;783:23–35.

[32] Revollo J, Pearce MG, Petibone DM, Mittelstaedt RA, Dobrovolsky VN. Confirmation of *Pig-a* mutation in flow cytometry-identified CD48-deficient T-lymphocytes from F344 rats. Mutagenesis 2015;30(3):315–24.

[33] Dobrovolsky VN, Revollo J, Pearce MG, Pacheco-Martinez MM, Lin H. CD48-deficient T-lymphocytes from DMBA-treated rats have de novo mutations in the endogenous *Pig-a* gene. Environ Mol Mutagen 2015;56(8):674–83.

[34] Masumura K, Toyoda-Hokaiwado N, Ukai A, Gondo Y, Honma M, Nohmi T. Dose-dependent *de novo* germline mutations detected by whole-exome sequencing in progeny of ENU-treated male *gpt* delta mice. Mutat Res 2016;810:30–9.

[35] Masumura K, Toyoda-Hokaiwado N, Ukai A, Gondo Y, Honma M, Nohmi T. Estimation of the frequency of inherited germline mutations by whole exome sequencing in ethyl nitrosourea-treated and untreated *gpt* delta mice. Genes Environ 2016;38:10.

[36] Picher AJ, Budeus B, Wafzig O, Kruger C, Garcia-Gomez S, Martinez-Jimenez MI, et al. TruePrime is a novel method for whole-genome amplification from single cells based on *Tth*PrimPol. Nat Commun 2016;7:13296.

[37] Sabina J, Leamon JH. Bias in whole genome amplification: causes and considerations. Methods Mol Biol 2015;1347:15–41.

[38] Schmitt MW, Kennedy SR, Salk JJ, Fox EJ, Hiatt JB, Loeb LA. Detection of ultra-rare mutations by next-generation sequencing. Proc Natl Acad Sci USA 2012;109(36): 14508–13.

[39] Gregory MT, Bertout JA, Ericson NG, Taylor SD, Mukherjee R, Robins HS, et al. Targeted single molecule mutation detection with massively parallel sequencing. Nucleic Acids Res 2016;44(3):e22.

[40] Travers KJ, Chin CS, Rank DR, Eid JS, Turner SW. A flexible and efficient template format for circular consensus sequencing and SNP detection. Nucleic Acids Res 2010;38(15):e159.

[41] Miura D, Dobrovolsky VN, Kasahara Y, Katsuura Y, Heflich RH. Development of an in vivo gene mutation assay using the endogenous *Pig-A* gene: I. Flow cytometric detection of CD59-negative peripheral red blood cells and CD48-negative spleen T-cells from the rat. Environ Mol Mutagen 2008;49(8):614–21.

[42] Phonethepswath S, Franklin D, Torous DK, Bryce SM, Bemis JC, Raja S, et al. *Pig-a* mutation: kinetics in rat erythrocytes following exposure to five prototypical mutagens. Toxicol Sci 2010;114(1):59–70.

[43] Bemis JC, Hall NE, Dertinger SD. Erythrocyte-based *Pig-a* gene mutation assay. Methods Mol Biol 2013;1044:51–77.

[44] Dobrovolsky VN, Boctor SY, Twaddle NC, Doerge DR, Bishop ME, Manjanatha MG, et al. Flow cytometric detection of *Pig-A* mutant red blood cells using an erythroid-specific antibody: application of the method for evaluating the in vivo genotoxicity of methylphenidate in adolescent rats. Environ Mol Mutagen 2010;51(2):138–45.

[45] Dobrovolsky VN, Cao X, Bhalli JA, Heflich RH. Detection of *Pig-a* mutant erythrocytes in the peripheral blood of rats and mice. Methods Mol Biol 2014;1105:205–21.

[46] Kimoto T, Chikura S, Suzuki K, Kobayashi X, Itano Y, Horibata K, et al. Further development of the rat *Pig-a* mutation assay: measuring rat *Pig-a* mutant bone marrow erythroids and a high throughput assay for mutant peripheral blood reticulocytes. Environ Mol Mutagen 2011;52(9):774–83.

[47] Phonethepswath S, Bryce SM, Bemis JC, Dertinger SD. Erythrocyte-based *Pig-a* gene mutation assay: demonstration of cross-species potential. Mutat Res 2008;657(2):122–6.

[48] Labash C, Avlasevich SL, Carlson K, Berg A, Torous DK, Bryce SM, et al. Mouse *Pig-a* and micronucleus assays respond to N-ethyl-N-nitrosourea, benzo[a]pyrene, and ethyl carbamate, but not pyrene or methyl carbamate. Environ Mol Mutagen 2016;57(1):28–40.

[49] Kimoto T, Suzuki K, Kobayashi XM, Dobrovolsky VN, Heflich RH, Miura D, et al. Manifestation of *Pig-a* mutant bone marrow erythroids and peripheral blood erythrocytes in mice treated with N-ethyl-N-nitrosourea: direct sequencing of *Pig-a* cDNA from bone marrow cells negative for GPI-anchored protein expression. Mutat Res 2011;723(1):36–42.

[50] Dobrovolsky VN, Elespuru RK, Bigger CA, Robison TW, Heflich RH. Monitoring humans for somatic mutation in the endogenous *PIG-a* gene using red blood cells. Environ Mol Mutagen 2011;52(9):784–94.

[51] Dertinger SD, Avlasevich SL, Bemis JC, Chen Y, MacGregor JT. Human erythrocyte *PIG-A* assay: an easily monitored index of gene mutation requiring low volume blood samples. Environ Mol Mutagen 2015;56(4):366–77.

[52] Rondelli T, Berardi M, Peruzzi B, Boni L, Caporale R, Dolara P, et al. The frequency of granulocytes with spontaneous somatic mutations: a wide distribution in a normal human population. PLoS One 2013;8(1):e54046.

CHAPTER TWELVE

An In Vitro Male Germ Cell Assay and Its Application for Detecting Phase Specificity of Genotoxins/ Mutagens

Khaled Habas, Martin H. Brinkworth, Diana Anderson
University of Bradford, Bradford, United Kingdom

1. INTRODUCTION

Germline mutations resulting from exposure to genotoxic and mutagenic agents are of great concern because they not only affect the exposed generation but also may be transmitted to their offspring [1].

Genetic toxicology and mutation fields in vitro are now focusing on determining how different germ cell types may contribute to possible genetic effects. To assess genotoxicity, different endpoints should be taken into consideration including beside point mutation, changes in chromosome number (polyploidy or aneuploidy) and structure (breaks, deletions, rearrangements, etc.). Characterization of gene mutations and understanding of these spermatogenic events may provide further insight into the role of environmental mutagens in male germ cells. Isolation of spermatogenic cells from testes is one of the critical steps to address these important issues [1]. In vivo and in vitro genotoxicity tests have been developed with different endpoints to detect genetic damage and its biological consequences in different germ cell types [2]. Generally, this field still lacks novel, more sensitive, less animal-intensive, and higher-throughput approaches to detect genotoxic effects in the male germ cells and associated genetic changes [3]. The ability to study them directly has also been hampered due to the lack of unique markers [4] and the difficulties to obtain testis tissues from humans [5]. There is a parallelogram approach, originally developed by Sobels in the late 1970s [6]. It is still used and applied to extrapolate these genotoxic consequences from in vitro to in vivo human germ cells [7]. This method includes the morphological specific locus test and the dominant lethal assay.

Mutagenicity: Assays and Applications
ISBN 978-0-12-809252-1
http://dx.doi.org/10.1016/B978-0-12-809252-1.00012-2

251

These assays have low sensitivity, use large numbers of progeny to provide the suggested sample size, and are very time-consuming and expensive [8].

Currently, there are many types of tests, which have been developed and could help in this parallelogram approach, because these new assays can detect mutagenic/genotoxic events in male germ cells directly. This new approach could also be suitable for extrapolating genetic changes in vitro to human populations, compared with the older classical approaches [8]. There has been a growing interest in the use of "omics" tools to monitor the development of biomarker panels to help predict whether a compound would cause a specific response under specific exposure conditions [9]. Recently, numbers of alternative animal methods for measuring the mutagenic/genotoxic of chemicals have been developed [10]. Male germ cells can be separated from whole testis tissues by different approaches; one method isolates cells using the velocity sedimentation (STA-PUT). Male germ cells in mice, including spermatogonia, spermatocytes, and spermatids, have been isolated by STA-PUT [11,12].

Our ability to isolate rodent spermatogonia, spermatocytes, and spermatids could provide novel targets for examination of the three phases of spermatogenesis for regulatory purposes, in the reduction in the number of animals, or offer new targets for gene therapy for the treatment of male infertility.

Here we describe a cost-effective and time-saving approach that uses a single protocol to enrich multiple testicular cell populations (spermatogonia, spermatocytes, and spermatids) from very few animals.

2. SPERMATOGENESIS

Spermatogenesis is a highly complex process characterized by mitotic, meiotic, and haploid differentiation phases. Spermatogenesis begins in the seminiferous tubules of the testes in which spermatogonial stem cells proliferate and differentiate into type A1 spermatogonia. Type A1 spermatogonia undergo a series of coordinated mitotic divisions, giving rise to type B spermatogonia, which enter the meiotic phase of spermatogenesis and then develop as primary spermatocytes [13]. Spermatocytes undergo two rounds of meiotic divisions, following a single DNA duplication, and by genetic exchange among homologous chromosomes produce four haploid spermatids. The spermatids then undergo extensive chromatin condensation and morphological changes to develop into mature spermatozoa Fig. 12.1 [14]. The mechanisms of regulation of this orchestrated developmental process,

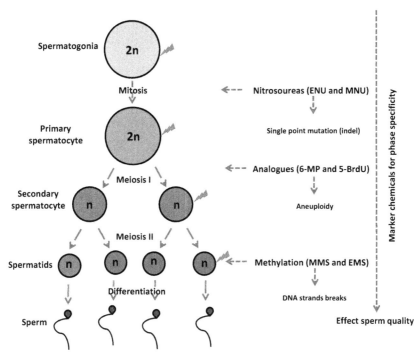

Figure 12.1 Diagram showing the cycle of the germ cells and different mutations that are predominant in the different phases of spermatogenesis.

including proliferation, differentiation at the intratesticular and extratesticular level, can be disturbed at every event of spermatogenesis [15]. Exposure to environmental genotoxicants can cause de novo mutations in spermatogenesis and lead to genetic health concerns for future generations [16]. Early phases of spermatogenesis have good repair mechanisms that allow detection and repair of damaged DNA.

However, mutagenicity and genotoxicity in male germ cells are of special concern because the later phase of spermatogenesis becomes more transcriptionally quiescent and the capability to repair damaged DNA is lost. Ultimately, unrepaired damage in spermatogenesis leads to the production of mutated spermatozoa and possible genetic defects in the offspring [17]. The molecular mechanisms causing these genetic defects as a result of exposure directly or indirectly to environmental toxicants can affect spermatogenesis and could lead to infertility in men [18]. Environmental exposures can affect male fertility by causing a reduction in sperm parameters such as sperm count, motility, etc. and damage the genetic integrity in the male germ cell. The interaction between genetic and environmental factors during male germ cell development is thought to be a possible determinant of germ cell mutation. It is known that spermatogenesis is controlled by many genetic and epigenetic factors.

Many studies have been trying to determine the molecular mechanism underlying spermatogenesis in rodents, but little is known about genetic and epigenetic regulation of spermatogonia, spermatocytes, and spermatids in humans due to the difficulties in obtaining human testicular tissues [5]. Further improvement of animal models and the combination of these with possible human investigation of sperm are required to provide better molecular understandings into the mechanisms of germline mutation [19].

In the following sections, we shall present the promising model system that has historically helped our understanding of germ cell differentiation and function. This may also circumvent some of the difficulties inherent to the study of germline mutation. We shall also discuss the different susceptibilities to the DNA damage response and the variability in genetic sensitivity among germ cell phases to various mutagens.

3. ASSESSMENT OF MUTAGENIC AGENTS CAUSING DNA DAMAGE IN MALE GERM CELLS IN THE STA-PUT SYSTEM

There has been a continuing concern about human health from exposures to environmental compounds. We have developed a new and sensitive method for evaluation of DNA damage relating to phase specificity of germ cell mutagens in vitro. However, assessment of germ cell damage plays a critical role in classifying and labeling compounds for mutagenic potential. The level of gene expression in response to DNA damage is dependent on both the gentoxicant and the type of germ cell in vitro [20]. DNA damage from specific double-strand breaks (DSBs) is difficult to repair and highly toxic [21]. DSBs can occur as a result of endogenous and exogenous insults, for example, genotoxic stress and replication fork defects [22]. Alkylating agents induce mutagenesis via transferring an ethyl group to nucleophilic oxygen or nitrogen sites on DNA, leading to base mismatch during DNA replication [23]. Also, a base or nucleoside analogue is a genotoxic compound that is incorporated into DNA, causing specific locus mutations and inhibition of cell proliferation [24]. Sega [25] has reported that the variability in genetic sensitivity among germ cell phases to several mutagenic agents can clarify how much chemical reaches a target and what molecular targets could be affected between different phases and also whether or not repair of DNA damage has occurred. However, various chemicals have been found to induce genetic mutation in early stages of germ cells (premeiotic). For example, Russell [26] showed in vivo that the N-ethyl-N-nitrosourea (ENU) produced the greatest increase in DNA damage in spermatogonia. Other chemicals incorporated into DNA in vivo have shown that the meiotic phase was most sensitive to base analogue agents such as 6-mercaptopurine (6-MP) and 5-bromo-2'-deoxy-uridine (5-BrdU) [27,28]. Several chemicals have been found to bind strongly to protamine in spermatid and early spermatozoa stages in vivo such as methyl methanesulfonate and ethyl methanesulfonate (postmeiotic phase) as shown in Table 12.1 [29]. These chemicals, which we have investigated, also yield their highest genetic damage in these same germ cell phases in vitro.

The conclusions from the experimental outcomes compiled above outline a model that has different cell types with different susceptibilities to these chemicals. Crucially, we have been able to validate this without the need for time-consuming, resource-intensive animal studies.

Table 12.1 Different responses for DNA damage in spermatogenic cells after treatment with ENU, MNU, 6-MP, 5-BrdU, MMS, and EMS in the Comet assay

		Treatment concentration 0.05 mM		
Compound	Species	Spermatogonia	Spermatocytes	Spermatids
N-ethyl-	Mouse [a]	*	ns	ns
N-nitrosourea	Rat[b]	*	ns	ns
N-methyl-	Mouse [a]	*	ns	ns
N-nitrosourea	Rat[b]	ND	ND	ND
Methyl	Mouse [a]	ns	ns	*
methanesulfonate	Rat[b]	ns	ns	*
Ethyl	Mouse [a]	ns	ns	*
methanesulfonate	Rat[b]	ND	ND	ND
6-Mercaptopurine	Mouse [a]	ns	*	ns
	Rat[b]	ns	*	ns
5-Bromo-	Mouse [a]	ns	*	ns
2'-deoxyuridine	Rat[b]	ns	ND	ns

*, statistically significant increase above untreated controls; *ND*, not done; *ns*, nonsignificant. *P* value of >0.05 was considered nonsignificant by comparison with untreated cells (negative control) and when a *P* value of <0.05 was considered significant.

5-BrdU, 5-bromo-2'-deoxy-uridine; *6-MP*, 6-mercaptopurine; *EMS*, ethyl methanesulfonate; *ENU*, N-ethyl-N-nitrosourea; *MMS*, methyl methanesulfonate; *MNU*, N-methyl-N-nitrosourea.
[a]NMRI mice.
[b]Sprague–Dawley rats.

4. METHODS EXISTING FOR DETECTION SIGNALS FROM POTENTIAL GERM CELL MUTAGENS

Over the last decade, a number of assays have been developed and many of these assays have been used widely to measure the DNA damage from germ cell mutagens. Among these assays, the comet assay and micronucleus assay are the most frequently used. Recent studies report that the combination of the comet and micronucleus assays is the best battery of tests, because both assays are rapid and highly sensitive tests and allow the detection of DNA-strand breaks, and/or genome mutations and chromosome changes, respectively [30,31]. Different forms of premutational and mutational alterations can be detected in male germ cells, including DNA breaks and abasic sites, which can be measured using the comet assay [32,33]. Our studies, with germ cells isolated through STA-PUT, have shown that the Comet assay is a usable, rapid, and sensitive assay to detect DNA damage in primary testis cell cultures in rodents and offers a useful system for genotoxicity assessment of effects in spermatogenic cells without relying solely on animal studies [34]. The advantage of the comet assay in vitro is the lack of the necessity for cell proliferation. However, other assays, ex vivo/in vitro, have possible access to all tissues for investigation [35]. Furthermore, the in vivo comet assay requires special expensive transgenic animals and has the possibility to be combined into subchronic toxicology studies united with the standard micronucleus assay compared with the transgenic rodent mutation assay [36]. Therefore this method is applied in clinical diagnosis and management of male infertility [3]. The micronucleus assay is also used for genotoxicity screening of new and promising applications in medicine and biology [37]. A chromosomal aberration was identified in spermatocytes, and micronuclei were detected in spermatids [38,39]. Recommendations and trial protocol guidelines on performing both alkaline and neutral versions of the comet assay and micronucleus assay have been reported by various experts [30,40]. In addition to the comet and micronucleus assays, the terminal deoxynucleotidyl transferase–mediated (TdT) deoxyuridine triphosphate (dUTP) nick end labeling assay (TUNEL) is the most widely used to measure sperm DNA integrity [3]. This assay assesses DNA breaks in situ as assessed via the incorporation of dUTP at the sites of breaks [41]. Similar results were found with the TUNEL assay, in primary testis cell cultures in rodents in the same way as for DNA damage [20].

Detection and identification of germline genetic mutations in spermatogenic cells is critical to the understanding of their molecular mechanisms and growth regulation during the spermatogenesis process. There are a number of genes that have been identified which, if mutated, are related to germline mutation. However, many methods exist for the detection of germ cell mutation. Real-time polymerase chain reaction (PCR) amplification methods and especially those with the capacity of high multiplexing have become universally utilized in laboratory practice [42]. These methods include the use of allele-specific competitive blocker PCR [43], real-time genotyping with locked nucleic acids [44], and allele-specific kinetic PCR in conjunction with modified polymerases [45]. However, a genetic mutation is a main key parameter used for the molecular characterization of germline mutation and cancer. The ability to detect these mutations on RNA and DNA during the spermatogenesis process requires the isolation and enrichment of these different cell types.

5. MUTAGENICITY AND GENOTOXICITY TESTING IN GERM CELLS USING STA-PUT VELOCITY SEDIMENTATION

Mammalian spermatogenesis is a very complex differentiation process that occurs in successive mitotic, meiotic, and postmeiotic phases in the seminiferous tubules of the testis. The possibility of obtaining homogeneous germ cells at defined stages of development is a prerequisite for the study of meiotic and postmeiotic germ cell function. Because of the complex process of spermatogenesis, which depends on special hormones and different molecular conditions in the testes, a dependable in vitro culture system for the full procedure of germ cells has not yet been fully developed because it is usually not possible to culture the cells for longer than 24 h [46]. Methods of cell culture have been established to create or to mimic cells, which are similar to primordial germ cells. However, to date these methods have been unable to produce large numbers of these cells and also failed to generate the later spermatogenic cell types in vitro [47]. It is well known that the spermatogenic cell types are significantly different in size; this difference allows for a single cell suspension gained from whole testes to be separated with a liquid gradient [48].

Several methods have been proposed for separation of animal cells based on velocity sedimentation [49,50], equilibrium density fractionation [51], magnetic activated cell sorting (MACS) [52], and fluorescence-activated cell sorting (FACS) [53].

The STA-PUT method is a unit gravity cell sedimentation system that allows the purification of specific cell types based on size and density through a linear bovine serum albumin (BSA) gradient. The optimal conditions for rodent cells have been reported [48]. The successful application of this technique to male germ cell fractions has been described [54]. The STA-PUT has numerous advantages compared with the two commonly used approaches for isolating germ cells types: FACS and elutriation. The STA-PUT method needs numerous pieces of specific glassware assembled in a cold room. Therefore, it is cheaper than using a cell sorter or an elutriator [55]. The STA-PUT method yields higher amounts of cells per cell type and the purity of each cell population is higher than those obtained with FACS [56]. Currently, cell sorting using magnetic beads, MACS, has been used for enrichment of spermatogonial cells from a mixed germ cell population; however, it is not appropriate for isolation of spermatocytes and spermatids because of the lack of information of suitable surface markers [57]. For a study that needs large number of yields of germ cell types at ~90% purity, the STA-PUT is a perfect technique [48].

In animal tests the rule of the 3R's, **reduction, refinement,** and **replacement**, has been established to be an area of common ground for researchers who use animals. This method used in these studies has achieved this standardization. This could lead to attempts to introduce this method of genotoxicity enrichment for a wide range of species.

The STA-PUT method, which requires very few animals compared with traditional methods, aids in the **reduction** of animal numbers. Since the animals are not treated and cells are treated in vitro, this **refines** toxicological approaches. Thus, because this approach combines these advantages, its use in a genotoxicity testing strategy could possibly replace in vivo testing and provide a suitable model in vitro to study DNA damage in different phases of spermatogenesis [12].

In vitro experiments have demonstrated that the STA-PUT technique could be a valuable method to make appropriate assessment of DNA, mRNA, and protein and investigate the biochemical and toxic effects of environmental and pharmacological agents on germ cells.

6. PROTOCOL OF THE STA-PUT SYSTEM

The protocol involves different steps: set up of the apparatus of the STA-PUT, preparation of cell suspension from testes, loading and sedimentation, and fraction collection.

7. SETTING UP THE STA-PUT APPARATUS

Good construction of the gradient is essential for effective cell separation. Extreme care should be taken to ensure that the gradient is both even and stable. All equipment especially the glassware and tubing will be washed and sterilized with 70% ethanol. The BSA gradient should be prepared using 250 mL of 4% and 250 mL of 2% BSA in Dulbecco's Minimum Eagle's Medium (DMEM). Both BSA solutions should be filtered before use with a 0.45-μm mounted filter. The fraction collector will be placed directly under the sedimentation chamber. All gradient chambers and the fraction tubes will be in place and covered with plastic wrap to prevent contamination. The cells have been grossly separated into Sertoli cells (adhering to the culture plates) and germ cells (in suspension in culture medium) fractions. As much as possible of the supernatant, enriched with germ cells, should be removed and transferred to fresh 50 mL tubes. The supernatant should be centrifuged for 10 min at $600 \times g$.

8. ISOLATING SPERMATOGENIC CELLS FROM WHOLE TESTES

Decapsulated tissue will be washed twice in a 50 mL conical tube in 30 mL of ice-cold DMEM. Minced tissues will be suspended in fresh DMEM and shaken vigorously for 1 min to disperse tubules. The tissue will be left to settle for 5 min on ice, and the supernatant will be discarded. This procedure will be repeated twice to mechanically remove red blood cells and free Leydig cells. The resulting pellet will be digested (I) in 10 mL of DMEM containing 5 mg/mL collagenase, 1 μg/mL DNase, and 1 mg/mL hyaluronidase. The sample will be gently swirled by hand and placed in a water bath at 32°C for 20 min with shaking at 5 min intervals. Samples will be removed from the water bath and the tubule tissue will be allowed to settle to the bottom of the tube. The supernatant in the tube containing the interstitial cells will be gently pipetted off with more care to avoid removing the partially digested tubules. Further digestion (II) with same enzyme will be performed for 20 min at 32°C [12]. The supernatant containing cells in suspension will be transferred without disturbing the larger chunks into a fresh Falcon tube containing 20 mL of DMEM + 10% fetal bovine serum (FBS) on ice to inhibit proteolytic activity. Three mL of 0.25% trypsin–EDTA and 300 μL DNase I will be added to the remaining cell pellet and swirled by hand and then placed in a water bath for 15 min incubation at 32°C. The tube will be shaken by hand every 5 min during this period to assist in dissociating the germ cells and the Sertoli cells. The supernatant will be removed gently and added to the previously prepared 50 mL Falcon tube containing 20 mL of DMEM + 10% FBS and the first supernatant removed from the digestion. This pool should now contain predominantly germ and Sertoli cells. The cell suspension will be filtered through a sterile 0.80 µm nylon filter to remove any undigested fragments or cell clumps. The filtered cells will be centrifuged at $600 \times g$ for 10 min at room temperature. The supernatant will be removed and the cell pellet resuspended in 10 mL DMEM.

9. CELL LOADING AND SEDIMENTATION

The BSA solutions will be transferred into the gradient maker apparatus. This is assembled as in Fig. 12.2 and a linear gradient will be generated from 250 mL of 2% BSA and 250 mL of 4% BSA solutions in the appropriate cell buffer chambers. Next, 50 mL of BSA will be loaded into the loading chamber. A cell suspension will be loaded and the stirrer will start to move under the tube, which makes it possible for the cells to flow through into the sedimentation chamber in about 5–7 min. The gradient valve between the two chambers will be opened to allow the two BSA solutions to mix, and then the release valve will be opened, thus allowing the BSA to flow into the syringe to generate the gradient. After 2.5 h at room temperature, cells will be collected in 12 mL fractions and eluted dropwise into 45 test tubes. Fractions will be placed on ice immediately after collection.

10. ANALYSIS OF FRACTIONS

The cells in each fraction will be examined under a phase contrast microscope, and consecutive fractions containing cells of a similar size and morphology will be spun down by low-speed centrifugation and resuspended in DMEM. The immunohistochemistry and Western blot assays should show that the purities of isolated rodent spermatogonia, spermatocytes, and spermatids should be 90%, and the viability of these isolated cells could be over 98% [20].

11. CONCLUDING REMARKS

Mutagenesis detection is a critical key for testing the genotoxic potential of compounds produced in our environment and medical applications. Both mouse and rat models are useful and complement

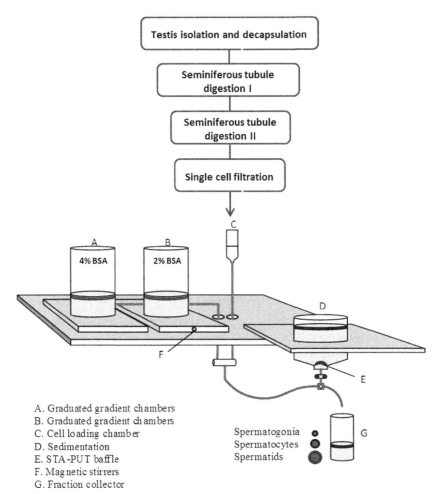

A. Graduated gradient chambers
B. Graduated gradient chambers
C. Cell loading chamber
D. Sedimentation
E. STA-PUT baffle
F. Magnetic stirrers
G. Fraction collector

Spermatogonia
Spermatocytes
Spermatids

Figure 12.2 Diagram shows the STA-PUT velocity sedimentation apparatus involving two graduated gradient chambers, a cell loading chamber, a magnetic stirrer, Teflon spin bar, a sedimentation chamber with an STA-PUT baffle, and a three way micro-metering valve. Silicone tubes connect the graduated gradient chambers, cell buffer loading, and the sedimentation chamber to each other.

each other. This chapter provides a cost-effective and time-saving protocol that is able to isolate multiple spermatogenic cell populations with high purities and viabilities from human sperm and from the testis tissues from rodents using STA-PUT velocity sedimentation. These purified cells can be successfully utilized for several genotoxicity/mutagenicity assays including the comet assay to detect DNA-strand breaks; chromosomal changes in the micronucleus assay, spermatogonia, spermatocytes, and spermatid-specific genes mRNA and DNA-expression quantified by qRT-PCR; detection of protein levels by Western blot; and specific proteins by immunohistochemistry staining. Thus, this in vitro method is a very useful technique for detecting genotoxic effects in male germ cells.

REFERENCES

[1] Chang YF, Lee-Chang JS, Panneerdoss S, MacLean 2nd JA, Rao MK. Isolation of Sertoli, Leydig, and spermatogenic cells from the mouse testis. BioTechniques 2011;51(5):341–2.

[2] Kang SH, Kwon JY, Lee JK, Seo YR. Recent advances in in vivo genotoxicity testing: prediction of carcinogenic potential using comet and micronucleus assay in animal models. J Cancer Prev 2013;18(4):277–88.

[3] Yauk CL, Aardema MJ, Benthem J, Bishop JB, Dearfield KL, DeMarini DM, et al. Approaches for identifying germ cell mutagens: report of the 2013 IWGT workshop on germ cell assays. Mutat Res Genet Toxicol Environ Mutagen 2015;783:36–54.

[4] Hofmann MC, Braydich-Stolle L, Dym M. Isolation of male germ-line stem cells; influence of GDNF. Dev Biol 2005;279(1):114–24.

[5] Liu Y, Niu M, Yao C, Hai Y, Yuan Q, Liu Y, et al. Fractionation of human spermatogenic cells using STA-PUT gravity sedimentation and their miRNA profiling. Sci Rep 2015;5:8084.

[6] Sobels FH. Some problems associated with the testing for environmental mutagens and a perspective for studies in "comparative mutagenesis". Mutat Res 1977;46(4): 245–60.

[7] Kienhuis AS, van de Poll MC, Wortelboer H, van Herwijnen M, Gottschalk R, Dejong CH, et al. Parallelogram approach using rat-human in vitro and rat in vivo toxicogenomics predicts acetaminophen-induced hepatotoxicity in humans. Toxicol Sci 2009;107(2):544–52.

[8] Verhofstad N, Linschooten JO, van Benthem J, Dubrova YE, van Steeg H, van Schooten FJ, et al. New methods for assessing male germ line mutations in humans and genetic risks in their offspring. Mutagenesis 2008;23(4):241–7.

[9] Wilson VS, Keshava N, Hester S, Segal D, Chiu W, Thompson CM, et al. Utilizing toxicogenomic data to understand chemical mechanism of action in risk assessment. Toxicol Appl Pharmacol 2013;271(3):299–308.

[10] Jung EM, Choi YU, Kang HS, Yang H, Hong EJ, An BS, et al. Evaluation of developmental toxicity using undifferentiated human embryonic stem cells. J Appl Toxicol 2015;35(2):205–18.

[11] Bellve AR, Cavicchia JC, Millette CF, O'Brien DA, Bhatnagar YM, Dym M. Spermatogenic cells of the prepuberal mouse. Isolation and morphological characterization. J Cell Biol 1977;74(1):68–85.

[12] Habas K, Anderson D, Brinkworth M. Development of an in vitro test system for assessment of male, reproductive toxicity. Toxicol Lett 2014;225(1):86–91.

[13] de Rooij DG. Proliferation and differentiation of spermatogonial stem cells. Reproduction 2001;121(3):347–54.

[14] Roeder GS. Meiotic chromosomes: it takes two to tango. Genes Dev 1997;11(20):2600–21.

[15] Sharlip ID, Jarow JP, Belker AM, Lipshultz LI, Sigman M, Thomas AJ, et al. Best practice policies for male infertility. Fertil Steril 2002;77(5):873–82.

[16] Demarini DM. Declaring the existence of human germ-cell mutagens. Environ Mol Mutagen 2012;53(3):166–72.

[17] Baarends WM, van der Laan R, Grootegoed JA. DNA repair mechanisms and gametogenesis. Reproduction 2001;121(1):31–9.

[18] Azenabor A, Ekun AO, Akinloye O. Impact of inflammation on male reproductive tract. J Reproduction Infertil 2015;16(3):123–9.
[19] Ferreira DW, Allard P. Models of germ cell development and their application for toxicity studies. Environ Mol Mutagen 2015;56(8):637–49.
[20] Habas K, Anderson D, Brinkworth M. Detection of phase specificity of in vivo germ cell mutagens in an in vitro germ cell system. Toxicology 2016;353–354:1–10.
[21] Jeruss JS, Woodruff TK. Preservation of fertility in patients with cancer. N Engl J Med 2009;360(9):902–11.
[22] Ward JF. DNA damage produced by ionizing radiation in mammalian cells: identities, mechanisms of formation, and reparability. Prog Nucleic Acid Res Mol Biol 1988;35:95–125.
[23] Imai Y, Feldman B, Schier AF, Talbot WS. Analysis of chromosomal rearrangements induced by postmeiotic mutagenesis with ethylnitrosourea in zebrafish. Genetics 2000;155(1):261–72.
[24] Morris SM. The genetic toxicology of 5-bromodeoxyuridine in mammalian cells. Mutat Res 1991;258(2):161–88.
[25] Sega GA. Molecular targets, DNA breakage, and DNA repair: their roles in mutation induction in mammalian germ cells, In: Allen JW, Bridges BA, Lyon MF, Moses MJ, Russell LB, editors. Banbury report 34: biology of mammalian germ cell mutagenesis. New York: Cold Spring Harbor Laboratory Press; 1990. p. 79–91.
[26] Russell LB, Russell WL. Frequency and nature of specific-locus mutations induced in female mice by radiations and chemicals: a review. Mutat Res 1992;296(1–2):107–27.
[27] Kanemitsu H, Yamauchi H, Komatsu M, Yamamoto S, Okazaki S, Nakayama H. Time-course changes in neural cell apoptosis in the rat fetal brain from dams treated with 6-mercaptopurine (6-MP). Histol Histopathol 2009;24(3):317–24.
[28] Levkoff LH, Marshall 2nd GP, Ross HH, Caldeira M, Reynolds BA, Cakiroglu M, et al. Bromodeoxyuridine inhibits cancer cell proliferation in vitro and in vivo. Neoplasia 2008;10(8):804–16.
[29] Anderson D, Hodge MC, Palmer S, Purchase IF. Comparison of dominant lethal and heritable translocation methodologies. Mutat Res 1981;85(6):417–29.
[30] ASSAY IVMAC. Oecd guideline for the testing of chemicals. 2014.
[31] Tice RR, Agurell E, Anderson D, Burlinson B, Hartmann A, Kobayashi H, et al. Single cell gel/comet assay: guidelines for in vitro and in vivo genetic toxicology testing. Environ Mol Mutagen 2000;35(3):206–21.
[32] Codrington AM, Hales BF, Robaire B. Spermiogenic germ cell phase-specific DNA damage following cyclophosphamide exposure. J Androl 2004;25(3):354–62.
[33] Simon L, Carrell DT. Sperm DNA damage measured by comet assay. Methods Mol Biol 2013;927:137–46.
[34] Habas K, Anderson D, Brinkworth MH. Germ cell responses to doxorubicin exposure in vitro. Toxicol Lett 2017;265.70–6.
[35] Ueno S, Kashimoto T, Susa N, Natsume H, Toya M, Ito N, et al. Assessment of DNA damage in multiple organs of mice after whole body X-irradiation using the comet assay. Mutat Res-gen Tox En 2007;634(1–2):135–45.
[36] Vasquez MZ. Combining the in vivo comet and micronucleus assays: a practical approach to genotoxicity testing and data interpretation. Mutagenesis 2010;25(2):187–99.
[37] Luzhna L, Kathiria P, Kovalchuk O. Micronuclei in genotoxicity assessment: from genetics to epigenetics and beyond. Front Genet 2013;4:131.
[38] Lahdetie J, Suutari A, Sjoblom T. The spermatid micronucleus test with the dissection technique detects the germ cell mutagenicity of acrylamide in rat meiotic cells. Mutat Res 1994;309(2):255–62.
[39] Xiao Y, Tates AD. Increased frequencies of micronuclei in early spermatids of rats following exposure of young primary spermatocytes to acrylamide. Mutat Res 1994;309(2):245–53.

[40] Azqueta A, Collins AR. The essential comet assay: a comprehensive guide to measuring DNA damage and repair. Arch Toxicol 2013;87(6):949–68.

[41] Gorczyca W, Gong J, Darzynkiewicz Z. Detection of DNA strand breaks in individual apoptotic cells by the in situ terminal deoxynucleotidyl transferase and nick translation assays. Cancer Res 1993;53(8):1945–51.

[42] Kurzawski G, Dymerska D, Serrano-Fernandez P, Trubicka J, Masojc B, Jakubowska A, et al. DNA and RNA analyses in detection of genetic predisposition to cancer. Hered Cancer Clin Pract 2012;10(1):17.

[43] Orou A, Fechner B, Utermann G, Menzel HJ. Allele-specific competitive blocker PCR: a one-step method with applicability to pool screening. Human Mutat 1995;6(2):163–9.

[44] Ugozzoli LA, Latorra D, Puckett R, Arar K, Hamby K. Real-time genotyping with oligo-nucleotide probes containing locked nucleic acids. Anal Biochem 2004;324(1):143–52.

[45] Germer S, Holland MJ, Higuchi R. High-throughput SNP allele-frequency determination in pooled DNA samples by kinetic PCR. Genome Res 2000;10(2):258–66.

[46] Dores C, Alpaugh W, Dobrinski I. From in vitro culture to in vivo models to study testis development and spermatogenesis. Cell Tissue Res 2012;349(3):691–702.

[47] Hayashi K, Ohta H, Kurimoto K, Aramaki S, Saitou M. Reconstitution of the mouse germ cell specification pathway in culture by pluripotent stem cells. Cell 2011;146(4):519–32.

[48] Bryant JM, Meyer-Ficca ML, Dang VM, Berger SL, Meyer RG. Separation of spermatogenic cell types using STA-PUT velocity sedimentation. J Vis Exp 2013;80.

[49] Bellve AR. Purification, culture, and fractionation of spermatogenic cells. Methods Enzymol 1993;225:84–113.

[50] Han SY, Zhou L, Upadhyaya A, Lee SH, Parker KL, DeJong J. TFIIAalpha/beta-like factor is encoded by a germ cell-specific gene whose expression is up-regulated with other general transcription factors during spermatogenesis in the mouse. Biol Reprod 2001;64(2):507–17.

[51] Meistrich ML, Trostle PK. Separation of mouse testis cells by equilibrium density centrifugation in renografin gradients. Exp Cell Res 1975;92(1):231–44.

[52] He Z, Kokkinaki M, Jiang J, Dobrinski I, Dym M. Isolation, characterization, and culture of human spermatogonia. Biol Reprod 2010;82(2):363–72.

[53] Mays-Hoopes LL, Bolen J, Riggs AD, Singer-Sam J. Preparation of spermatogonia, spermatocytes, and round spermatids for analysis of gene expression using fluorescence-activated cell sorting. Biol Reprod 1995;53(5):1003–11.

[54] Boucheron C, Baxendale V. Isolation and purification of murine male germ cells. Methods Mol Biol 2012;825:59–66.

[55] Bastos H, Lassalle B, Chicheportiche A, Riou L, Testart J, Allemand I, et al. Flow cytometric characterization of viable meiotic and postmeiotic cells by Hoechst 33342 in mouse spermatogenesis. Cytometry Part A 2005;65(1):40–9.

[56] Getun IV, Torres B, Bois PR. Flow cytometry purification of mouse meiotic cells. J Vis Exp 2011;50.

[57] Gassei K, Ehmcke J, Schlatt S. Efficient enrichment of undifferentiated GFR alpha 1+ spermatogonia from immature rat testis by magnetic activated cell sorting. Cell Tissue Res 2009;337(1):177–83.

CHAPTER THIRTEEN

Fluorescence In Situ Hybridization in Genotoxicity Testing

Dayton M. Petibone, Wei Ding

National Center for Toxicological Research, US Food and Drug Administration, Jefferson, AR, United States

1. INTRODUCTION

1.1 Levels of DNA Damage

Numerous endogenous and exogenous stresses can produce DNA mutations that might alter the function or change gene expression patterns of encoded proteins, which could potentially result in carcinogenesis. In terms of mutation size, there are effectively three levels. At the DNA base level, gene mutations might involve single or multiple base changes. The next level is structural chromosome aberrations, such as translocations, dicentric chromosomes, and acentric fragments. Finally, there are numerical chromosome changes, where there is a gain or loss of whole chromosomes, resulting in aneuploidies. These types of mutations are characteristic of many diseases, such as cancer, and their detection the basis for genotoxicity evaluations or exposure assessments in humans.

1.2 Approaches to Measuring Genotoxicity

The initial application of cytogenetic methods was to study chromosomes and their structure to identify inherited aberrations correlated with human disease. As a crucial means for studying inherited chromosome aberrations, the use of cytogenetic assessments to evaluate induced DNA damage was only natural. In addition to chromosome structural analysis, techniques to assess acquired DNA damage include the comet assay for studying DNA strand breaks and the micronucleus (MN) assay for studying chromatin loss. Certain advantages of these assays are that they are rapid, inexpensive, and permit analysis of DNA damage with single-cell resolution. The Organization for Economic Cooperation and Development (OECD) has developed in vitro and/or in vivo test guidelines for performing the comet

Mutagenicity: Assays and Applications
ISBN 978-0-12-809252-1
http://dx.doi.org/10.1016/B978-0-12-809252-1.00013-4

265

assay [1], MN assay [2,3], and chromosomal aberration test [4,5] in regulatory genotoxicity assessments.

Nevertheless, these conventional genotoxicity assays have a limited ability to detect important types of DNA damage. The comet assay is a rapid and straightforward means used for detecting DNA strand breaks following genotoxic exposures. Still, the comet assay is an indicator assay, as it detects DNA lesions that might be repaired or fixed into DNA mutations. The MN test detects chromatin loss, but cannot readily distinguish between clastogenic events involving segmental chromosome loss, aneugenic events involving whole chromosome loss, or asymmetrical nuclear division. Chromosome translocations are important DNA mutations, given their persistence and association with human cancers. However, solid staining methods such as Giemsa staining cannot detect chromosome translocations, and while chromosome banding detects translocations, the method is labor intensive and requires highly specialized training to perform. Consequently, although chromosome banding provides reliable chromosome structural assessment data when evaluating small samples in clinical settings, it is not particulary well suited for analyzing the large numbers of metaphase cells needed for genotoxicity evaluations or adverse exposure assessments in humans. Integrating fluorescence in situ hybridization (FISH) into comet, MN, and chromosome aberration assessments can provide detection of DNA damage related to carcinogenicity. This is accomplished through identification of cancer gene–specific damage in the comet assay, distinguishing aneugenic from clastogenic events in the MN assay, and identification of chromosome translocations during chromosome analysis.

1.3 FISH History and Overview

The advent of FISH technologies (Fig. 13.1) brought cytogenetics into the molecular biology era, engendering the field of molecular cytogenetics. FISH, simply stated, is the use of labeled polynucleotide probes that hybridize to their complementary target sequence in situ and, indirectly or directly, report a fluorescent signal. Visualization of FISH signals can determine the localization, arrangement, and number of the targeted sequences in metaphase and interphase cells alike. The first reports describing in situ hybridization used radioactively labeled RNA probes in combination with autoradiography [6–8]. Later, development of fluorescence detection methods proved useful in labeling chromosomes in metaphase cells [9–13], and since then FISH has steadily replaced the use of radiolabeled probes. Strategies employing FISH probes provide several distinct

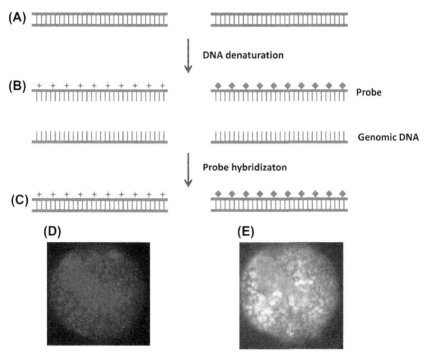

Figure 13.1 Simplified illustration of FISH basic elements. (A) Genomic DNA containing the target sequences is denatured. (B) Following genomic DNA denaturation, it is exposed to FISH probes complementary to the genomic DNA target of interest, shown directly labeled with a red (light gray in print versions) star or green (dark gray in print versions) diamond fluorochrome. (C) The red (light gray stars in print versions) and green (dark gray diamonds in print versions) FISH probes hybridize to their complementary target sequences in the genomic DNA. Images of human interphase nuclei that are labeled with whole chromosome red (light gray in print versions) and green (dark gray in print versions) FISH probes, counterstained with DAPI (blue (gray in print versions)) and viewed through a (D) DAPI single-band pass filter, and (E) a triple-band pass filter to reveal the nuclear distribution of the labeled chromosomes during interphase.

advantages over their radiolabeled predecessors. For example, the FISH probes are safer, more stable, and easier to work with than radioisotope-labeled probes. Additionally, the FISH probes have the advantage of faster throughput, including shorter hybridization times and rapid microscopic analysis. Finally, multiplexing FISH probes has the advantage of simultaneous multicolor analysis of multiple targets in a single specimen.

FISH probes come with a variety of different labeling and detection options. A common FISH probe labeling strategy is to incorporate nucleotide bases coupled to haptens such as digoxigenin or biotin. Antibodies or avidin (a biotin-binding protein) conjugated to a fluorochrome then bind the haptens and produce a FISH signal. Alternatively, FISH probes labeled with nucleotide bases directly linked to fluorochromes permit straightforward DNA target detection without involving the antibody-labeling step. Initially, to generate whole chromosome FISH probes, chromosomes were typically flow sorted or isolated by chromosome microdissection [14,15]. DNA amplification using degenerate oligonucleotide-primed PCR (DOP-PCR) was a commonly used method to generate FISH probes from the isolated chromosomes [16,17]. However, most current practices use cloned DNA in plasmids or artificial chromosome vectors to create labeled FISH probes by random priming, nick translation, or PCR.

1.4 FISH-Based Techniques Used in Genetic Toxicology

This review addresses how the integration of FISH into conventional genotoxicity tests improve their predictive value when evaluating induced genetic damage in somatic cells from human studies, rodent models, and mammalian cell cultures. The present chapter reviews using FISH combined with the comet assay to investigate damage to genomic DNA regions surrounding genes involved in carcinogenesis. Also reviewed is how FISH in combination with the MN assay can discern aneuploidy from clastogenicity. Finally discussed is how whole chromosome FISH probes used for "painting" metaphase cell chromosomes can rapidly identify structural chromosome damage, such as chromosome translocations, that are important due to their persistence and implications in cancer.

2. INTEGRATION OF FISH INTO THE COMET ASSAY

The comet assay (i.e., single cell gel electrophoresis) is a method widely used for measuring DNA damage following potentially genotoxic treatments [18,19]. In the comet assay, cells are embedded in low melting point (LMP) agarose for immobilization onto microscope slides, followed by lysis to remove the cellular and nuclear membranes. Exposure to strong alkaline (pH ≥ 13) buffer relaxes DNA supercoiling and unwinds the DNA. After unwinding, electrophoretic separation of nuclear DNA results in migration of the fragmented DNA loops toward the anode more quickly than the intact DNA loops that remain associated with the nucleus. Thus, following fluorescent dye staining, the nuclear DNA appearance resembles a comet with a head containing intact DNA loops and a tail containing fragmented DNA. Assessment of DNA damage with the comet images uses the percent DNA in the comet tail to reflect the degree of DNA damage [1,18,20]. Recently, the OECD developed a test guideline (Test no. 489) for in vivo mammalian alkaline comet assay use in regulatory genotoxicity testing of pharmaceuticals. The in vivo comet assay is an optional test of the genotoxicity test battery cited in the S2R1 guidance of the International Council on Harmonization (ICH) [1,22].

The first description of combining the comet assay with FISH (comet–FISH) was published in 1997 [23]. FISH probes were used in combination with the comet assay to investigate nuclear organization of telomeres and centromeres and FISH signal migration into comet tails in electrophoresed comet preparations of human peripheral blood lymphocytes (PBLs) [23]. Gene-specific FISH probes in combination with the comet assay can evaluate damage to regions of DNA that surround the genes of interest, based on whether the FISH signals are present in the comet tail or in the head. For example, if using a FISH probe detecting a somatic gene existing as two copies in a diploid somatic cell, observing two fluorescent signals in the comet head of a damaged cell indicate that the gene is in an undamaged region of DNA. However, if one or both FISH signals appear in the comet tail, this indicates that a DNA break or breaks have occurred in the proximity of the probed gene [24,25]. The level of damage to specific DNA regions probed can be expressed as the percentage of FISH signals present in the comet head versus the comet tail [26]. Conversely, the comet–FISH technique can also monitor DNA repair following genotoxic treatment. Time-dependent changes in the position of gene-specific FISH signals from the comet tail to the comet head, that occur during a recovery interval following a genotoxic treatment, can indicate repair of the DNA lesions within and around the gene of interest [27].

2.1 Application of Comet–FISH in Genetic Toxicology

Comet–FISH provides a tool to study the sensitivity of DNA regions surrounding cancer-relevant genes toward genotoxic agents, to better understand the mechanisms of carcinogenesis [28]. There are several examples of how, following genotoxic treatments, comet–FISH was used for evaluating damage to DNA regions where cancer-relevant genes localize. After treating primary human colon cells with hydrogen peroxide (H_2O_2), and the lipid oxidation products trans-2-hexenal and 4-hydroxy-2-nonenal (HNE), comet–FISH was used for evaluating damage to nuclear DNA regions containing the *APC, KRAS*, and *p53* cancer-relevant genes [29]. The investigators reported cell type–specific susceptibility to DNA damage in the regions surrounding these genes. This study revealed that DNA surrounding the *p53* gene FISH signal was sensitive to H_2O_2, trans-2-hexenal, and HNE in primary human colon cells, but that DNA surrounding the *APC* and *KRAS* genes were only sensitive to trans-2-hexenal. While in a human colon cell line, *p53*, but not *APC* and *KRAS*, exhibited greater sensitivity only to HNE [29].

The relationship between repair to DNA surrounding specific genes and repair to the overall chromatin was studied in Chinese hamster ovary (CHO) cells and human lymphocytes treated with H_2O_2, or a photosensitizer plus light exposure to induce oxidative DNA damage [30]. Following exposures, the CHO cells displayed preferential repair of oxidative damage in the area surrounding the *MGMT* gene, whereas in the human lymphocytes there was evidence indicating that DNA strand breaks surrounding the *p53* gene were repaired much more quickly when compared with repair to total DNA damage [30]. In this way, comet–FISH provided a means to study the repair rates of damaged DNA surrounding specific genes in relationship to the overall nuclear chromatin structure.

Comet–FISH has been applied to evaluating the ability of genotoxicants to damage telomeric DNA [23,31,32]. Telomeres are repetitive DNA sequences on the ends of each chromosome that protect them from degradation and prohibit fusion with other chromosomes. Comet–FISH was used for investigating whether chemotherapeutic agents could bring about telomere instability in human noncancerous cells. After exposing PBLs isolated from healthy humans to the chemotherapeutic drugs, bleomycin (BLM) and mitomycin C (MMC), telomere-specific FISH probes quantitatively detected the drugs effect on telomere DNA damage. This study revealed that after BLM and MMC treatment, appearance of telomere FISH signals in comet tails paralleled total DNA migration, thus showing no evidence of preferential damage to telomeric DNA [31]. Furthermore, a follow-up study revealed cisplatin reduced BLM-induced migration of telomeric DNA to comet tails to a greater extent than it did for migration of total DNA, due to the telomere-specific cross-linking effect of cisplatin [32]. These studies illustrate the utility of comet–FISH for assessing damage to the telomeric sequences of chromosomes following genotoxic treatments.

Comet–FISH can also evaluate damage to a specific chromosomal region as compared to overall DNA damage [33]. Acute myeloid leukemia (AML) is associated with previous exposures to benzene, or to chemotherapeutic drugs such as melphalan (alkylating agent) and etoposide (topoisomerase II inhibitor). Deletion of a region on the long arm of chromosome 5 (5q31) is frequently observed with benzene exposures and in AML patients exposed to alkylating agents. In addition, rearrangements involving chromosome region 11q23 encompassing the *MLL* gene, are associated with distinct clinical features and a poor prognosis in AML patients exposed to topoisomerase inhibitors. After exposing human TK6 B-lymphoblastoid cells to melphalan, etoposide, or the alkylating benzene metabolite, hydroquinone (HQ), comet–FISH was used for assessing breakage at the chromosome 5q31 and the 11q23 regions. The study revealed that HQ, melphalan, and etoposide all induced DNA breaks in a dose-dependent manner at both chromosome 5q31 and 11q23 regions [33]. HQ produced significantly more DNA damage at the 5q31 region than at the 11q23 region, while the effect of melphalan treatment was similar to HQ treatment, it was not significant; and etoposide produced slightly more DNA damage at the 11q23 region [33]. Therefore, the comet–FISH findings support observations of benzene-exposed populations or AML patients previously treated with alkylating agents, having deletions to chromosome region 5q31. In addition, the comet–FISH analysis supported evidence for translocations of the *MLL* gene in the 11q23 region, that is observed in AML patients following treatment with topoisomerase II inhibitors. Thus, these findings illustrate the ability of comet–FISH to identify specific chromosome region biomarkers related to exposures that might induce leukemia in human populations.

The combination of comet–FISH offers the possibility to detect total genomic DNA damage simultaneously with detection of damage to specific DNA regions within the same specimen [25]. As discussed, comet–FISH has been applied to identify damage to DNA surrounding cancer relevant genes in response to different genotoxic agents [29] and to study the repair of specific genes in relationship overall chromatin repair [30]. Comet–FISH also facilitates analysis of telomeric DNA sensitivity to chemotherapy drugs [31,32] and the study of specific chromosome regions related to biomarker research in humans [33]. Application of comet–FISH in genetic toxicology studies provides an opportunity to collect valuable information for further interpreting and understanding the impact of DNA damage.

2.2 Factors to Consider When Using Comet–FISH During Genotoxicity Evaluations

Several method papers are available that detail applications of comet–FISH that are adaptable to genotoxicity studies [28,34–36]. When using the comet–FISH technique to assess DNA damage, the ability of FISH probes to migrate from the comet head depends on the gel concentration, and is a factor to consider during experimental design [37]. In addition, when using comet–FISH for investigating DNA damage in nondividing cells, two signals are often present if the gene appears as a single copy on both somatic chromosomes. However, when using comet–FISH in heavily damaged cells or in cultured cells, where the cells might be dividing or the cell lines are genetically unstable, the number of FISH signals detected might be inconsistent, as was observed with experiments in CHO cells [30]. Finally, applying FISH probes to comet slides after unwinding and electrophoresis creates a three-dimensional (3D) configuration of hybridized regions. This 3D configuration creates an important difference between whole chromosome FISH experiments in two-dimensional (2D) space and comet–FISH experiments because this 3D structure makes imaging and scoring comet–FISH slides somewhat more difficult and time-consuming.

3. INTEGRATION OF FISH WITH THE MICRONUCLEUS ASSAY

The MN test is widely employed for genotoxicity testing and a biomarker assay for adverse exposures in humans [2,3]. Chromosome breaks and mitotic spindle disruptions resulting in lagging chromosomes might produce chromatin structures excluded from the main nuclei, referred to as micronuclei, that can be used as a measure of aneuploidy and clastogenicity. The MN assay rapidly gained popularity as a genetic toxicology endpoint because scoring micronuclei requires little training and can rapidly assess thousands of cells, giving it an increased sensitivity and statistical power compared with the chromosomal aberration assay. The MN assay was initially developed to measure genetic damage in erythroblasts from rat bone marrow [38]. However, a significant advancement to MN test procedures occurred when cytochalasin B was added to cultured cells to block cell division [39], resulting in what is now commonly referred to as the cytokinesis-blocked micronucleus (CBMN) test. With the CBMN assay, it is possible to identify first division cells as binucleated cells and evaluate them for MN content within the cytoplasm, providing single cell resolution.

FISH techniques integrated into the CBMN assay (CBMN–FISH) increase its utility, chiefly through the ability to reveal clastogenic events, aneugenic events, or asymmetrical nuclear division (Fig. 13.2). Using FISH probes for pancentromeric DNA that label all centromeres or FISH probes for chromosome-specific centromeres, it is possible to distinguish exposures that result in MN containing an acentric chromosome fragment from MN containing a whole chromosome [40]. This is possible because a MN containing an acentric chromosome fragment will not have a centromeric FISH probe signal, whereas a MN containing a chromosome will be positive for a centromeric FISH probe signal. The effectiveness of FISH to reveal information on mechanisms of chromosome damage and MN composition

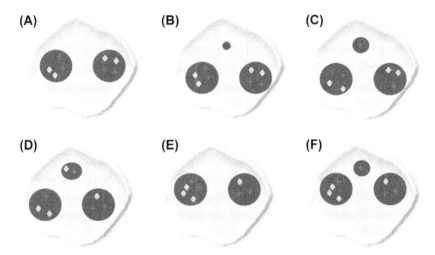

Figure 13.2 Schematic of CBMN–FISH with chromosome-specific centromeric probes labeling one chromosome pair in red (gray star in print versions) and one chromosome pair in green (gray diamond in print versions), within the nuclei (blue (light gray in print versions)) of cytokinesis-blocked binucleated cells to illustrate simple C– MN, C+ MN, and asymmetrical nuclear division events. (A) A binucleated cell with two daughter nuclei containing two nonhomologous chromosome signals. (B) A binucleated cell with two daughter nuclei and a C– MN containing chromatin from an acentric chromosome fragment resulting from a clastogenic exposure. (C) A binucleated cell with two daughter nuclei and a C+ MN containing a chromosome centromeric signal resulting from an aneugenic exposure. (D) A binucleated cell with two daughter nuclei and a C+ MN containing multiple chromosome centromeric signals. (E) A binucleated cell with aneuploidy resulting from asymmetrical nuclear division, where one hyperdiploid daughter nuclei contains an extra green (gray diamond in print versions) signal for a chromosome centromere and one hypodiploid daughter nuclei is missing a green (gray diamond in print versions) signal for a chromosome centromere. (F) A binucleated cell with aneuploidy resulting from asymmetrical nuclear division with an accompanying C+ MN.

is recognized in the OECD test guideline (TG 487) for conducting the in vitro MN assay [2].

The ability to distinguish between centromere negative MN (C− MN) and centromere positive MN (C+ MN) with FISH is informative, given that a suspected adverse exposure often will predominately induce only one type of these MN. It is also possible to analyze aneuploidy as an asymmetrical nuclear division, where daughter nuclei have unequal numbers of centromeric FISH probes [41–43]. Discernment of C− MN from C+ MN as a biomarker might reveal important differences induced by an exposure, such as elevated aneuploidy, when the overall MN frequencies between exposed subjects and unexposed subjects might be insignificant. In this way, FISH evaluation of the C− MN or C+ MN frequencies might considerably improve the sensitivity for detecting an exposure effect for a suspected genotoxic agent.

3.1 Application of CBMN–FISH in Genetic Toxicology

Numerous genotoxicity studies have incorporated pancentromeric or chromosome-specific centromeric FISH probes into CBMN–FISH. As previously stated, the CBMN–FISH combination has made it possible to resolve C− MN, C+ MN, and aneuploidy resulting from asymmetrical cell division (Fig. 13.2). As it is known that most radiation-induced MN contain primarily acentric fragments while spontaneous MN originate from whole chromosomes, CBMN–FISH applied to radiation studies to evaluate C− MN frequencies decreased the detectable radiation absorption threshold from 0.2 Gy to 0.1 Gy [44]. Therefore, using pancentromeric probes in CBMN–FISH can increase the assay sensitivity to reveal a dose effect when analyzing C− MN frequencies in populations accidentally exposed to radiation [44–46]. Paclitaxel, a chemotherapeutic drug that disrupts tubulin depolymerization, was analyzed for genotoxic effect by CBMN–FISH using pancentromeric probes in human primary T-lymphocytes [47]. This study revealed paclitaxel-induced dose-dependent MN formation, with 85% of the MN positive for centromeric signal, and many C+ MN containing more than one centromeric signal [47]. Investigation into the effect caffeine has on cell division and MN formation in human PBLs utilized CBMN–FISH with chromosome 11 and chromosome 17 centromere-specific probes [43]. The results indicated that low concentrations of caffeine are capable of inducing aneuploidy in a dose-dependent manner and that the increased MN levels were unrelated to chromatid breaks in human PBLs. The investigators attributed the aneuploidy to asymmetrical cell division due to misalignment of chromosomes following caffeine treatment [43]. The CBMN–FISH assay is also an effective tool for investigating DNA damage resulting from occupational exposures to the industrial chemical styrene. The CBMN–FISH assay identified significantly elevated MN levels in styrene exposed subjects compared to the control subjects. The elevated MN frequencies appeared to involve both C− MN and C+ MN, with C+ MN frequencies correlated with urinary styrene metabolite levels [48]. These examples incorporating CBMN–FISH into genotoxicity assessments illustrate its power to determine whether clastogenicity, aneuploidy, or both result from genotoxic exposures to predict carcinogenic effects with greater accuracy.

3.2 Factors to Consider When Using CBMN–FISH During Genotoxicity Evaluations

The MN assay remains an important cytogenetic tool for evaluating the genotoxicity of suspected agents due to its simplicity, accuracy, and in vivo and in vitro applicability. Moreover, integration of CBMN–FISH into genotoxicity assessments and adverse exposure studies improves the conventional MN assay sensitivity. Both pancentromeric and chromosome-specific centromeric probes for individual chromosomes, or chromosomes in combination, are commercially available. Additionally, numerous studies have utilized CBMN–FISH, and several methods are available that are adaptable to genotoxicity studies [49–51]. While incorporating pancentromeric or chromosome-specific centromere FISH labeling into MN tests can distinguish clastogenicity from aneugenicity, several considerations need to be made when using them to conduct genotoxic assessments. Using pancentromeric FISH probes for determining chromosome nondisjunction or asymmetrical nuclear division is not recommended due to the difficulty in enumerating all the centromeres within nuclei. Also, age-related increases in the frequency and composition of spontaneous MN is an important factor when conducting adverse exposure assessments in humans. Aged subjects have elevated MN levels and different MN composition, particularly involving sex chromosomes, than do younger subjects [52–54]. Utilization of FISH probes for X chromosome–specific centromeric DNA identified an overrepresentation of X chromosome signal in lymphocyte MNs from women with increasing rates correlated with aging [53]. Using Y chromosome FISH probes revealed that in males there was an age-dependent increase in MN containing Y chromosomes [55]. Another consideration is that in vitro studies have revealed that MN frequencies increase in human lymphocytes following extended culture duration with and without cytochalasin B, primarily due to increasing C+ MN frequencies [56,57]. Yet, it is still unclear how extended culturing influences MN frequencies in lymphocytes, possibly due to elevated X chromosome MN frequencies or increased expansion of cells with a higher MN baseline.

4. APPLICATION OF FISH TO DETECT STRUCTURAL CHROMOSOME ABERRATIONS

Chemical or physical exposures producing DNA double strand breaks (DSBs) could result in cell death or cancer if left unrepaired or are misrepaired. Repair of DNA DSBs occurs primarily through the homologous repair (HR) and nonhomologous end joining (NHEJ) pathways. DNA DSBs in chromatin of somatic cells in G1 phase of the cell cycle (Fig. 13.3A) can result is several outcomes following repair. Restitution of the two break sites in a chromosome through HR might occur with (or without) removal or addition of DNA bases at the break site that are not detectable at the chromosome level (Fig. 13.3B). The DNA DSBs could also result in exchange of genetic material between nonhomologous chromosomes through NHEJ, which might produce a chromosome aberration detectable with whole chromosome FISH in metaphase cells.

Acquired chromosome aberrations, particularly when involving exchanges of genetic material, are a prominent trait of cancer and therefore are used to assess potential genotoxicity to predict carcinogenicity. Following DNA DSBs, there are different possible types of simple chromosome aberrations produced, depending on how the broken DNA ends rejoin (Fig. 13.3C–H). Translocations are the most stable and therefore most persistent type of chromosome aberration. The simplest chromosome translocations could be reciprocal (balanced) exchanges (Fig. 13.3C), or nonreciprocal (unbalanced) exchanges (Fig. 13.3D), that produce linear chromosomes with a single centromere. Generally, it is regarded that most translocations are reciprocal, but that the appearance of a nonreciprocal translocation is the result of a segment of translocated DNA that is below the limit of visual detection [59,60]. An insertion involves three breaks in two chromosomes that rejoin to form at least one linear chromosome with an interrupting interstitial segment from another chromosome (Fig. 13.3E). Chromosome insertions are also stable aberrations, and their persistence is similar to translocations, but insertions occur at much lower frequencies and thus are infrequently observed.

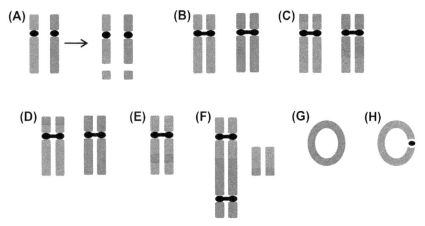

Figure 13.3 Commonly observed simple aberrations when evaluating FISH painted human metaphase cells. Shown are two nonhomologous chromosomes in different colors. (A) An example of breaks in two nonhomologous chromosomes in G1 phase, where several outcomes are possible that can be visualized in the G2/M phase chromosomes (B–H). (B) The nonhomologous chromosomes are restituted, leaving no visible aberration detected. (C) A reciprocal (balanced) translocation involving an exchange from a single break in two chromosomes. (D) A nonreciprocal translocation (unbalanced) where only one exchange is detected. (E) An insertion event arising from an exchange following two interstitial breaks in one chromosome and a single break in the other chromosome. (F) A dicentric chromosome with two centromeres and an accompanying compound acentric fragment. (G) An acentric ring chromosome, and (H) a centric ring.

A number of simple chromosome aberrations are considered unstable because they result in cell death or are lost during mitosis. These unstable aberrations include acentric fragments, dicentric chromosomes, and centric rings. Dicentric chromosomes involve the joining of two broken chromosomes to form a derivative chromosome with two centromeres (Fig. 13.3F). Metaphase cells with dicentric chromosomes will frequently contain a compound acentric fragment consisting of the sections from broken chromosomes (Fig. 13.3F). Dicentric chromosomes should occur at the same frequency as chromosome translocations following acute clastogenic exposures due to the randomness involved in joining two breaks in two broken chromosomes, but due to their unstable nature, dicentric frequencies will decrease over time. Acentric fragments are DNA segments associated with a metaphase cell that are lacking a centromere, and can be simple (derived from a single break in a single chromosome), compound (derived from acentric segments from two broken chromosomes), and either in a linear or a ring form (Fig. 13.3G). When DNA DSBs occur in both arms of a chromosome, they can then join to form a centric ring (Fig. 13.3H). Unstable chromosome aberrations are considered as such because they lack a centromere, have more than one functional centromere, or in the case of ring chromosomes, odd numbers of sister chromatid exchanges lead to interlocking rings that cannot be resolved during mitosis. Thus, these aberrations result in loss of the unstable chromosomes during mitosis. More complex types of chromosome exchanges are also possible, including chromosomes with three or more centromeres and multiple translocation and insertion events within a single derivative chromosome [61].

A majority of the studies utilizing whole chromosome FISH are dedicated to translocation enumeration in radiation biology, while the use of FISH in testing chemical induced chromosome aberrations has yet to receive widespread implementation. This is primarily due to the fact that the chromosomal aberrations induced by genotoxic chemicals are unstable, and quantifying them using the chromosomal aberration test remains a better suited method [62]. However, given the link between stable chromosome aberrations such as translocations and insertions with cancer, it seems appropriate that whole chromosome FISH techniques should be applicable to chemical genotoxicity studies.

4.1 Application of FISH in Analysis of Chromosome Damage

Multicolor whole chromosome FISH utilizes probes conjugated with different fluorochromes to allow for simultaneously labeling whole chromosomes with different colors, a process referred to as chromosome "painting." Multicolor FISH is possible for any combination of all human chromosomes, because paints are available that will uniquely paint each chromosome a different color [63,64]. One advantageous painting scheme to assess chromosome damage in humans is to simultaneously paint the six largest chromosome pairs. When chromosome pairs 1, 2, and 4 are labeled in red, chromosome pairs 3, 5, and 6 are labeled in green, and all chromosomes counterstained blue, this painting scheme can detect rearrangements between three colors (Fig. 13.4) [65,66]. Application of this three-color FISH technique revealed increased chromosome translocation frequencies correlated with radiation exposure dose in retrospective biodosimetry studies among hospital radiological technologists [67]. In a related study, the three-color FISH chromosome analysis revealed increased translocation frequencies associated with cumulative radiation dose from personal diagnostic X-ray examinations in a cohort of radiation technologists [68]. In addition, this FISH painting technique was used for evaluating chromosome aberration frequencies in PBLs from Gulf War veterans, who were

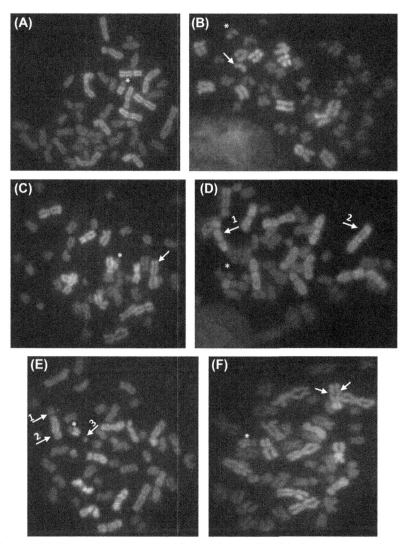

Figure 13.4 Human B-lymphoblastoid metaphase cells with chromosome pairs 1, 2, and 4 painted red and chromosome pairs 3, 5, and 6 painted green with blue counterstain. Indicated are types of chromosome aberrations that are identifiable when using whole chromosome FISH painting. (A) Undamaged metaphase cell containing a known stable translocation (*) where a segment from chromosome 3 translocated onto chromosome 21, t(3; 21) [81], present in all metaphase cells from this cell line. (B) Acentric fragment (*arrow*) from a red chromosome. (C) Nonreciprocal (unbalanced) translocation involving a segment of a green chromosome on a red chromosome. (D) Reciprocal (balanced) translocation with (1) a segment of red chromosome translocated onto a green chromosome and (2) a segment of green chromosome translocated onto a red chromosome. (E) Dicentric chromosome involving green and blue chromosomes. *Arrows indicate* (1) the blue chromosome centromere, (2) the green chromosome centromere of the dicentric chromosome, and (3) an acentric compound fragment consisting of a green segment and a blue segment. (F) An insertion of a red chromosome segment in a green chromosome.

long-term exposed to depleted uranium [69]. In the Gulf War veterans study, Bakhmutsky et al. [69] revealed that there was no correlation between uranium urine levels and detectable increases in chromosome aberration frequencies. Application of human FISH probes to the rhesus monkey (*Macaca mulatta*), an important biomedical and behavioral model, is possible due to its genetic similarity to humans. The human whole chromosome FISH probes that label the six largest chromosome pairs in humans, label seven chromosome pairs when used for painting *M. mulatta* metaphase cells, with the following chromosomes painted for human (*M. mulatta*): 1(1), 2q(12), 2p(13), 4(5) pairs in red, and 3(2), 5(6), and 6(4) pairs in green [70]. The whole chromosome FISH painting of metaphase cells from *M. mulatta* treated with methylphenidate hydrochloride (MPH), a drug prescribed in treatment of the pediatric attention-deficit/hyperactivity disorder (ADHD), indicated that MPH did not induce chromosome aberrations [71]. Recently, this three-color whole chromosome FISH was used to evaluate the ability of an oxygen-functionalized graphene, a nanomaterial investigated for numerous biomedical applications, to induce chromosome aberrations in B-lymphoblastoid cell lines with either a *p53* wild type or *p53* null status [72]. Whole chromosome FISH painting indicated there were significant differences in the overall frequencies of chromosome aberrations, including translocations, in *p53* null cells but not *p53* wild type cells treated with oxygen-functionalized graphene. This is an important consideration for the use of oxidized graphene nanomaterials in treatment of cancers lacking functional p53. Although by no means an exhaustive account for applications of FISH painting in adverse exposure studies, these examples provide an illustration of the utility whole chromosome FISH has in genetic toxicology.*p53p53*

4.2 Factors to Consider When Using Whole Chromosome FISH in Genotoxicity Evaluations

Several commercial options are available for ready-to-use FISH probes or for kits to synthesize FISH probes, depending on the investigator's needs. There are numerous studies that have used whole chromosome FISH painting in genotoxicity studies, and methods are available that detail FISH techniques applicable to both human and animal models [70,73,74]. Several confounding variables need to be considered when using whole chromosome FISH paints to study the genotoxicity of suspected agents, particularly when conducting human population studies. The relevance that chromosome translocations have on human health makes them of notable significance among all types of morphological chromosome aberrations. Albeit, detection of chromosome translocations must be done through chromosome banding or painting, making them the most difficult chromosome aberration to detect. While banded chromosome analysis is inexpensive and useful in clinical studies to determine if a patient has a constitutional chromosome alteration, it is time-consuming and requires specialized training. For applications such as exposure assessments in populations or genotoxicity evaluations, where large numbers of cells must be analyzed, the use of FISH molecular methods is clearly preferable.

Of particular importance when using FISH to assess the genotoxicity of a suspected agent in human populations are the effects of age, and environmental and lifestyle factors. Because chromosome aberrations can arise spontaneously, and translocations accumulate with age, a clear link has been established between age and the background frequencies of cytogenetic damage observed in numerous populations [75,76]. Cigarette smoking is the largest confounding lifestyle factor, yet it is a weak contributor to baseline cytogenetic damage, as compared with the effect of age on translocation frequencies [76]. Additional confounding factors in human population studies include environmental exposures to asbestos or coal, or having a previous disease. It is prudent to include assessment of these confounding factors in any human population study.

As the use of whole chromosome FISH in cytogenetic studies continued to expand, a need was realized for a common nomenclature that investigators could use to communicate their findings. Therefore, the nomenclature system PAINT (protocol for aberration identification and nomenclature terminology) for aberration type was developed [77,78]. The PAINT system provides investigators with a straightforward, descriptive nomenclature for reporting chromosome aberration findings when conducting genotoxicity studies. Likewise, different investigators use different FISH probe sets that simultaneously label between three to six pairs of chromosomes, and are thus analyzing different subsets of the whole genome. This means that detection of chromosome translocations is limited to those that occur between chromosomes of a different color, a factor that varies between studies depending on the FISH probe combinations used. Therefore, to standardize reporting of aberration frequencies, conversion formulas enable investigators using different whole chromosome FISH combinations to compare findings between studies [79,80]. This information can be conveyed as the whole genome equivalents or the number of metaphase cells that need be scored to be equivalent to one banded cell. Tucker (2010) gives a thorough description of how to make use of one such conversion equation to determine banded cell equivalents when using whole chromosome FISH [60].

The final consideration when using whole chromosome FISH paints in genotoxicity studies or to conduct exposure assessments in humans is the treatment of clonal cells containing aberrations. Particularly those clonal cells containing chromosome translocations, as these cells can survive mitosis. Clonal cells might confound the findings of human population studies using PBLs, as the frequency of clones from immune stem cells might undergo clonal expansion, these cells might increase in frequency with age. The same clonal events are possible in studies using immortalized culture cells, where we can predict the clonal expansion of cells with translocations should occur at near the same frequencies as normal cells. Therefore, characterizing the aberration type and quantifying cells that are clones in human population studies and genetic toxicology studies should be a routine part of analysis to determine if there are any confounding effects of the clonal cells during data analysis.

5. CONCLUSIONS

Assessing an agent's genotoxicity represents an essential component of toxicity evaluations. Incorporating FISH techniques into the comet, MN, and chromosome aberration endpoints has proved it can increase the predictive value of these conventional assays used in genetic toxicology. Further, the integration of FISH techniques into genotoxicity endpoints discussed here can be used for quantifying adverse human exposures, construct exposure-response curves, and collect in vivo translocation persistence data for biodosimetry. As automated scoring and FISH probe production technologies advance, it is anticipated that FISH techniques will have a promising future for further improving the predictive value of these widely used genotoxicity tests.

DISCLAIMER

The opinions presented in this article are those of the authors and do not necessarily reflect the views or policies of the US Food and Drug Administration.

REFERENCES

[1] OECD. Test no. 489: in vivo mammalian alkaline comet assay. 2016.
[2] OECD. Test no. 487: in vitro mammalian cell micronucleus test. OECD Publishing; 2016.
[3] OECD. Test no. 474: mammalian erythrocyte micronucleus test. OECD Publishing; 2016.

[4] OECD. Test no. 473: in vitro mammalian chromosomal aberration test. OECD Publishing; 2016.

[5] OECD. Test no. 475: mammalian bone marrow chromosomal aberration test. OECD Publishing; 2016.

[6] Gall JG, Pardue ML. Formation and detection of rna-DNA hybrid molecules in cytological preparations. Proc Natl Acad Sci USA 1969;63:378–83.

[7] John HA, Birnstiel ML, Jones KW. Rna-DNA hybrids at the cytological level. Nature 1969;223:582–7.

[8] Pardue ML, Gall JG. Molecular hybridization of radioactive DNA to the DNA of cytological preparations. Proc Natl Acad Sci USA 1969;64:600–4.

[9] Landegent JE, Baan RA, Hoeijmakers JH, Raap AK, Van der Ploeg M. Hybridocytochemistry with 2-acetylaminofluorene-modified probes. Cell Biol Int Rep 1984;8:186.

[10] Landegent JE, Jasen in de Wal N, Baan RA, Hoeijmakers JH, Van der Ploeg M. 2-acetylaminofluorene-modified probes for the indirect hybridocytochemical detection of specific nucleic acid sequences. Exp Cell Res 1984;153:61–72.

[11] Pinkel D, Gray JW, Trask B, van den Engh G, Fuscoe J, van Dekken H. Cytogenetic analysis by in situ hybridization with fluorescently labeled nucleic acid probes. Cold Spring Harb Symp Quant Biol 1986;51(Pt .1):151–7.

[12] Pinkel D, Landegent J, Collins C, Fuscoe J, Segraves R, Lucas J, et al. Fluorescence in situ hybridization with human chromosome-specific libraries: detection of trisomy 21 and translocations of chromosome 4. Proc Natl Acad Sci USA 1988;85:9138–42.

[13] Van Prooijen-Knegt AC, Van Hoek JF, Bauman JG, Van Duijn P, Wool IG, Van der Ploeg M. In situ hybridization of DNA sequences in human metaphase chromosomes visualized by an indirect fluorescent immunocytochemical procedure. Exp Cell Res 1982;141:397–407.

[14] Meltzer PS, Guan XY, Burgess A, Trent JM. Rapid generation of region specific probes by chromosome microdissection and their application. Nat Genet 1992;1:24–8.

[15] Telenius H, Ponder BA, Tunnacliffe A, Pelmear AH, Carter NP, Ferguson-Smith MA, et al. Cytogenetic analysis by chromosome painting using dop-pcr amplified flow-sorted chromosomes. Genes Chromosomes Cancer 1992;4:257–63.

[16] Carter NP, Bebb CE, Nordenskjo M, Ponder BA, Tunnacliffe A. Degenerate oligonucleotide-primed pcr: general amplification of target DNA by a single degenerate primer. Genomics 1992;13:718–25.

[17] Aubele M, Smida J. Degenerate oligonucleotide-primed pcr. PCR Protoc 2003:315–8.

[18] Collins AR. The comet assay for DNA damage and repair: principles, applications, and limitations. Mol Biotechnol 2004;26:249–61.

[19] Tice RR, Agurell E, Anderson D, Burlinson B, Hartmann A, Kobayashi H, et al. Single cell gel/comet assay: guidelines for in vitro and in vivo genetic toxicology testing. Environ Mol Mutagen 2000;35.206–21.

[20] Klaassen CD. Casarett & Doull's toxicology: the basic science of poisons. 8th ed. 2013. p. 445–80.

[21] ICH. ICH s2(r1): guidance on genotoxicity testing and data interpretation for pharmaceuticals intended for human use. 2011. http://www.Ich.Org/fileadmin/public_web_site/ich_products/guidelines/safety/s2_r1/step4/s2r1_step4.Pdf.

[22] Deleted in review.

[23] Santos SJ, Singh NP, Natarajan AT. Fluorescence in situ hybridization with comets. Exp Cell Res 1997;232:407–11.

[24] Glei M, Hovhannisyan G, Pool-Zobel BL. Use of comet-fish in the study of DNA damage and repair: review. Mutat Res 2009;681:33–43.

[25] Hovhannisyan GG. Fluorescence in situ hybridization in combination with the comet assay and micronucleus test in genetic toxicology. Mol Cytogenet 2010;3:17.

[26] Spivak G, Cox RA, Hanawalt PC. New applications of the comet assay: comet-fish and transcription-coupled DNA repair. Mutat Res 2009;681:44–50.

[27] Shaposhnikov S, Frengen E, Collins AR. Increasing the resolution of the comet assay using fluorescent in situ hybridization–a review. Mutagenesis 2009;24:383–9.

[28] Schlormann W, Glei M. Detection of DNA damage by comet fluorescence in situ hybridization. Methods Mol Biol 2012;920:91–100.

[29] Glei M, Schaeferhenrich A, Claussen U, Kuechler A, Liehr T, Weise A, et al. Comet fluorescence in situ hybridization analysis for oxidative stress-induced DNA damage in colon cancer relevant genes. Toxicol Sci 2007;96:279–84.

[30] Horvathova E, Dusinska M, Shaposhnikov S, Collins AR. DNA damage and repair measured in different genomic regions using the comet assay with fluorescent in situ hybridization. Mutagenesis 2004;19:269–76.

[31] Arutyunyan R, Gebhart E, Hovhannisyan G, Greulich KO, Rapp A. Comet-fish using peptide nucleic acid probes detects telomeric repeats in DNA damaged by bleomycin and mitomycin c proportional to general DNA damage. Mutagenesis 2004;19:403–8.

[32] Arutyunyan R, Rapp A, Greulich KO, Hovhannisyan G, Haroutiunian S, Gebhart E. Fragility of telomeres after bleomycin and cisplatin combined treatment measured in human leukocytes with the comet-fish technique. Exp Oncol 2005;27:38–42.

[33] Escobar PA, Smith MT, Vasishta A, Hubbard AE, Zhang L. Leukaemia-specific chromosome damage detected by comet with fluorescence in situ hybridization (comet-fish). Mutagenesis 2007;22:321–7.

[34] Shaposhnikov S, El Yamani N, Collins AR. Fluorescent in situ hybridization on comets: fish comet. Methods Mol Biol 2015;1288:363–73.

[35] Shaposhnikov S, Thomsen PD, Collins AR. Combining fluorescent in situ hybridization with the comet assay for targeted examination of DNA damage and repair. Methods Mol Biol 2011;682:115–32.

[36] Glei M, Schlormann W. Analysis of DNA damage and repair by comet fluorescence in situ hybridization (comet-fish). Methods Mol Biol 2014;1094:39–48.

[37] Rapp A, Bock C, Dittmar H, Greulich K-O. Uv-a breakage sensitivity of human chromosomes as measured by comet-fish depends on gene density and not on the chromosome size. J Photochem Photobiol B 2000;56:109–17.

[38] Heddle JA, Carrano AV. The DNA content of micronuclei induced in mouse bone marrow by gamma-irradiation: evidence that micronuclei arise from acentric chromosomal fragments. Mutat Res 1977;44:63–9.

[39] Fenech M, Morley A. Solutions to the kinetic problem in the micronucleus assay. Cytobios 1985;43:233–46.

[40] Kirsch-Volders M, Elhajouji A, Cundari E, Van Hummelen P. The in vitro micronucleus test: a multi-endpoint assay to detect simultaneously mitotic delay, apoptosis, chromosome breakage, chromosome loss and non-disjunction. Mutat Res 1997;392:19–30.

[41] Fenech M. Cytokinesis-block micronucleus assay evolves into a "cytome" assay of chromosomal instability, mitotic dysfunction and cell death. Mutat Res 2006;600:58–66.

[42] Fenech M, Kirsch-Volders M, Natarajan AT, Surralles J, Crott JW, Parry J, et al. Molecular mechanisms of micronucleus, nucleoplasmic bridge and nuclear bud formation in mammalian and human cells. Mutagenesis 2011;26:125–32.

[43] Hatzi VI, Karakosta M, Barszczewska K, Karachristou I, Pantelias G, Terzoudi GI. Low concentrations of caffeine induce asymmetric cell division as observed in vitro by means of the cbmn-assay and iFISH. Mutat Res 2015;793:71–8.

[44] Thierens H, Vral A. The micronucleus assay in radiation accidents. Ann Ist Super Sanita 2009;45:260–4

[45] Cho YH, Kim YJ, An YS, Woo HD, Choi SY, Kang CM, et al. Micronucleus-centromere assay and DNA repair gene polymorphism in lymphocytes of industrial radiographers. Mutat Res 2009;680:17–24.

[46] Thierens H, Vral A, Barbe M, Aousalah B, De Ridder L. A cytogenetic study of nuclear power plant workers using the micronucleus-centromere assay. Mutat Res 1999;445:105–11.

[47] Digue L, Orsiere T, De Meo M, Mattei MG, Depetris D, Duffaud F, et al. Evaluation of the genotoxic activity of paclitaxel by the in vitro micronucleus test in combination with fluorescent in situ hybridization of a DNA centromeric probe and the alkaline single cell gel electrophoresis technique (comet assay) in human t-lymphocytes. Environ Mol Mutagen 1999;34:269–78.

[48] Migliore L, Naccarati A, Coppede F, Bergamaschi E, De Palma G, Voho A, et al. Cytogenetic biomarkers, urinary metabolites and metabolic gene polymorphisms in workers exposed to styrene. Pharmacogenet Genomics 2006;16:87–99.

[49] Decordier I, Kirsch-Volders M. Fluorescence in situ hybridization (fish) technique for the micronucleus test. Genotoxicity Assess Methods Protoc 2013:237–44.

[50] Migliore L, Di Bucchianico S, Uboldi C. The in vitro micronucleus assay and fish analysis. Genotoxicity DNA Repair A Pract Approach 2014:73–102.

[51] Fenech M. Cytokinesis-block micronucleus cytome assay. Nat Protoc 2007;2:1084–104.

[52] Bakou K, Stephanou G, Andrianopoulos C, Demopoulos NA. Spontaneous and spindle poison-induced micronuclei and chromosome non-disjunction in cytokinesis-blocked lymphocytes from two age groups of women. Mutagenesis 2002;17:233–9.

[53] Hando JC, Nath J, Tucker JD. Sex chromosomes, micronuclei and aging in women. Chromosoma 1994;103:186–92.

[54] Tucker JD, Nath J, Hando JC. Activation status of the x chromosome in human micronucleated lymphocytes. Hum Genet 1996;97:471–5.

[55] Nath J, Tucker JD, Hando JC. Y chromosome aneuploidy, micronuclei, kinetochores and aging in men. Chromosoma 1995;103:725–31.

[56] Falck G, Catalan J, Norppa H. Influence of culture time on the frequency and contents of human lymphocyte micronuclei with and without cytochalasin b. Mutat Res 1997;392:71–9.

[57] Sgura A, Antoccia A, Ramirez MJ, Marcos R, Tanzarella C, Degrassi F. Micronuclei, centromere-positive micronuclei and chromosome nondisjunction in cytokinesis blocked human lymphocytes following mitomycin c or vincristine treatment. Mutat Res 1997;392:97–107.

[58] Deleted in review.

[59] Natarajan AT. Fluorescence in situ hybridization (fish) in genetic toxicology. J Environ Pathol Toxicol Oncol 2001;20:293–8.

[60] Tucker JD. Chromosome translocations and assessing human exposure to adverse environmental agents. Environ Mol Mutagen 2010;51:815–24.

[61] Savage JR, Simpson PJ. Fish "painting" patterns resulting from complex exchanges. Mutat Res 1994;312:51–60.

[62] Kirkland D. Chromosome aberration testing in genetic toxicology-past, present and future. Mutat Res 1998;404:173–85.

[63] Schrock E, du Manoir S, Veldman T, Schoell B, Wienberg J, Ferguson-Smith MA, et al. Multicolor spectral karyotyping of human chromosomes. Science 1996;273: 494–7.

[64] Speicher MR, Gwyn Ballard S, Ward DC. Karyotyping human chromosomes by combinatorial multi-fluor fish. Nat Genet 1996;12:368–75.

[65] Tucker JD, Cofield J, Matsumoto K, Ramsey MJ, Freeman DC. Persistence of chromosome aberrations following acute radiation: I, paint translocations, dicentrics, rings, fragments, and insertions. Environ Mol Mutagen 2005;45:229–48.

[66] Matsumoto K, Ramsey MJ, Nelson DO, Tucker JD. Persistence of radiation-induced translocations in human peripheral blood determined by chromosome painting. Radiat Res 1998;149:602–13.

[67] Bhatti P, Preston DL, Doody MM, Hauptmann M, Kampa D, Alexander BH, et al. Retrospective biodosimetry among united states radiologic technologists. Radiat Res 2007;167:727–34.

[68] Sigurdson AJ, Bhatti P, Preston DL, Doody MM, Kampa D, Alexander BH, et al. Routine diagnostic X-ray examinations and increased frequency of chromosome translocations among U.S. Radiologic technologists. Cancer Res 2008;68:8825–31.

[69] Bakhmutsky MV, Squibb K, McDiarmid M, Oliver M, Tucker JD. Long-term exposure to depleted uranium in gulf-war veterans does not induce chromosome aberrations in peripheral blood lymphocytes. Mutat Res 2013;757:132–9.

[70] Petibone DM, Morris SM, Hotchkiss CE, Mattison DR, Tucker JD. Technique for culturing macaca mulatta peripheral blood lymphocytes for fluorescence in situ hybridization of whole chromosome paints. Mutat Res 2008;653:76–81.

[71] Morris SM, Dobrovolsky VN, Shaddock JG, Mittelstaedt RA, Bishop ME, Manjanatha MG, et al. The genetic toxicology of methylphenidate hydrochloride in non-human primates. Mutat Res 2009;673:59–66.

[72] Petibone DM, Mustafa T, Bourdo SE, Lafont A, Ding W, Karmakar A, et al. P53-competent cells and p53-deficient cells display different susceptibility to oxygen functionalized graphene cytotoxicity and genotoxicity. J Appl Toxicol 2017 (Epub ahead of print).

[73] Pathak R, Koturbash I, Hauer-Jensen M. Detection of inter-chromosomal stable aberrations by multiple fluorescence in situ hybridization (mFISH) and spectral karyotyping (SKY) in irradiated mice. J Vis Exp 2017:e55162.

[74] Petibone DM, Tucker JD, Morris SM. Chromosome painting of mouse peripheral blood and spleen tissues. In: Sierra LM, Gaivão I, editors. Genotoxicity and DNA repair: a practical approach. New York, NY: Springer New York; 2014. p. 141–58.

[75] Ramsey MJ, Moore 2nd DH, Briner JF, Lee DA, Olsen L, Senft JR, et al. The effects of age and lifestyle factors on the accumulation of cytogenetic damage as measured by chromosome painting. Mutat Res 1995;338:95–106.

[76] Sigurdson AJ, Ha M, Hauptmann M, Bhatti P, Sram RJ, Beskid O, et al. International study of factors affecting human chromosome translocations. Mutat Res 2008;652:112–21.

[77] Tucker JD, Morgan WF, Awa AA, Bauchinger M, Blakey D, Cornforth MN, et al. A proposed system for scoring structural aberrations detected by chromosome painting. Cytogenet Cell Genet 1995;68:211–21.

[78] Tucker JD, Morgan WF, Awa AA, Bauchinger M, Blakey D, Cornforth MN, et al. Paint: A proposed nomenclature for structural aberrations detected by whole chromosome painting. Mutat Res 1995;347:21–4.

[79] Garcia Sagredo JM, Vallcorba I, Lopez Y, Sanchez-Hombre MD, Resino M, Ferro MT. Chromosome painting in biological dosimetry: assessment of the ability to score stable chromosome aberrations using different pairs of paint probes. Environ Health Perspect 1996;104(Suppl .3):475–7.

[80] Lucas JN, Awa A, Straume T, Poggensee M, Kodama Y, Nakano M, et al. Rapid translocation frequency analysis in humans decades after exposure to ionizing radiation. Int J Radiat Biol 1992;62:53–63.

[81] Honma M. Generation of loss of heterozygosity and its dependency on p53 status in human lymphoblastoid cells. Environ Mol Mutagen 2005;45:162–76.

CHAPTER FOURTEEN

DNA Damage, Repair, and Maintenance of Telomere Length: Role of Nutritional Supplements

Krupa Kansara, Souvik Sen Gupta
Ahmedabad University, Ahmedabad, India

1. INTRODUCTION: DNA REPAIR AND TELOMERE

The DNA is synonymous to the hard drive of a computer that stores all the data. DNA contains all the information (data) essential for life. It is a chemical entity, which is subjected to continuous assault and if not repaired, may lead to mutations and many diseases. The assault may be external (environmental; like UV radiation) or internal (reactive oxygen species formed as a by-product of metabolism, collision of replication and transcription machinery). The mechanism in which cells detect their DNA damage, send signals for damage presence and work to repair the damage is collectively known as DNA damage response (DDR) [1,2]. Cells have developed a number of mechanisms to detect and repair the various types of DNA damages irrespective of the damage caused by internal or external insult. Mammalian cells utilize a minimum of five major DNA repair mechanisms. Base excision repair (BER) is initiated typically by lesion-specific DNA glycosylases that remove the damaged base leading to production of abasic site. BER does not distort the overall structure of the DNA helix. The mismatch repair (MMR) pathway is important for postreplication repair of misincorporated bases that escape the proofreading activity of replication polymerases. The MMR pathway also takes care of the insertion–deletion loops, which may result from the slippage of polymerase during repetitive DNA replication. The highly versatile nucleotide excision pathway (NER) can recognize and remove a wide variety of bulky, helix-distorting lesions from DNA. Though similar to BER, NER is more complex and requires about 30 different proteins for recognition of damaged DNA, local opening of the DNA helix around the lesion, excision of a short single-strand

Mutagenicity: Assays and Applications
ISBN 978-0-12-809252-1
http://dx.doi.org/10.1016/B978-0-12-809252-1.00014-6

287

segment of DNA spanning the lesion, and sequential repair synthesis followed by strand ligation. Global genome NER (GG-NER) (eliminates DNA lesions throughout the genome) and transcription coupled NER (TC-NER) (repairs lesions located on the coding strand of actively transcribed genes) are two related subpathways of the NER system. Other than the initial DNA damage recognition step, both these pathways are similar mechanistically. The most hazardous biological DNA damage is DNA double strand breaks (DSBs). There are two main pathways in mammalian cells for the repair of DSBs—homologous recombination (HR) and non-homologous end joining (NHEJ). HR is an error-free mechanism as it utilizes the genetic information contained in the undamaged sister chromatid as a template. However, NHEJ involves elimination of DSBs by direct ligation of the broken ends and thus is error prone. NHEJ operates in all phases of the mammalian cell cycle, whereas HR is restricted to the late S and G2 phases of the cell cycle [126].

Telomeres are DNA protein complexes, which are present at the chromosome ends and function to protect the genome from degradation and aberrant chromosomal fusion. This implies that they play a pivotal role in maintaining genomic integrity. Normally, a small portion of telomeric DNA is lost with each cell division cycle and when a critical limit is reached, the cell undergoes senescence or apoptosis. The ability to extend telomeres is conferred by the enzyme telomerase, which is present in germline and certain hematopoietic cells but at a very low or undetectable level in somatic cells [3]. Thus in somatic cells, telomeres undergo progressive shortening with each cell division cycle, and this property may be used to determine the life span of the cell or organism. During the replication of linear chromosomes, DNA polymerase replicates the DNA termini from 5′ to 3′ direction using a RNA primer for initiation. When this RNA primer is removed after DNA replication, telomeric DNA sequence is lost from the ends [4–7]. In germ cells, stem cells, and many cancer cells, the reverse transcriptase enzyme telomerase adds telomeric sequence to the ends of the newly synthesized DNA, maintaining the telomere length [8,9]. A complex of proteins, termed shelterin, associate with telomeric repeats when cells are not undergoing division and replication and thus prevent undesired lengthening of telomeres by restricting access to telomerase [10]. Telomere length is epigenetically regulated by DNA and histone methylation. Lack of normal DNA or histone methylation marks at elongated telomeres may result in increased telomeric recombination [11–13]. Although, the structure of chromatin rather than the telomere length is the determinant of recombination, longer telomeres undergo more frequent recombination than the shorter ones [13]. Telomere dysfunction is linked to the development of age-related pathologies including Parkinson disease, Alzheimer disease, cardiovascular disease, and cancer [14–18]. Telomere length has been shown to be associated with nutritional status and thus is of significant interest to nutritionists across the world. Healthy lifestyles and diets are positively correlated with telomere length. Reports indicate that the telomerase activity in peripheral blood mononuclear cells is modulated by changes in diet and lifestyle [19]. However, it is not clear whether this modulation translates into changes in telomere length. This chapter has been attempted to discuss the current knowledge on the interplay of nutrition, DNA repair and telomere length, and elaborate on how various nutrients may influence telomere length.

2. VITAMINS: DNA REPAIR AND TELOMERE LENGTH

2.1 Vitamin A

Vitamin A is a fat soluble vitamin known to have powerful antioxidant property. It is involved in reducing inflammation by fighting free radical damage. This property in turn will reduce DNA damage in the cells. There is no evidence about the direct involvement of vitamin A in the DNA repair pathways; the role of this vitamin is more toward the prevention of cells from oxidative DNA damage. Indeed, vitamin A plays a protective role in tumor progression by inhibition of N-myc gene expression [20]. Dietary intake of vitamin A is positively associated with telomere length [21]. Vitamin A plays an important role in the immune response [22], and its deficiency may lead to infections in individuals [23], which may lead to shortening of the telomeres. Supplementation with vitamin A has been shown to reduce the plasma concentration of the inflammatory cytokine tumor necrosis factor alpha and increased the concentration of the antiinflammatory cytokine interleukin-10 in vitamin A–deficient individuals [23]. However, supplementation with vitamin A beyond the dietary requirement, as might be the case in individuals who take multivitamins, does not appear to have a dose-dependent effect on telomere length [24].

2.2 Folate

Folate is a water-soluble vitamin B. It is the generic term used for naturally occurring food folate as well as folic acid. Folate functions as coenzyme or cosubstrate in synthesis of nucleic acids and metabolism of amino acids [127].

Low levels of folate are associated with colorectal cancer. Due to its pivotal role in one carbon metabolism, folate is crucial for DNA synthesis, repair, and methylation [25]. There are many folate-metabolizing enzymes in the cell. 5,10-Methylenetetrahydrofolate is converted to 5-methyltetrahydrofolate by the key enzyme 5,10-methylenetetrahydrofolate reductase, which in turn controls whether folate is employed for DNA synthesis or DNA methylation. Reports show that in subjects homozygous for a common variant (C677T) of the gene encoding for this enzyme have a decreased risk for colorectal cancer, which suggests that DNA synthesis and repair may be increased in these individuals [25]. Folate is also reported to play an important role in the prevention of uracil incorporation into DNA and hypomethylation of DNA. Indeed, folate deficiency causes increase in the levels of plasma homocysteine (Hcy) which in turn is significantly correlated with increased micronucleus formation and reduced telomere length [26]. Folate deficiency causes thymidylate stress, which is associated with increased uracil incorporation into DNA. This leads to the activation of BER pathway in which uracil DNA glycosylase removes the genomic uracil in the cells, and evidence indicates that HR responds to these persistent BER strand breaks [27].

The plasma concentration of folate has also been reported to be associated with telomere length in both men and women [28,29]. Methylene–tetrahydrofolate (THF) is used for the synthesis of pyrimidine thymidylate and purines, thus providing precursors for DNA synthesis. Methyl–THF provides methyl groups for the methylation of homocysteine (Hcy) to methionine, the precursor of S-adenosylmethionine (SAM), which is the universal methyl donor in biological methylation reactions including those of DNA and histones. Folate acts as a precursor for thymidylate synthesis, and thus low levels of folate induces misincorporation of uracil in place of thymidine in DNA. When these uracil bases are removed by excision repair enzymes, strand breaks are formed in DNA [30]. The thymidine in the telomeric sequence may be replaced by uracil under folate deficiency, leading to telomeric DNA damage. Low nutritional status of folate results in short telomeres, possibly due to DNA damage [28,29]. Folate deficiency would result in imbalance of nucleotide pools in the cells, which can destabilize replication forks and cause shortening of telomeres in a process independent of DNA damage due to uracil misincorporation [31]. Availability of folate is also related to DNA methylation due to its role in generating the methyl donor SAM. Folate deficiency has been shown to be associated with genomic DNA hypomethylation [32]. DNA methylation and expression of epigenetically regulated genes can be modified by folate supplementation [33]. Loss of epigenetic regulation of telomere length or loss of DNA and histone methylation may lead to elongated telomeres in cells [11,12,34]. Low folate status has been shown to be associated with longer telomeres in men [29] probably due to DNA hypomethylation. However, similar association has not been found in a population that consists of predominantly females [29]. This suggests that folate may have differential effect on telomere length based on gender. Chromosome rearrangements as indicated by nucleoplasmic bridges have also been reported under low folate conditions in cell culture [35,36].

Telomere dysfunction–induced genomic instability could be one of the mechanisms by which folate affects the risk for diseases. Impairment of remethylation of Hcy to methionine due to inadequate levels of folate will result in elevated Hcy concentrations [37]. Thus, plasma total Hcy (tHcy) functions as a marker of folate status. Elevated tHcy is linked to a decrease in telomere length [28,38]. However, very high tHcy concentration has been shown to be related with longer telomeres possibly due to loss of epigenetic regulation resulting from DNA hypomethylation [29].

2.3 Vitamin B12

Vitamin B12 is a water-soluble vitamin, which plays a key role towards the normal functioning of the brain and nervous system. Vitamin B12 exists in several forms and contains the mineral cobalt and thus compounds with vitamin B12 are collectively known as cobalamines. This vitamin acts as an antioxidant [39] and thus significantly protects cells from oxidative DNA damage. It has been reported that vitamin B12 supplementation decreases the frequency of micronuclei formation [40] and also causes a significant decrease in the amount of arsenic-induced free radicals [41].

Methylation of homocysteine to form methionine, the precursor of SAM, is catalyzed in a vitamin B12–dependent reaction. Despite the role of vitamin B12 in generation of methyl groups for methylation reactions, plasma concentration of vitamin B12 or intake of vitamin B12 has not shown an association with telomere length [24,29]. However, women who take vitamin B12 supplements have been found to have longer telomeres than nonusers [24]. Supraphysiological doses of vitamin B12 derived from supplements can inhibit nitric oxide synthase [42] and potentially reduce inflammation. The reduced oxidative stress and inflammation due to the high dose of vitamin B12 may explain the longer telomeres in individuals who use vitamin B12 supplements.

2.4 Vitamin D

Vitamin D consists of a group of fat-soluble steroids, which have multiple biological effects. They are responsible for intestinal absorption of many minerals such as calcium and zinc. In humans, vitamin D3 (cholecalciferol) and vitamin D2 (ergocalciferol) are the two most important vitamin D compounds. Vitamin D is produced in the human body by direct exposure to sunlight. However, exposure to sunlight is also the major cause of skin cancer [43]. Sunlight consists of ultraviolet radiation (UVR), which enhances accumulation of p53 in skin cells and in turn upregulates DNA repair genes. However, cells, which are irreparably damaged are destined to death [43]. Literature indicates that vitamin D, which is produced in the skin, can serve to minimize UVR-induced tumor formation. It is suggested that the vitamin D receptor (VDR), either in the presence or absence of its ligand 1, 25-dihydroxyvitamin D, can limit the cancer forming potential following UVR [44]. NER is the main pathway by which UVR-induced DNA damage is repaired. It has been shown that VDR null mice exhibit impaired DDR [44] suggesting that vitamin D and VDR play a pivotal role in DNA repair particularly in the context of skin cells.

Concentration of vitamin D in serum is associated positively with the telomere length in women [45]. 1, 25-Dihydroxyvitamin D, the biological active form of vitamin D, possesses immunosuppressive properties [22]. This is reflected in the inverse relationship between plasma concentration of vitamin D and the inflammatory marker CRP [46]. Telomere length is negatively correlated to the plasma concentration of CRP [47]. Addition of 1, 25-dihydroxyvitamin D to the cell culture medium reduces expression of the proliferation factor granulocyte–macrocyte colony stimulating factor that is important for the proliferation of all lineages of hematopoietic cells [48] and hence reduces proliferation of lymphocytes [49].

In addition, vitamin D also reduces the expression of inflammation mediators interleukin-2 [50] and interferon gamma [51]. The turnover of cells is limited by the antiinflammatory and antiproliferative properties of vitamin D, causing significant decrease in the attrition of telomere length.

2.5 Vitamins C and E

Vitamin C is a water-soluble vitamin, chemically known as ascorbic acid, which is naturally present in citrus fruits. Vitamin E, on the other hand, is a fat-soluble vitamin chemically known as tocopherols. The antioxidant properties of vitamins C and E are widely acknowledged [52].

Vitamin C is well known to decrease oxidative damage to human cells, and recent studies also indicate that it regulates the expression of certain genes participating in the repair process [53]. At plasma levels greater than 50 µmol/L, vitamin C decreases the frequency of chromosomal aberrations in groups of individuals who had insufficient dietary intake of vitamin C and were occasionally exposed to mutagens. High plasma level of vitamin C also counteracts the damage induced by air pollution and decreases the levels of micronuclei formation in smokers challenged with γ-irradiation [54]. Kiwifruit, a rich source of vitamin C, has been shown to stimulate BER pathway in vitro [55]. Vitamin E has been shown to exert antigenotoxic effect in *Danio rerio* [56]. The vitamin E isoform γ-tocotrienol reduces DSB formation and also lowers the level of chromosomal aberrations when HUVEC cells pretreated with γ-tocotrienol are exposed to ionizing radiation. Moreover, this also increases the expression levels of the DNA repair gene RAD50 [57] suggesting that vitamin E plays an important role in DNA DSB repair pathway. Vitamin E supplementation also plays a role in reducing Dnmt1 gene expression and thus methylation in obese mouse model [58].

Intake of vitamins C and E either from diet or from multivitamins is positively associated with longer telomeres in a dose-dependent manner in women [24]. Age-dependent shortening of telomeres as well as decrease in telomerase activity in cell cultures can be slowed down and life span can be increased by addition of physiological concentrations of vitamin C or vitamin E to the culture medium [59–61]. In cells treated with vitamin E (6-O-phosphorylated form of α-tocopherol), a reduction was observed in the amount of reactive oxygen species (ROS) due to scavenging by the vitamin [61]. This process may limit oxidative damage to telomeric DNA that would otherwise cause shortening of telomere length. Vitamin C has been shown to have antiaging effects on human pluripotent stem cell–derived cardiomyocytes specifically due to modulation of the expression of telomere related genes [62]. Another cross-sectional cohort study showed that high plasma concentration of vitamin C is associated with longer telomere length in elderly individuals [63] suggesting that it plays a protective role in telomere maintenance.

2.6 Nicotinamide

Nicotinamide is also known as niacinamide, which is a vitamin found in food and also used as a dietary supplement. Green vegetables, meat, milk, and yeasts are common sources of this vitamin. Nicotinamide is used by the human body to form two important coenzymes: nicotinamide adenine dinucleotide (NAD) and nicotinamide adenine dinucleotide phosphate (NADP). Both NAD and NADP are required for many essential reactions including energy production, DNA repair, and regulation of cell death. Thus nicotinamide is a very important vitamin, particularly in the context of humans. The role of poly ADP ribose polymerase (PARP) in DNA repair is well documented. NAD is used by PARP to transfer ADP ribose units to acceptor proteins [64] thus highlighting the importance of nicotinamide in DNA repair. There are numerous in vitro and in vivo studies indicating that PARP-1 and NAD^+ status influence cellular responses to genotoxicity, which can lead to mutagenesis and cancer formation. Nicotinamide acts as a precursor for NAD^+ and also as a substrate for PARP-1 and thus is tightly related to DNA repair and genomic integrity. PARP-1 is a nuclear protein, which binds to DNA DSBs, forms ADP ribose polymers on acceptor proteins, and ultimately recruits DNA repair complexes to the DSBs [65]. PARP-1 plays key role in NER [66] and BER [67], and thus nicotinamide indirectly plays a pivotal role in NER and BER pathways. Nicotinamide increases intracellular NAD^+ and enhances the repair of damaged DNA induced by N-methyl-nitro-N-nitrosoguanidine (MNNG) in primary human mammary epithelial cells [68]. NAD^+ depletion increases spontaneous DNA damage in human HaCaT keratinocytes in the absence of genotoxic stress, which is reversible with the addition of nicotinamide [69]. It has also been suggested that high dose (5 mM or more) of nicotinamide inhibits DNA repair through PARP-1 inhibition, whereas low dose nicotinamide increases the rejoining of DNA strand breaks through the provision of NAD^+ [70]. This implies that the cellular concentration of nicotinamide is very important in the context of DNA repair.

In response to DNA damage, the activity of PARPs and synthesis of poly(ADP-ribose) increase with corresponding utilization of nicotinamide adenine dinucleotide [71–73]. Incidence of micronuclei in lymphocytes, which is an indicator of genomic instability, is inversely related to the dietary intake of nicotinamide [74]. Nicotinamide also has the potential to influence telomere length due to the role of PARPs in telomere elongation. Telomere length is positively regulated by the PARPs tankyrase 1 and 2 and by ADP-ribosylation of telomere repeat binding factor-1, which is a negative regulator of telomere elongation [75,76]. It has also been shown that in vitro addition of nicotinamide in human fibroblast cell culture decreases the shortening of telomeres but increases the replicative capacity of the cells [77]. While ADP-ribosylation of telomere repeat binding factor-2 does not modulate telomere length or telomerase activity, it is important for the maintenance of telomere integrity [78]. All these findings suggest that nicotinamide may influence telomere length through multiple mechanisms.

3. MINERALS: DNA REPAIR AND TELOMERE LENGTH

3.1 Magnesium

Magnesium is the second most abundant element and is involved in basically all metabolic pathways. It is required for the catalytic activity of a wide array of enzymes including those involved in DNA replication, DNA repair, and RNA synthesis [79]. Magnesium is itself not genotoxic but is an absolute requirement for maintaining genomic stability. Dietary magnesium intake has been shown to be positively related to telomere length in women [24]. Magnesium deficiency has been shown to cause shortening of telomeres both in vivo and in vitro [80,81]. This decrease in telomere length due to magnesium deficiency is also accompanied by an increase in oxidative stress [80], which is one of the factors that result in telomere attrition.

Low serum magnesium concentration is associated with high concentration of the inflammatory marker CRP [82]. Reduced availability of magnesium ions negatively influences genome integrity. Magnesium ion binding is necessary for efficiency and fidelity of DNA polymerase [83,84]. Furthermore, magnesium is essential for the functioning of endonucleases involved in BER of DNA and for maintaining chromosome structure by binding to the phosphate of nucleotides [79,85]. Insufficiency of magnesium reduces DNA repair capacity [86] and induces chromosomal abnormalities [87,88]. Thus, magnesium affects telomere length by modulating DNA integrity and repair in addition to its role in oxidative stress and inflammation.

3.2 Zinc

Zinc (Zn) is chemically similar to magnesium in some respect. Zn is perceived as of exceptional biologic and public health importance. Zinc-dependent enzymes in the cell include DNA polymerases, RNA polymerases, and reverse transcriptases [89–91]. Zn is an essential component of many proteins involved in biological defense against oxidative stress, and its depletion may enhance DNA damage by impairing DNA repair mechanisms [92]. High dietary and plasma Zn is inversely related with the risk for developing breast cancer [93]. Providing additional zinc in the cell culture medium increases activity of telomerase, which is a reverse transcriptase [94]. Zinc is also necessary for activation of poly (ADP-ribose) polymerase involved in DNA repair at DNA damage sites [95].

In humans, deficiency of dietary zinc has been shown to cause DNA damage [96,99]. In older subjects, percentage of cells with critically short telomeres or decrease in telomere length are associated with decrease in concentration of intracellular labile zinc and the zinc-binding protein, metallothionein, in peripheral blood mononuclear cells [97]. Maternal diet deficient in zinc causes chromosomal abnormalities including fusion between chromosomes in the offspring of rats [87]. One of the principal reasons for fusion between chromosomes is the loss of telomere cap by attrition, which in this case could be due to DNA damage arising from zinc deficiency. Zinc also has a protective role in oxidative stress [98]. While a direct role for zinc in removing ROS or free radicals has not been proven, it has been shown that dietary zinc deficiency is associated with oxidative damage [96,99]. Supplementation with zinc can reduce oxidative stress and inflammation [100,101]. Binding of zinc decreases the susceptibility of oxidation of the sulfhydryl groups in purified systems [102], and it has been proposed that zinc competes with other prooxidant metals such as iron for binding with cysteine, which prevents the formation of free radicals [98]. Zinc supplementation reduces the incidence of infection [100,103], which is another factor that leads to telomere attrition by higher turnover of cells. All the above facts suggest that zinc can affect the length of telomeres by a variety of mechanisms, which include influencing the activity of telomerase, affecting DNA integrity, oxidative stress, and modulating the susceptibility to infection.

3.3 Iron

Iron (Fe) is a metal in the first transition series and is the most common element on earth. It exists in a wide range of oxidation states, $+2$ and $+3$ being the most common. It plays a very important role in biology as it forms complex with oxygen in hemoglobin and myoglobin and is pivotal for oxygen transport inside the body. Iron is necessary for facilitating assembly of Fe–S cluster proteins, heme-binding proteins, and ribonucleotide reductases in most eukaryotic cells [104,105]. Many DNA replication and repair proteins such as Pol α, Rad 3/XPD, and Dna2 require iron as a cofactor [106]. Scientists have found that BER glycosylases that remove a wide variety of damaged bases contain the Fe–S cluster cofactor [107]. Also, XPD which is classified as an SF2-DNA helicase plays an important role in NER [106]. The Dna2 protein has been implicated in DSB repair. It has been showed that mutations in the Fe–S domain of Dna2 affect the ability of protein complexes to bind broken DNA and thus impairs DNA replication, which indicates the essential function of Fe–S in this process [108]. Mutations of iron-requiring proteins are associated with diseases characterized by defects in DNA repair and poor response to replication stress in mammals [106], which proves the potential role of Fe in DNA repair.

In contrast to the effect of other nutrients, use of iron supplements is associated with shorter telomeres [24,109]. Iron is a prooxidant that can bind to cysteine residues of proteins and result in formation of hydroxyl free radicals. Iron supplement intake has been shown to increase free radical excretion in feces in healthy individuals [110]. The shorter telomeres observed in iron supplement users could be due to the free radical generating capacity of iron and resultant oxidative stress [109]. Iron intake from diet or multivitamins, which may contain less iron than iron supplements, is not negatively associated with telomere length [24].

4. OTHER BIOACTIVE DIETARY COMPONENTS AND TELOMERE LENGTH

4.1 Omega-3 Fatty Acids

Plasma concentration of marine omega-3 fatty acids, docosahexanoic acid and eicosapentaenoic acid, has been shown to be positively associated with reduced attrition of telomere length in study subjects over a period of 5–8 years [111]. Individuals who have higher baseline omega-3 fatty acid concentration showed the maximum effect of omega-3 fatty acids on telomere attrition. Omega-3 fatty acid concentration in plasma is associated with low proinflammatory markers and high antiinflammatory markers [112]. In mice, diets enriched in marine omega-3 fatty acids have been shown to enhance the activities of antioxidant enzymes superoxide dismutase, catalase, and glutathione peroxidase, and this also increased life span [113,114]. The changes in the antiinflammatory and antioxidant properties induced by omega-3 fatty acids decrease the cell turnover and oxidative DNA damage and thus may reduce telomere shortening.

4.2 Polyphenols

Polyphenols in grape seed and tea have been reported to have antioxidant and antiinflammatory properties [115,116]. Tea consumption is negatively associated with biomarkers of inflammation [117]. Possibly, due to this antiinflammatory effect, habitual tea drinkers have longer telomeres in peripheral blood cells than those who drink tea less often [118]. In mice, administration of grape seed polyphenols through diet resulted in a trend for longer telomeres when compared to controls [119].

4.3 Curcumin

Curcumin, an active ingredient of the spice turmeric, is an important dietary component. It possesses properties similar to polyphenols. It induces synthesis of the antioxidant glutathione and inhibits the release of the chemokine interleukin-8. It also activates the nuclear transcription factor NF-kappa B, which mediates the inflammatory response [120]. A decreased DNA damage has been shown in mice that were fed with diets containing curcumin. These mice also contained longer telomeres compared to mice that were fed with a control diet [119].

5. DIET, LIFESTYLE, AND TELOMERE LENGTH

Diet and lifestyle can influence inflammation, oxidative stress, and psychological stress all of which cause telomere attrition, and thus these could also influence telomere length. A healthy lifestyle with a diet high in fruits and vegetables combined with exercise, lower body mass index, and no smoking is associated with longer telomeres [121]. Use of multivitamins has been associated with longer telomeres in women. However, when analyzed for individual vitamin intake, only vitamins C and E are associated with telomere length [24]. Most multivitamin users follow a healthy lifestyle, which is another factor that could influence their telomere length [109].

A small pilot study in men showed that comprehensive changes in lifestyle including a change in diet to include more low fat, unrefined plant-based food supplemented with omega-3 fatty acids (from fish oil), soy and vitamins C and E for a period of 3 months increases the telomerase activity of peripheral blood mononuclear cells. Increase in telomerase activity was correlated to the decrease in plasma concentration of low-density lipoprotein and decrease in psychological distress [19]. The telomere length was not measured in this study because the short duration of the study was not considered sufficient to show any changes in telomere length. Consumption of whole grains and other plant-based foods ameliorates inflammation [122]. This is the probable reason why dietary fiber intake specifically from cereals and whole grains is positively associated with telomere length [123]. In a multiethnic cohort, it has been shown that the intake of processed meat is inversely related to the telomere length [124]. Many studies have shown that maternal diet exerts a strong influence on the health and disease risks of the adult offspring. Maternal diet low in proteins during gestation can lead to increased DNA damage and accelerated shortening of aortic telomere length of the offspring in rats [125].

6. CONCLUDING REMARKS

Figs. 14.1 and 14.2 present a schematic summary of the role of the various nutritional supplements in telomere length maintenance and DNA repair. Any external or internal insult to the cellular DNA activates

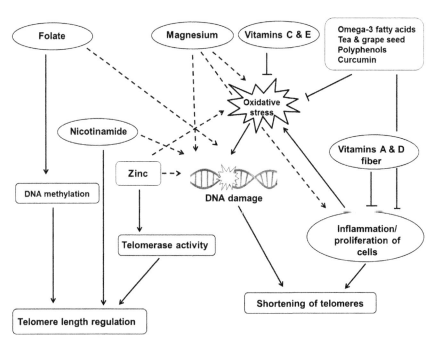

Figure 14.1 Role of nutritional supplements in genomic damage and telomere length maintenance. *Solid arrow* indicates direct role, *dashed arrow* indicates indirect role, and ⊥ *indicates* inhibitory role.

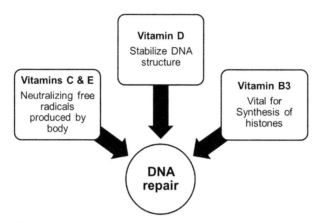

Figure 14.2 Role of nutritional supplements in DNA repair.

the DNA repair pathways. Most of the nutritional supplements discussed in this chapter affect the DNA repair pathways indirectly by modulating the oxidative stress inside the cells. However, folate has been shown to cause thymidylate stress and thus activates the BER pathway. Also, the levels of vitamin D can indirectly facilitate the NER pathway. Vitamin C stimulates the BER pathway in the cells and vitamin D plays an important role on the DSB repair of DNA. Nicotinamide indirectly plays an important role in both BER and NER pathways and also in DSB repair. Both magnesium and zinc are indirectly related to DNA repair. Magnesium is required for proper functioning of DNA polymerases and zinc depletion causes impairment in DNA repair pathways. Iron is particularly important because Fe–S cluster cofactor is an important factor of many DNA repair enzymes. Such enzymes have been found to play major role in BER and NER pathways.

Length of telomeres is an indicator of biological aging. Due to the incomplete replication of linear chromosomes by DNA polymerase, telomeric repeats at the ends are lost with each cell division. Dysfunction of telomeres is associated with the development of many age–related diseases. Telomere length and attrition of telomeric repeats are influenced by nutrition in human and animal models. Damage to telomeric DNA due to either oxidative stress or reduced availability of nucleotide precursors results in shorter telomeres. Antiinflammatory and antioxidant nutrients can reduce the erosion of telomeres. Nutrients also have the potential to influence the regulation of telomere length. For example, folate can regulate telomere length via its role in epigenetic status of DNA and histones, whereas nicotinamide can regulate the length of telomeres through its role as a substrate for posttranslational modification of telomere-associated proteins.

REFERENCES

[1] Hoeijmakers JH. DNA damage, aging, and cancer. N Engl J Med 2009;361:1475–85.

[2] Stracker TH, Usui T, Petrini JH. Taking the time to make important decisions: the checkpoint effector kinases Chk1 and Chk2 and the DNA damage response. DNA Repair 2009;8:1047–54.

[3] Shammas MA. Telomeres, lifestyle, cancer, and aging. Curr Opin Clin Nutr Metab Care 2011;14:28–34.

[4] Olovnikov AM. Principle of marginotomy in template synthesis of polynucleotides. Dokl Akad Nauk SSSR 1971;201:1496–9.

[5] Harley CB, Futcher AB, Greider CW. Telomeres shorten during ageing of human fibroblasts. Nature 1990;345:458–60.

[6] Olovnikov AM. Telomeres, telomerase, and aging: origin of the theory. Exp Gerontol 1990;31:443–8.

[7] Olovnikov AM. A theory of marginotomy. The incomplete copying of template margin in enzymic synthesis of polynucleotides and biological significance of the phenomenon. J Theor Biol 1973;41:181–90.

[8] Shippen-Lentz D, Blackburn EH. Functional evidence for an RNA template in telomerase. Science (New York, NY) 1990;247:546–52.

[9] Harley CB. Telomere loss: mitotic clock or genetic time bomb? Mutat Res 1990;256:271–82.

[10] De Lange T. Shelterin: the protein complex that shapes and safeguards human telomeres. Gene Dev 2005;19:2100–10.

[11] Gonzalo S, Jaco I, Fraga MF, Chen T, Li E, Esteller M, Blasco MA. DNA methyltransferases control telomere length and telomere recombination in mammalian cells. Nat Cell Biol 2006;8:416–24.

[12] Benetti R, Gonzalo S, Jaco I, Schotta G, Klatt P, Jenuwein T, Blasco MA. Suv4-20h deficiency results in telomere elongation and derepression of telomere recombination. J Cell Biol 2007;178:925–36.

[13] Slijepcevic P, Hande MP, Bouffler SD, Lansdorp P, Bryant PE. Telomere length, chromatin structure and chromosome fusigenic potential. Chromosoma 1997;106:413–21.

[14] Guan JZ, Maeda T, Sugano M, Oyama J, Higuchi Y, Suzuki T, Makino N. A percentage analysis of the telomere length in Parkinson's disease patients. J Gerontol 2008;63:467–73.

[15] Minamino T, Miyauchi H, Yoshida T, Ishida Y, Yoshida H, Komuro I. Endothelial cell senescence in human atherosclerosis: role of telomere in endothelial dysfunction. Circulation 2002;105:1541–4.

[16] Panossian LA, Porter VR, Valenzuela HF, Zhu X, Reback E, Masterman D, Cummings JL, Effros RB. Telomere shortening in T cells correlates with Alzheimer's disease status. Neurobiol Aging 2003;24:77–84.

[17] Wu X, Amos CI, Zhu Y, Zhao H, Grossman BH, Shay JW, Luo S, Hong WK, Spitz MR. Telomere dysfunction: a potential cancer predisposition factor. J Natl Canc Inst 2003;95:1211–8.

[18] Oh BK, Kim H, Park YN, Yoo JE, Choi J, Kim KS, Lee JJ, Park C. High telomerase activity and long telomeres in advanced hepatocellular carcinomas with poor prognosis. Lab Investig 2008;88:144–52.

[19] Ornish D, Lin J, Daubenmier J, Weidner G, Epel E, Kemp C, Magbanua MJ, Marlin R, Yglecias L, Carroll PR, Blackburn EH. Increased telomerase activity and comprehensive lifestyle changes: a pilot study. Lancet 2008;9:1048–57.

[20] De Flora S, Bagnasco M, Vainio H. Modulation of genotoxic and related effects by carotenoids and vitamin A in experimental models: mechanistic issues. Mutagenesis 1999;14:153–72.

[21] Nomura SJ, Robien K, Zota AR. Serum folate, vitamin B-12, vitamin a, gamma-tocopherol, alpha-tocopherol, and carotenoids do not modify associations between cadmium exposure and leukocyte telomere length in the general US adult population. J Nutr 2017;147:538–48.

[22] Mora JR, Iwata M, von Andrian UH. Vitamin effects on the immune system: vitamins A and D take centre stage. Nat Rev 2008;8:685–98.

[23] Aukrust P, Muller F, Ueland T, Svardal AM, Berge RK, Froland SS. Decreased vitamin A levels in common variable immunodeficiency: vitamin A supplementation in vivo enhances immunoglobulin production and downregulates inflammatory responses. Eur J Clin Investig 2000;30:252–9.

[24] Xu Q, Parks CG, DeRoo LA, Cawthon RM, Sandler DP, Chen H. Multivitamin use and telomere length in women. Am J Clin Nutr 2009;89:1857–63.

[25] Duthie SJ, Narayanan S, Sharp L, Little J, Basten G, Powers H. Folate, DNA stability and colo-rectal neoplasia. Proc Nutr Soc 2004;63:571–8.

[26] Fenech M. Folate (vitamin B9) and vitamin B12 and their function in the maintenance of nuclear and mitochondrial genome integrity. Mutat Res 2012;733:21–33.

[27] Berger SH, Pittman DL, Wyatt MD. Uracil in DNA: consequences for carcinogenesis and chemotherapy. Biochem Pharmacol 2008;76:697–706.

[28] Richards JB, Valdes AM, Gardner JP, Kato BS, Siva A, Kimura M, Lu X, Brown MJ, Aviv A, Spector TD. Homocysteine levels and leukocyte telomere length. Atherosclerosis 2008;200:271–7.

[29] Paul L, Cattaneo M, D'Angelo A, Sampietro F, Fermo I, Razzari C, Fontana G, Eugene N, Jacques PF, Selhub J. Telomere length in peripheral blood mononuclear cells is associated with folate status in men. J Nutr 2009;139:1273–8.

[30] Blount BC, Mack MM, Wehr CM, MacGregor JT, Hiatt RA, Wang G, Wickramasinghe SN, Everson RB, Ames BN. Folate deficiency causes uracil misincorporation into human DNA and chromosome breakage: implications for cancer and neuronal damage. Proc Natl Acad Sci USA 1997;94:3290–5.

[31] Toussaint M, Dionne I, Wellinger RJ. Limited TTP supply affects telomere length regulation in a telomerase-independent fashion. Nucleic Acids Res 2005;33:704–13.

[32] Friso S, Choi SW, Girelli D, Mason JB, Dolnikowski GG, Bagley PJ, Olivieri O, Jacques PF, Rosenberg IH, Corrocher R, Selhub J. A common mutation in the 5,10-methylenetetrahydrofolate reductase gene affects genomic DNA methylation through an interaction with folate status. Proc Natl Acad Sci USA 2002;99:5606–11.

[33] Ingrosso D, Cimmino A, Perna AF, Masella L, De Santo NG, De Bonis ML, Vacca M, D'Esposito M, D'Urso M, Galletti P, Zappia V. Folate treatment and unbalanced methylation and changes of allelic expression induced by hyperhomocysteinaemia in patients with uraemia. Lancet (London, England) 2003;361:1693–9.

[34] Garcia-Cao M, O'Sullivan R, Peters AH, Jenuwein T, Blasco MA. Epigenetic regulation of telomere length in mammalian cells by the Suv39h1 and Suv39h2 histone methyltransferases. Nat Genet 2004;36:94–9.

[35] Crott JW, Mashiyama ST, Ames BN, Fenech M. The effect of folic acid deficiency and MTHFR C677T polymorphism on chromosome damage in human lymphocytes in vitro. Cancer Epidemiol Biomarkers Prev 2001;10:1089–96.

[36] Leopardi P, Marcon F, Caiola S, Cafolla A, Siniscalchi E, Zijno A, Crebelli R. Effects of folic acid deficiency and MTHFR C677T polymorphism on spontaneous and radiation-induced micronuclei in human lymphocytes. Mutagenesis 2006;21:327–33.

[37] Selhub J, Jacques PF, Wilson PW, Rush D, Rosenberg IH. Vitamin status and intake as primary determinants of homocysteinemia in an elderly population. Jama 1993;270:2693–8.

[38] Bull CF, O'Callaghan NJ, Mayrhofer G, Fenech MF. Telomere length in lymphocytes of older South Australian men may be inversely associated with plasma homocysteine. Rejuvenation Res 2009;12:341–9.

[39] Birch CS, Brasch NE, McCaddon A, Williams JH. A novel role for vitamin B(12): cobalamins are intracellular antioxidants in vitro. Free Radic Biol Med 2009;47:184–8.

[40] Joksic I, Leskovac A, Petrovic S, Joksic G. Vitamin B12 reduces ribavirin-induced genotoxicity in phytohemaglutinin-stimulated human lymphocytes. Tohoku J Exp Med 2006;209:347–54.

[41] Majumdar S, Mukherjee S, Maiti A, Karmakar S, Das AS, Mukherjee M, Nanda A, Mitra C. Folic acid or combination of folic acid and vitamin B(12) prevents short-term arsenic trioxide-induced systemic and mitochondrial dysfunction and DNA damage. Environ Toxicol 2009;24:377–87.

[42] Weinberg JB, Chen Y, Jiang N, Beasley BE, Salerno JC, Ghosh DK. Inhibition of nitric oxide synthase by cobalamins and cobinamides. Free Radic Biol Med 2009;46:1626–32.

[43] Dixon KM, Tongkao-On W, Sequeira VB, Carter SE, Song EJ, Rybchyn MS, Gordon-Thomson C, Mason RS. Vitamin D and death by sunshine. Int J Mol Sci 2013;14:1964–77.

[44] Bikle DD. Protective actions of vitamin D in UVB induced skin cancer. Photochem Photobiol Sci 2012;11:1808–16.

[45] Richards JB, Valdes AM, Gardner JP, Paximadas D, Kimura M, Nessa A, Lu X, Surdulescu GL, Swaminathan R, Spector TD, Aviv A. Higher serum vitamin D concentrations are associated with longer leukocyte telomere length in women. Am J Clin Nutr 2007;86:1420–5.

[46] Oelzner P, Muller A, Deschner F, Huller M, Abendroth K, Hein G, Stein G. Relationship between disease activity and serum levels of vitamin D metabolites and PTH in rheumatoid arthritis. Calcif Tissue Int 1998;62:193–8.

[47] Aviv A, Valdes A, Gardner JP, Swaminathan R, Kimura M, Spector TD. Menopause modifies the association of leukocyte telomere length with insulin resistance and inflammation. J Clin Endocrinol Metab 2006;91:635–40.

[48] Tobler A, Gasson J, Reichel H, Norman AW, Koeffler HP. Granulocyte-macrophage colony-stimulating factor. Sensitive and receptor-mediated regulation by 1,25-dihydroxyvitamin D_3 in normal human peripheral blood lymphocytes. J Clin Investig 1987;79:1700–5.

[49] Lemire JM, Adams JS, Sakai R, Jordan SC. 1 α,25-dihydroxyvitamin D_3 suppresses proliferation and immunoglobulin production by normal human peripheral blood mononuclear cells. J Clin Investig 1984;74:657–61.

[50] Lemire JM, Adams JS, Kermani-Arab V, Bakke AC, Sakai R, Jordan SC. 1,25-Dihydroxyvitamin D3 suppresses human T helper/inducer lymphocyte activity in vitro. J Immunol 1985;134:3032–5.

[51] Reichel H, Koeffler HP, Tobler A, Norman AW. 1 α,25-Dihydroxyvitamin D_3 inhibits gamma-interferon synthesis by normal human peripheral blood lymphocytes. Proc Natl Acad Sci USA 1987;84:3385–9.

[52] Honarbakhsh S, Schachter M. Vitamins and cardiovascular disease. Br J Nutr 2009;101:1113–31.

[53] Konopacka M. Role of vitamin C in oxidative DNA damage. Postepy Hig Med Dosw (Online) 2004;58:343–8.

[54] Sram RJ, Binkova B, Rossner Jr P. Vitamin C for DNA damage prevention. Mutat Res 2012;733:39–49.

[55] Collins AR. Kiwifruit as a modulator of DNA damage and DNA repair. Adv Food Nutrit Res 2013;68:283–99.

[56] Rocco L, Mottola F, Santonastaso M, Saputo V, Cusano E, Costagliola D, Suero T, Pacifico S, Stingo V. Anti-genotoxic ability of alpha-tocopherol and Anthocyanin to counteract fish DNA damage induced by musk xylene. Ecotoxicology (London, England) 2015;24:2026–35.

[57] Pathak R, Bachri A, Ghosh SP, Koturbash I, Boerma M, Binz RK, Sawyer JR, Hauer-Jensen M. The vitamin E analog gamma-tocotrienol (GT3) suppresses radiation-induced cytogenetic damage. Pharmaceut Res 2016;33:2117–25.

[58] Remely M, Ferk F, Sterneder S, Setayesh T, Kepcija T, Roth S, Noorizadeh R, Greunz M, Rebhan I, Wagner KH, Knasmuller S, Haslberger A. Vitamin E modifies high-fat diet-induced increase of DNA strand breaks, and changes in expression and DNA methylation of Dnmt1 and MLH1 in C57BL/6J male mice. Nutrients 2017;9.

[59] Furumoto K, Inoue E, Nagao N, Hiyama E, Miwa N. Age-dependent telomere shortening is slowed down by enrichment of intracellular vitamin C via suppression of oxidative stress. Life Sci 1998;63:935–48.

[60] Yokoo S, Furumoto K, Hiyama E, Miwa N. Slow-down of age-dependent telomere shortening is executed in human skin keratinocytes by hormesis-like-effects of trace hydrogen peroxide or by anti-oxidative effects of pro-vitamin C in common concurrently with reduction of intracellular oxidative stress. J Cell Biochem 2004;93:588–97.

[61] Tanaka Y, Moritoh Y, Miwa N. Age-dependent telomere-shortening is repressed by phosphorylated alpha-tocopherol together with cellular longevity and intracellular oxidative-stress reduction in human brain microvascular endotheliocytes. J Cell Biochem 2007;102:689–703.

[62] Kim YY, Ku SY, Huh Y, Liu HC, Kim SH, Choi YM, Moon SY. Anti-aging effects of vitamin C on human pluripotent stem cell-derived cardiomyocytes. Age (Dordrecht, Netherlands) 2013;35:1545–57.

[63] Sen A, Marsche G, Freudenberger P, Schallert M, Toeglhofer AM, Nagl C, Schmidt R, Launer LJ, Schmidt H. Association between higher plasma lutein, zeaxanthin, and vitamin C concentrations and longer telomere length: results of the Austrian Stroke Prevention Study. J Am Geriatr Soc 2014;62:222–9.

[64] Sousa FG, Matuo R, Soares DG, Escargueil AE, Henriques JA, Larsen AK, Saffi J. PARPs and the DNA damage response. Carcinogenesis 2012;33:1433–40.

[65] Surjana D, Halliday GM, Damian DL. Role of nicotinamide in DNA damage, mutagenesis, and DNA repair. J Nucleic Acids 2010;2010:157591.

[66] Stierum RH, van Herwijnen MH, Hageman GJ, Kleinjans JC. Increased poly(ADP-ribose) polymerase activity during repair of (+/-)-anti-benzo[a]pyrene diolepoxide-induced DNA damage in human peripheral-blood lymphocytes in vitro. Carcinogenesis 1994;15:745–51.

[67] Trucco C, Oliver FJ, de Murcia G, Menissier-de Murcia J. DNA repair defect in poly(ADP-ribose) polymerase-deficient cell lines. Nucleic Acids Res 1998;26:2644–9.

[68] Jacobson EL, Shieh WM, Huang AC. Mapping the role of NAD metabolism in prevention and treatment of carcinogenesis. Mol Cell Biochem 1999;193:69–74.

[69] Benavente CA, Jacobson EL. Niacin restriction upregulates NADPH oxidase and reactive oxygen species (ROS) in human keratinocytes. Free Radic Biol Med 2008;44:527–37.

[70] Riklis E, Kol R, Marko R. Trends and developments in radioprotection: the effect of nicotinamide on DNA repair. Int J Radiat Biol 1990;57:699–708.

[71] Kreimeyer A, Wielckens K, Adamietz P, Hilz H. DNA repair-associated ADP-ribosylation in vivo. Modification of histone H1 differs from that of the principal acceptor proteins. J Biol Chem 1984;259:890–6.

[72] Halldorsson H, Gray DA, Shall S. Poly (ADP-ribose) polymerase activity in nucleotide permeable cells. FEBS Lett 1978;85:349–52.

[73] Berger NA, Sikorski GW, Petzold SJ, Kurohara KK. Association of poly(adenosine diphosphoribose) synthesis with DNA damage and repair in normal human lymphocytes. J Clin Investig 1979;63:1164–71.

[74] Fenech M, Baghurst P, Luderer W, Turner J, Record S, Ceppi M, Bonassi S. Low intake of calcium, folate, nicotinic acid, vitamin E, retinol, beta-carotene and high intake of pantothenic acid, biotin and riboflavin are significantly associated with increased genome instability–results from a dietary intake and micronucleus index survey in South Australia. Carcinogenesis 2005;26:991–9.

[75] Smith S, Giriat I, Schmitt A, de Lange T. Tankyrase, a poly(ADP-ribose) polymerase at human telomeres. Science (New York, NY) 1998;282:1484–7.

[76] Cook BD, Dynek JN, Chang W, Shostak G, Smith S. Role for the related poly(ADP-Ribose) polymerases tankyrase 1 and 2 at human telomeres. Mol Cell Biol 2002;22:332–42.

[77] Kang HT, Lee HI, Hwang ES. Nicotinamide extends replicative lifespan of human cells. Aging Cell 2006;5:423–36.

[78] Dantzer F, Giraud-Panis MJ, Jaco I, Ame JC, Schultz I, Blasco M, Koering CE, Gilson E, Menissier-de Murcia J, de Murcia G, Schreiber V. Functional interaction between poly(ADP-Ribose) polymerase 2 (PARP-2) and TRF2: PARP activity negatively regulates TRF2. Mol Cell Biol 2004;24:1595–607.

[79] Hartwig A. Role of magnesium in genomic stability. Mutat Res 2001;475:113–21.

[80] Martin H, Uring-Lambert B, Adrian M, Lahlou A, Bonet A, Demougeot C, Devaux S, Laurant P, Richert L, Berthelot A. Effects of long-term dietary intake of magnesium on oxidative stress, apoptosis and ageing in rat liver. Magne Res 2008;21:124–30.

[81] Killilea DW, Ames BN. Magnesium deficiency accelerates cellular senescence in cultured human fibroblasts. Proc Natl Acad Sci USA 2008;105:5768–73.

[82] Guerrero-Romero F, Rodriguez-Moran M. Relationship between serum magnesium levels and C-reactive protein concentration, in non-diabetic, non-hypertensive obese subjects. Int J Obes Relat Metab Disord 2002;26:469–74.

[83] Batra VK, Beard WA, Shock DD, Krahn JM, Pedersen LC, Wilson SH. Magnesium-induced assembly of a complete DNA polymerase catalytic complex. Structure 2006;14:757–66.

[84] Sirover MA, Loeb LA. Metal activation of DNA synthesis. Biochem Biophys Res Comm 1976;70:812–7.

[85] Mazia D. The particulate organization of the chromosome. Proc Natl Acad Sci USA 1954;40:521–7.

[86] Mahabir S, Wei Q, Barrera SL, Dong YQ, Etzel CJ, Spitz MR, Forman MR. Dietary magnesium and DNA repair capacity as risk factors for lung cancer. Carcinogenesis 2008;29:949–56.

[87] Bell LT, Branstrator M, Roux C, Hurley LS. Chromosomal abnormalities in maternal and fetal tissues of magnesium- or zinc-deficient rats. Teratology 1975;12:221–6.

[88] Jayson GG. Bivalent metal ions as the coupling factor between cell metabolism and the rate of cell mutation. Nature 1961;190:144–6.

[89] Springgate CF, Mildvan AS, Abramson R, Engle JL, Loeb LA. Escherichia coli deoxyribonucleic acid polymerase I, a zinc metalloenzyme. Nuclear quadrupolar relaxation studies of the role of bound zinc. J Biol Chem 1973;248:5987–93.

[90] Poiesz BJ, Seal G, Loeb LA. Reverse transcriptase: correlation of zinc content with activity. Proc Natl Acad Sci USA 1974;71:4892–6.

[91] Terhune MW, Sandstead HH. Decreased RNA polymerase activity in mammalian zinc deficiency. Science (New York, NY) 1972;177:68–9.

[92] Valko M, Jomova K, Rhodes CJ, Kuca K, Musilek K. Redox- and non-redox-metal-induced formation of free radicals and their role in human disease. Arch Toxicol 2016;90:1–37.

[93] Alam S, Kelleher SL. Cellular mechanisms of zinc dysregulation: a perspective on zinc homeostasis as an etiological factor in the development and progression of breast cancer. Nutrients 2012;4:875–903.

[94] Nemoto K, Kondo Y, Himeno S, Suzuki Y, Hara S, Akimoto M, Imura N. Modulation of telomerase activity by zinc in human prostatic and renal cancer cells. Biochem Pharmacol 2000;59:401–5.

[95] Ikejima M, Noguchi S, Yamashita R, Ogura T, Sugimura T, Gill DM, Miwa M. The zinc fingers of human poly(ADP-ribose) polymerase are differentially required for the recognition of DNA breaks and nicks and the consequent enzyme activation. Other structures recognize intact DNA. J Biol Chem 1990;265:21907–13.

[96] Song Y, Chung CS, Bruno RS, Traber MG, Brown KH, King JC, Ho E. Dietary zinc restriction and repletion affects DNA integrity in healthy men. Am J Clin Nutr 2009a;90:321–8.

[97] Cipriano C, Tesei S, Malavolta M, Giacconi R, Muti E, Costarelli L, Piacenza F, Pierpaoli S, Galeazzi R, Blasco M, Vera E, Canela A, Lattanzio F, Mocchegiani E. Accumulation of cells with short telomeres is associated with impaired zinc homeostasis and inflammation in old hypertensive participants. J Gerontol 2009;64:745–51.

[98] Bray TM, Bettger WJ. The physiological role of zinc as an antioxidant. Free Radic Biol Med 1990;8:281–91.

[99] Song Y, Leonard SW, Traber MG, Ho E. Zinc deficiency affects DNA damage, oxidative stress, antioxidant defenses, and DNA repair in rats. J Nutr 2009b;139:1626–31.

[100] Bao B, Prasad AS, Beck FW, Snell D, Suneja A, Sarkar FH, Doshi N, Fitzgerald JT, Swerdlow P. Zinc supplementation decreases oxidative stress, incidence of infection, and generation of inflammatory cytokines in sickle cell disease patients. Transl Res 2008;152:67–80.

[101] Hennig B, Meerarani P, Toborek M, McClain CJ. Antioxidant-like properties of zinc in activated endothelial cells. J Am Coll Nutr 1999;18:152–8.

[102] Gibbs PN, Gore MG, Jordan PM. Investigation of the effect of metal ions on the reactivity of thiol groups in human 5-aminolaevulinate dehydratase. Biochem J 1985;225:573–80.

[103] Meydani SN, Barnett JB, Dallal GE, Fine BC, Jacques PF, Leka LS, Hamer DH. Serum zinc and pneumonia in nursing home elderly. Am J Clin Nutr 2007;86:1167–73.

[104] Dlouhy AC, Outten CE. The iron metallome in eukaryotic organisms. Met Ions Life Sci 2013;12:241–78.

[105] Heath JL, Weiss JM, Lavau CP, Wechsler DS. Iron deprivation in cancer–potential therapeutic implications. Nutrients 2013;5:2836–59.

[106] Zhang C. Essential functions of iron-requiring proteins in DNA replication, repair and cell cycle control. Protein Cell 2014;5:750–60.

[107] Lukianova OA, David SS. A role for iron-sulfur clusters in DNA repair. Curr Opin Chem Biol 2005;9:145–51.

[108] Wu Y, Brosh Jr RM. DNA helicase and helicase-nuclease enzymes with a conserved iron-sulfur cluster. Nucleic Acids Res 2012;40:4247–60.

[109] Aviv A. Leukocyte telomere length: the telomere tale continues. Am J Clin Nutr 2009;89:1721–2.

[110] Lund EK, Wharf SG, Fairweather-Tait SJ, Johnson IT. Oral ferrous sulfate supplements increase the free radical-generating capacity of feces from healthy volunteers. Am J Clin Nutr 1999;69:250–5.

[111] Farzaneh-Far R, Lin J, Epel ES, Harris WS, Blackburn EH, Whooley MA. Association of marine omega-3 fatty acid levels with telomeric aging in patients with coronary heart disease. Jama 2010;303:250–7.

[112] Ferrucci L, Cherubini A, Bandinelli S, Bartali B, Corsi A, Lauretani F, Martin A, Andres-Lacueva C, Senin U, Guralnik JM. Relationship of plasma polyunsaturated fatty acids to circulating inflammatory markers. J Clin Endocrinol Metab 2006;91:439–46.

[113] Jolly CA, Muthukumar A, Avula CP, Troyer D, Fernandes G. Life span is prolonged in food-restricted autoimmune-prone (NZB x NZW)F(1) mice fed a diet enriched with (n-3) fatty acids. J Nutr 2001;131:2753–60.

[114] Kesavulu MM, Kameswararao B, Apparao C, Kumar EG, Harinarayan CV. Effect of omega-3 fatty acids on lipid peroxidation and antioxidant enzyme status in type 2 diabetic patients. Diabetes Metabol 2002;28:20–6.

[115] Vitseva O, Varghese S, Chakrabarti S, Folts JD, Freedman JE. Grape seed and skin extracts inhibit platelet function and release of reactive oxygen intermediates. J Cardiovasc Pharmacol 2005;46:445–51.

[116] Frei B, Higdon JV. Antioxidant activity of tea polyphenols in vivo: evidence from animal studies. J Nutr 2003;133:3275S–84S.

[117] De Bacquer D, Clays E, Delanghe J, De Backer G. Epidemiological evidence for an association between habitual tea consumption and markers of chronic inflammation. Atherosclerosis 2006;189:428–35.

[118] Chan R, Woo J, Suen E, Leung J, Tang N. Chinese tea consumption is associated with longer telomere length in elderly Chinese men. Br J Nutr 2010;103:107–13.

[119] Thomas P, Wang YJ, Zhong JH, Kosaraju S, O'Callaghan NJ, Zhou XF, Fenech M. Grape seed polyphenols and curcumin reduce genomic instability events in a transgenic mouse model for Alzheimer's disease. Mutat Res 2009;661:25–34.

[120] Biswas SK, McClure D, Jimenez LA, Megson IL, Rahman I. Curcumin induces glutathione biosynthesis and inhibits NF-kappaB activation and interleukin-8 release in alveolar epithelial cells: mechanism of free radical scavenging activity. Antioxidants Redox Signal 2005;7:32–41.

[121] Mirabello L, Huang WY, Wong JY, Chatterjee N, Reding D, Crawford ED, De Vivo I, Hayes RB, Savage SA. The association between leukocyte telomere length and cigarette smoking, dietary and physical variables, and risk of prostate cancer. Aging Cell 2009;8:405–13.

[122] Lopez-Garcia E, Schulze MB, Fung TT, Meigs JB, Rifai N, Manson JE, Hu FB. Major dietary patterns are related to plasma concentrations of markers of inflammation and endothelial dysfunction. Am J Clin Nutr 2004;80:1029–35.

[123] Cassidy A, De Vivo I, Liu Y, Han J, Prescott J, Hunter DJ, Rimm EB. Associations between diet, lifestyle factors, and telomere length in women. Am J Clin Nutr 2010;91:1273–80.

[124] Nettleton JA, Diez-Roux A, Jenny NS, Fitzpatrick AL, Jacobs Jr DR. Dietary patterns, food groups, and telomere length in the Multi-Ethnic Study of Atherosclerosis (MESA). Am J Clin Nutr 2008;88:1405–12.

[125] Tarry-Adkins JL, Martin-Gronert MS, Chen JH, Cripps RL, Ozanne SE. Maternal diet influences DNA damage, aortic telomere length, oxidative stress, and antioxidant defense capacity in rats. Faseb J 2008;22:2037–44.

[126] Chengzhuo G, Robert EH, Elaine MH. DNA repair pathways and mechanisms. In: Lesley AM, Stephanie MC, Elaine MH, editors. DNA repair of cancer stem cells. 2013. [Chapter 2].

[127] Warzyszynska JE, Kim Y-IJ. Folate in Human Health and Disease. Chichester: eLS. John Wiley & Sons Ltd; 2014. http://dx.doi.org/10.1002/9780470015902.a0002268.pub2. http://www.els.net.

INDEX

'*Note*: Page numbers followed by "f" indicate figures, "t" indicate tables.'